Barron's Review Course Series

Let's Review:

Math A

Second Edition

Lawrence S. Leff
Former Assistant Principal, Mathematics Supervision
Franklin D. Roosevelt High School
Brooklyn, New York

BARRON'S

Dedication

To Rhona
For the understanding,
for the love,
and with love.

All inquiries should be addressed to:
Barron's Educational Series, Inc.
250 Wireless Boulevard
Hauppauge, New York 11788
www.barronseduc.com

Library of Congress Catalog Card No. 2002074779

International Standard Book No. 0-7641-2296-7

Library of Congress Cataloging-in-Publication Data

Leff, Lawrence S.
    Let's review. Math A / Lawrence S. Leff.–2nd ed.
        p.   cm.—(Barron's review course series)
    Includes index.
    Summary: Provides a review covering mathematical reasoning; geometry; algebraic methods; area, volume, and measurement; graphs and transformations; and probability and statistics. Includes two sample Regents examinations.
    ISBN 0-7641-2296-7
    1. Mathematics.   2. Mathematics—Study and teaching (Secondary)—New York (State)   [1. Mathematics.   2. Mathematics—Examinations—Study guides.]
    I. Title: Math A.   II. Title.   III. Series.

QA39.3 .L43 2003
510—dc21
                                                                    2002074779

PRINTED IN THE UNITED STATES OF AMERICA
9 8 7 6 5 4 3 2 1

TABLE OF CONTENTS

PREFACE
Second Edition

The second edition of *Let's Review: Math A* collects and organizes *all* of the various topics, concepts, and mathematical reasoning skills required by the most recent Mathematics A Regents Examinations in a way that is easy for students to understand and convenient for teachers to use when planning their lessons.

Who Should Use This Book?

- *Students Who Wish to Raise Their Grades*
  Students enrolled in Regents-level mathematics classes preparing for the Mathematics A Regents Examination will find this book helpful if they need either additional explanation and practice on a troublesome topic being studied in class or review on specific topics.
- *Teachers Who Wish Additional Help in Planning Their Lessons*
  Since this book is designed to be compatible with all styles of instruction and curriculum organization, classroom teachers will want to include *Let's Review: Math A* in their personal, departmental, and school libraries. Teachers will find it a valuable lesson-planning aid as well as a helpful source of classroom exercises, homework problems, and test questions.
- *School Districts and Mathematics Departments*
  School districts and mathematics departments who want to bring their mathematics curricula and classroom instruction into closer alignment with the Mathematics A Regents Examination will find this book particularly useful. The table of contents indicates how the wide range of topics and diverse skills required by the Mathematics A Regents Exam can be organized into a logical sequence of lessons. The accompanying chart shows how the book can be used as the foundation of a comprehensive $1\frac{1}{2}$- or 2-year course that leads to the Mathematics A exam.

Term	3-Term Course	4-Term Course
1	Chapters: 1 to 5	Chapters: 1 to 4
2	Chapters: 6 to 11	Chapters: 5 to 8
3	Chapters: 12 to 16, Appendix I	Chapters: 9 to 11, 13
4		Chapters: 12, 14 to 16, Appendix I

School districts may wish to adopt this book for their Mathematics A-level classes as a cost-effective way of supplementing existing materials with up-to-date coverage of Mathematics A topics, TI-83 graphing calculator instruction, and thorough Regents examination preparation in a *single* book.

What's New in the Second Edition?

The second edition of *Let's Review: Math A* places greater emphasis on the role of problem-solving and the graphing calculator in the development of the subject.

- Chapter 3 (Problem Solving and Logical Reasoning) has been rewritten and expanded to include more work with motion problems, as well as a new section titled "Solving Problems in Different Ways." This section shows how the same problem can be approached in different ways by capitalizing on some of the key features of the TI-83 graphing calculator.
- Various sections have been enhanced with additional contextualized problems and simplified explanations.
- To ensure that the book continues to reflect recent trends in the Mathematics A Regents Examination, minor additions and deletions in content have been made throughout the book.

What Special Features Does This Book Have?

- *Core Curriculum*
  The second edition continues to offer *full* topic coverage of the New York State *Core Curriculum*, on which the Mathematics A Regents Examination is based. Thus, this book is recommended for use in any college preparatory mathematics class for students who will take the New York State Mathematics A Regents Exam.
- *Mathematics A Topics and Question Types*
  This book not only covers standard Mathematics A Regents topics, but also reflects the increased emphasis on mathematical reasoning, contextualized word problems, and the use of graphing calculator technology. Because almost half of the Mathematics A Regents Examination consists of standard multiple-choice questions, each chapter includes a generous sampling of multiple-choice questions organized by topic. Practice questions that require more sustained work, creative analysis, and writing are also featured.
- *A Compact Format Designed for Self-Study and Rapid Learning*
  For easy reference, the major topics of the book are grouped by their branches of mathematics. The logical arrangement of topics is further

enhanced by a clear writing style that quickly identifies essential ideas while avoiding unnecessary details. Helpful diagrams, convenient summaries, and numerous step-by-step demonstration examples will be appreciated by students who need an easy-to-read book that provides complete and systematic preparation for the Mathematics A Regents Examination.

- *Mathematics A Regents Examination Preparation*
  Traditional Mathematics Regents Examination topics as well as the new topics and types of questions introduced by the Mathematics A Regents Examination are covered fully. Actual Mathematics A Regents Exams with answers are included in Appendix II.

- *Graphing Calculator Solutions*
  The main graphing tools of the Texas Instruments TI-83 family of graphing calculators are introduced when needed and actual screen shots are included.

- *Answers to Practice Exercises*
  Practice exercises include questions at different levels of difficulty that are designed to build understanding, skill, and confidence. Answers to the practice exercises, except the open-ended writing questions and some of the graphing questions, are provided to give students valuable feedback that will lead to greater understanding and mastery of the material.

LAWRENCE S. LEFF
December 2002

Unit One

MATHEMATICAL REASONING

CHAPTER 1

WORKING WITH NUMBERS AND VARIABLES

1.1 ALGEBRAIC TERMS AND SYMBOLS

KEY IDEAS

Algebra is arithmetic that uses both letters and numbers. These letters, called **variables**, are placeholders for unknown numbers. The language of algebra allows general statements to be made about numbers. We know, for example, that $2 + 3 = 3 + 2$. To generalize that the order in which *any* two numbers are added together does not matter, we can write $a + b = b + a$, where *a* and *b* are *variables* that stand for *any* two numbers.

Variables and Constants

Suppose you pick any number from the set $\{1, 2, 3, 4, 5\}$ and then add 9 to that number. The sum will depend on the number you pick. If you pick 3, the sum is $3 + 9$ or 12. If you pick 4, the sum is $4 + 9$ or 13. If the number you pick is not known, the sum can be represented by $x + 9$, where x is a place-holder for the actual number you will pick from the set. The letter x is a *variable*, while the number 9 is a *constant*.

- A **variable** is a symbol, usually a single letter of the alphabet, that stands for any member of a given set of possible replacement values.
- A **constant** is a fixed number. Sometimes a letter can represent a constant; an example is the constant π (read as "pi"), which is approximately equal to $\frac{22}{7}$.

Equal and Comparison Symbols

The symbol $=$ is read as "is equal to," while the symbol $\neq$ means "is not equal to." For example, $1 + 1 = 2$ and $1 + 1 \neq 3$. If two numbers are not equal, then one is greater than the other. The symbol for "is greater than" is $>$, and the symbol for "is less than" is $<$.

For example:

- $8 > 7$ is read as "8 is greater than 7."
- $1 < 4$ is read as "1 is less than 4."
- $x \le 3$ is read as "x is less than *or* equal to 3." Hence x is *at most* 3.
- $y \ge 6$ is read as "y is greater than *or* equal to 6." Hence y is *at least* 6.

The symbol $\approx$ is read as "is approximately equal to." For example, $\dfrac{1}{3} \approx 0.33$.

Indicating Multiplication

In algebra, multiplication is often represented in ways that avoid using the symbol $\times$. Multiplication is indicated by:

- Centering a dot between two quantities. For example, $4 \cdot y$ means 4 times y.
- Placing parentheses around quantities written next to each other. For example, $(4)(y)$ means 4 times y.
- Writing two or more variables, or a number and one or more variables, next to each other. For example, ab means a times b, and $3n$ means 3 times n. When a number and a variable are written next to each other, the number is called the **coefficient** of the variable. Thus, in $3n$, 3 is the *coefficient* of n. If $n = 4$, then $3n = 3 \times 4 = 12$. Hence, $3n$ means $n + n + n$.

When a number is not written next to a variable, as in x, the coefficient of the variable is understood to be 1. Hence, x and $1 \cdot x$ mean the same thing.

Number Opposites

Positive 3 and negative 3 are *opposites* in the same sense that winning 3 games (+3) and losing 3 games (−3) are opposite situations. Keep in mind that:

- The sum of a number and its opposite is always 0. For example, $(+3) + (-3) = 0$.
- A negative sign in front of a number that is enclosed by parentheses indicates the opposite of that number. For example:

$$-(+4) = -4 \qquad \text{and} \qquad -(-4) = +4 \text{ or } 4$$

- The opposite of 0 is 0.

Numbers preceded by positive (+) or negative (−) signs are called **signed numbers**.

Number Line

The size relationship between positive and negative numbers can be represented by a rulerlike diagram called a **number line**. A number line extends indefinitely in opposite directions on either side of the point labeled 0, which is called the **origin**. On a horizontal number line, positive numbers are labeled in increasing size order to the right of 0, and negative numbers appear in decreasing size order to the left of 0. The opposite of each positive number lies on the opposite side of the origin, the same distance from 0 as its positive counterpart, as shown in Figure 1.1.

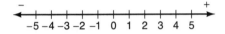

Figure 1.1 A Number Line

As you move along the number line from left to right, the numbers increase in value. For example, −2 is greater than −3, −1 is greater than −2, 0 is greater than −1, 1 is greater than 0, and so forth. In general:

- Any number on the number line is greater than any number to its left.
- The set $\{\ldots, -3, -2, -1, 0, 1, 2, 3, \ldots\}$ is called the set of **integers**.
- **Positive integers** are the integers to the *right* of 0 on the number line.
- **Negative integers** are the integers to the *left* of 0 on the number line.
- If a nonzero integer has no sign in front of it, you can assume the number is positive. Thus, 5 and +5 are the same number.
- A number that can be written as a fraction with an integer in the numerator and a nonzero integer in the denominator, such as $\frac{-3}{5}$, is called a **rational number**. The fraction $\frac{-3}{5}$ can be written in more than one way:

$$\frac{-3}{5} = \frac{3}{-5} = -\frac{3}{5} = -0.6$$

Absolute Value

On the number line −2 and +2 are each at a distance of 2 units from the origin. Since the **absolute value** of a number is its distance from the origin, the absolute value of −2 is 2, which is written as $|-2| = 2$. Also, $|+2| = 2$. In general, the absolute value of a number x is written as $|x|$ and is equal to x without an attached plus or minus sign.

3

Multiplying and Dividing Signed Numbers

To multiply or divide signed numbers, first multiply or divide the numbers without their signs. Then:

- If the original numbers have the same sign, make the answer positive. For example,

$$(+2) \times (+7) = +14 \quad \text{and} \quad \frac{+24}{+3} = +8$$

$$(-2) \times (-7) = +14 \quad \text{and} \quad \frac{-24}{-3} = +8$$

- If the original numbers have opposite signs, make the answer negative. For example,

$$(-2) \times (+7) = -14 \quad \text{and} \quad \frac{-24}{+3} = -8$$

Adding Signed Numbers

- To add signed numbers that have the same sign, add the numbers without their signs. Then write the sum using the common sign. For example,

$$(+2) + (+7) = +9 \quad \text{and} \quad (-2) + (-7) = -9$$

- To add signed numbers that have different signs, *subtract* the numbers while ignoring their signs. Then write the difference using the sign of the number with the larger absolute value. For example,

$$(-2) + (+7) = +5 \quad \text{and} \quad (+2) + (-7) = -5$$

Subtracting Signed Numbers

Rewrite a subtraction problem as an addition problem by changing the sign of the number that appears after the subtraction sign to its opposite. For example,

$$(+2) - (-7) = (+2) + (+7) = +9 \quad \text{and} \quad (+2) - (+7) = (+2) + (-7) = -5$$

Divisibility

Here are some facts about division that you should know:

- In the division example $12 \div 4 = 3$, the answer, 3, is called the **quotient**, 4 is called the **divisor**, and 12 is the **dividend**.
- A quotient or division example may also be written as a fraction. For example, $12 \div 4$ means $\frac{12}{4}$.
- Dividing by 0 is *not* allowed. An expression such as $3 \div 0$ (or $\frac{3}{0}$) is *undefined*. Also, the fraction $\frac{2}{x+4}$ is not defined when $x = -4$ since, if x is replaced by -4, the denominator becomes $-4 + 4 = 0$, making the original fraction $\frac{2}{0}$, which is not defined.
- An integer is **divisible** by a nonzero integer if the quotient has a remainder of 0. For example, 12 is *divisible* by 4 (or 4 divides 12 evenly) since the quotient is 3 with a remainder of 0. However, since $12 \div 7 = 1$ with a remainder of 5, 12 is *not* divisible by 7.
- **Even integers**, for example, -2, 0, 2, 4, and 6, are divisible by 2, while **odd integers**, for example, -1, 1, 3, and 5, are *not* divisible by 2.
- Any nonzero number divided by itself is 1. For example, $\frac{3}{3} = 1$ and $\frac{x}{x} = 1$ (provided that $x \neq 0$).

Example

In a softball league, 92 players are on 8 different teams. If each team has at least 11 players, what is the largest possible number of players on any one team?

Solution: **15**

Since $92 \div 11 = 8$ teams with a remainder of 4 players, there are 4 extra players in the softball league. If all 4 players are on the same team, that team has $11 + 4 = 15$ players, which is, therefore, the largest possible number of players any one team can have.p

Prime and Composite Numbers

A prime number is an integer greater than 1 that is divisible only by itself and 1. The numbers 3, 5, 7, 11, 13, 17, 19, and 23 are examples of prime numbers. The number 2 is the only even number that is also a prime number. Positive integers greater than 1 that are *not* prime are called **composite numbers**.

Check Your Understanding of Section 1.1

A. *Multiple Choice*

1. Which inequality could be used to represent the statement "*x* is *at least* 4"?

   (1) $x < 4$ (2) $x \le 4$ (3) $x > 4$ (4) $x \ge 4$

2. Which inequality could be used to represent the statement "*x* is *at most* 7"?

   (1) $x < 7$ (2) $x \le 7$ (3) $x > 7$ (4) $x \ge 7$

3. A class of 43 students is to be divided into committees in such a way that each student serves on exactly one committee. Each committee must have at least 3 members and at most 5 members. The possible number of committees that can be formed is

   (1) at least 8 and at most 14   (3) at least 9 and at most 14
   (2) at least 8 and at most 15   (4) at least 9 and at most 15

4. A jar contains between 40 and 50 marbles. If the marbles are taken out of the jar 3 at a time, 2 marbles will be left in the jar. If the marbles are taken out of the jar 5 at a time, 4 marbles will be left in the jar. How many marbles are in the jar?

   (1) 41   (2) 43   (3) 44   (4) 47

5. For a class trip to a museum, 461 students and 20 teachers will be taking buses. Each bus can seat a maximum of 52 persons. What is the *least* number of buses needed for the trip?

   (1) 8   (2) 9   (3) 10   (4) 11

B. *In each case, show how you arrived at your answer by clearly indicating all of the necessary steps, formula substitutions, diagrams, graphs, charts, etc.*

6–9. Find each sum.

6. $(-6) + (-1)$ 8. $-6.4 + (+3.9)$

7. $-3 + 7$ 9. $\left(+\dfrac{1}{2} \right) + \left(-\dfrac{1}{3} \right)$

10–13. Find each difference.

10. $(-7) - (-3)$ 12. $5 - 7$
11. $(-2) - (-6)$ 13. $-(-8) - (-5)$

14–17. Find each product.

14. $(-4)(-3)$

16. $(+1.5)(-0.3)$

15. $(-5)(+4)(-3)$

17. $\left(-\dfrac{3}{5}\right)\left(-\dfrac{10}{3}\right)$

18–21. Find each quotient.

18. $\dfrac{-21}{+7}$

20. $\dfrac{2.4}{-0.6}$

19. $\dfrac{-18}{-6}$

21. $\dfrac{(-14)-(-2)}{-3}$

22. The product of two integers is −24. If the same two integers are added together, the sum is −23. Find the smaller of the two integers.

23. The product of two integers is −40. If the same two integers are added together, the sum is −3. Find the larger of the two integers.

24. If all of the books on a shelf holding fewer than 45 books were put into piles of 5 books each, no books would remain. If the same set of books were put into piles of 7 books each, 2 books would remain. What is the greatest number of books that could be on the shelf?

1.2 PROPERTIES OF REAL NUMBERS

KEY IDEAS

The set of all points on the number line represents the set of **real numbers**. Real numbers behave in predictable ways. For example, we know that the order in which any two real numbers are added or multiplied does not matter.

Using Parentheses as a Grouping Symbol

When parentheses enclose two numbers connected by an arithmetic operation, then this operation is performed first. For example:

$$2+(1+5) = 2+6 = 8 \quad \text{and} \quad 2(3 \times 4) = 2 \times 12 = 24$$

Parentheses may be inserted simply to make an expression easier to read. For example, rather than writing $5x + 2x + 8 + 3$, we can write $(5x + 2x) + (8 + 3)$, thus grouping the variable terms together and grouping the constant terms together.

Working with Real Numbers

To be able to work with real numbers, some basic assumptions must be made about how these numbers behave. The set of real numbers has the following properties for addition and multiplication:

- **Closure Property**
  A set of numbers is **closed** for an arithmetic operation if the answer is always a number that belongs to the same set. For example, adding, subtracting, or multiplying two integers always results in another integer. Dividing two integers, however, does not necessarily result in another integer, as in 3 ÷ 4. Therefore, the set of integers is closed under addition, subtraction, and multiplication, but is *not* closed under division. The set of real numbers is closed under addition and multiplication.
- **Commutative Properties**
  The order in which *two* real numbers are added or multiplied does not matter. For instance, 2 + 3 = 3 + 2 = 5 and 2·3 = 3·2 = 6. Thus, the set of real numbers is **commutative** for both addition and multiplication. The set of real number is *not* commutative for division or subtraction. For example, 3 ÷ 4 is not the same as 4 ÷ 3 and 4 − 3 is not the same as 3 − 4.
- **Associative Properties**
  When *three* real numbers are added or multiplied together, it does not matter how the numbers are grouped, or *associated*, when performing the operation. Thus:

$$(2+3)+4 = 2+(3+4) = 9 \quad \text{and} \quad (2\cdot3)4 = 2(3\cdot4) = 24$$

- **Identity Properties**
  When 0 is *added* to a real number, the sum is that same real number. For example, 4 + 0 = 0 and 0 + 4 = 0. As a result, 0 is called the **identity element for addition.**
  When any real number is *multiplied* by 1, the product is that same real number. For example, 4 × 1 = 4 and 1 × 4 = 4. Hence, 1 is the **identity element for multiplication.**
- **Inverse Properties**
  Each real number has an opposite, called its **additive inverse**, so that the sum of that number and its additive inverse is 0. For example, the additive inverse of 4 is −4 since (4) + (−4) = 0.
  Each nonzero real number a has a reciprocal $\frac{1}{a}$, called its **multiplicative inverse**, so that the product of that number and its multiplicative inverse is always 1. For example, the multiplicative inverse (reciprocal) of 4 is $\frac{1}{4}$ and $4 \times \frac{1}{4} = 1.$

8

- **Distributive Property**
  The **distributive** property links the operations of multiplication and addition so that an expression like $3(2 + 4)$ can be evaluated as follows:

$$3(2+4) = (3 \times 2) + (3 \times 4) = 6 + 12 = 18$$

Defining a Binary Operation

Addition is an example of a **binary operation** since it works with exactly *two* numbers of a set at a time. Subtraction, multiplication, and division are also binary operations.

In the example below, a new binary operation symbolized as Δ is defined by using a table. The table summarizes the result of performing operation Δ on every possible pair of members of the set $\{A,B,C,D\}$. Read the table in much the same way that you would read an ordinary times or multiplication table. For example, to find the value of $B \Delta D$, locate B in the first column under the operation symbol Δ. Follow the same row across to the right until it meets the table entry under the column headed by D. According to the table, $B \Delta D = C$.

Example

The binary operation Δ is defined for the elements of $\{A,B,C,D\}$ using the accompanying table. Determine the:

(a) identity element
(b) inverse of B

Δ	A	B	C	D
A	C	D	A	B
B	D	A	B	C
C	A	B	C	D
D	B	C	D	A

Solution: (a) C

To determine the identity element, find the element under the operation symbol in the first column that produces the same element as it operates on. Notice that:

$C \Delta A = A$
$C \Delta B = B$
$C \Delta C = C$
$C \Delta D = D$

Δ	A	B	C	D
A	C	D	A	B
B	D	A	B	C
C	A	B	C	D
D	B	C	D	A

It is also the case that $A \Delta C = A$, $B \Delta C = B$, and $D \Delta C = D$. Whatever element C operates on, regardless of the order, the result is that same element. Hence, the identity element is C.

(b) D

To find the inverse of B, locate the element that, when it operates on B, results in C, the identity element. Since $B \Delta D = C$ and $D \Delta B = C$, the inverse of B is D.

Check Your Understanding of Section 1.2

Multiple Choice

1. Which set of numbers is *not* closed under multiplication?
  (1) {odd integers}   (3) {prime numbers}
  (2) {even integers}   (4) {rational numbers}

2. The statement $2x + 2y = 2(x + y)$ uses which property of real numbers?
  (1) commutative property   (3) distributive property
  (2) associative property   (4) closure property

3. Excluding 0, which set does *not* have a multiplicative inverse for each of its members?

  (1) {integers}   (4) $\left\{-1, \dfrac{1}{2}, 1, 2\right\}$
  (2) {rational numbers}
  (3) {positive real numbers}

4. The statement $(x + y) + z = z + (x + y)$ illustrates which property of real numbers?
  (1) commutative property   (3) distributive property
  (2) associative property   (4) closure property

5. If a, b, and c are real numbers, which statement is *always* true?
  (1) $a \div b = b \div a$ (3) $a(b \times c) = a \times b + a \times c$
  (2) $a(b + c) = (a + b) \times (a + c)$ (4) $a \times 0 = a + (-a)$

6. Which equation illustrates the additive inverse property for real numbers?

  (1) $a + (-a) = 0$ (4) $a \cdot \dfrac{1}{a} = 1$
  (2) $a + 0 = a$
  (3) $a + (-a) = -1$

7. Which equation illustrates the multiplicative identity property for real numbers?

(1) $x + 0 = x$ (2) $x \cdot 1 = x$ (3) $x \cdot \dfrac{1}{x} = 1$ (4) $x \cdot 0 = 0$

8. Which equation illustrates the associative property for addition?
(1) $a + (b + c) = a + (c + b)$ (3) $(a + b) + c = a + (b + c)$
(2) $a + (b + c) = (b + c) + a$ (4) $(a + b)c = ac + bc$

9. Which equation illustrates the multiplicative inverse property?
(1) $b \cdot 0 = 0$ (3) $b + 0 = b$

(2) $b + (-b) = 0$ (4) $b \cdot \dfrac{1}{b} = 1$

10. Which set is closed under subtraction?
(1) {odd integers}   (3) {counting numbers}
(2) {integers}   (4) {prime numbers}

11. What is the additive inverse of $-4a$?

(1) $\dfrac{a}{4}$ (2) $4a$ (3) $-\dfrac{4}{a}$ (4) $-\dfrac{1}{4a}$

12. What is the multiplicative inverse of $\dfrac{x}{2}(x \neq 0)$?

(1) 1 (2) $\dfrac{2}{x}$ (3) $-\dfrac{x}{2}$ (4) $2x$

13. To add $\dfrac{1}{2} + \dfrac{1}{3}$, the first step is to change each fraction into an equivalent fraction that has the least common denominator as its denominator by writing

$$\frac{1}{2} + \frac{1}{3} = \frac{1}{2} \times \left(\frac{3}{3}\right) + \frac{1}{3} \times \left(\frac{2}{2}\right)$$

This step can be justified by which property of real numbers?
(1) commutative property
(2) existence of an additive identity
(3) existence of a multiplicative identity
(4) distributive property

14. The operation ♣ for the set {C,L,U,B} is defined in the accompanying table. What is the identity element for ♣?

(1) C   (2) L   (3) U   (4) B

♣	C	L	U	B
C	B	U	C	L
L	U	B	L	C
U	C	L	U	B
B	L	C	B	U

15. What is the inverse of *b* in the system defined in the accompanying table?

(1) *a*   (2) *b*   (3) *c*   (4) *d*

∗	a	b	c	d
a	d	c	a	b
b	c	d	b	a
c	a	b	c	d
d	b	a	d	c

16–17. In the accompanying tables, the elements *S, A, L, E* and the operations ∗ and # are defined.

∗	S	A	L	E
S	A	S	E	L
A	S	A	L	E
L	E	L	A	S
E	L	E	S	A

#	S	A	L	E
S	A	L	E	S
A	L	E	S	A
L	E	S	A	L
E	S	A	L	E

16. What is the inverse of *L* under #?

(1) S   (2) A   (3) L   (4) E

17. What is the value of (A ∗ L) # (E ∗ S)?

(1) S   (2) A   (3) L   (4) E

1.3 EXPONENTS AND SCIENTIFIC NOTATION

KEY IDEAS

When numbers are multiplied together, the answer is called the **product**. Each of the numbers that are multiplied together to form a product is called a **factor** of that product. Thus, since $4 \times 3 = 12$, 4 and 3 are factors of 12. The product $2 \times 2 \times 2 \times 2 \times 2$ can be written using the shorthand notation 2^5, in which 2, the factor that is repeated, is called the **base**. The **exponent** (5) is the number of times the base (2) is used as a factor.

$$\underbrace{2 \times 2 \times 2 \times 2 \times 2}_{\text{2 is used as a factor 5 times}} = 2^{\overset{\text{Exponent}}{\downarrow}5}_{\uparrow\text{Base}} = 2 \text{ raised to the fifth power}$$

When an exponent is not written, it is assumed to be 1. Thus, $4 = 4^1$ and $x = x^1$.

Laws of Exponents

Powers of the *same* nonzero base can be *multiplied* by *adding* their exponents. For example, $5^2 \times 5^6 = 5^{2+6} = 5^8$ because

$$5^2 \times 5^6 = \underbrace{(5 \times 5) \times (5 \times 5 \times 5 \times 5 \times 5 \times 5) = 5^8}_{\text{8 factors of } 5 = 5^8}$$

Powers of the *same* nonzero base can be *divided* by *subtracting* their exponents. For example, $5^8 \div 5^6 = 5^{8-6} = 5^2$ because

$$\frac{5^8}{5^6} = \frac{5 \times 5 \times 5 \times 5 \times 5 \times 5 \times 5 \times 5}{5 \times 5 \times 5 \times 5 \times 5 \times 5}$$

$$= \frac{5 \times 5 \times 5 \times 5 \times \cancel{5 \times 5}}{\cancel{5 \times 5} \times 5 \times 5 \times 5 \times 5} \times \underbrace{5 \times 5}_{\substack{\text{2 factors} \\ \text{remain}}}$$

$$= 1 \times 5^2$$

$$= 5^2$$

Example 1

Evaluate:

(a) $3^4 - 4^3$ (b) $(-10)^3$ (c) $(-2)^4$ (d) -2^4

Solution:

(a) **17**

$$3^4 = 3 \times 3 \times 3 \times 3 = 81$$
$$4^3 = \quad 4 \times 4 \times 4 \ = 64$$
$$3^4 - 4^3 = \quad 81 - 64 \quad = 17$$

(b) **−1000**

$$(-10)^3 = (-10) \ \times (-10) \times (-10)$$
$$= (+100) \times (-10)$$
$$= -1000$$

(c) **+16**

$$(-2)^4 = \underbrace{(-2) \times (-2)} \times (-2) \times (-2)$$
$$= \underbrace{\quad +4 \quad \times (-2)} \times (-2)$$
$$= \quad\quad -8 \quad\quad \times (-2)$$
$$= +16$$

(d) **−16**

Since -2^4 means $-(2^4)$, do the power first:

$$-2^4 = -(2 \times 2 \times 2 \times 2) = -16$$

=== **MATH FACTS** ===

- When the base is negative and the exponent is an *odd* integer, the product will be a *negative* number. For example, $(-1)^{99} = -1$.
- When the base is negative and the exponent is an *even* integer, the product will be a *positive* number. For example, $(-1)^{100} = 1$.

Example 2

Multiple Choice:

The expression $2^4 \cdot 4^3$ is equivalent to

(1) 8^{12} (2) 8^7 (3) 2^{10} (4) 2^9

Solution: **(3)**

$$2^4 \cdot 4^3 = 2^4 \cdot \left(2^2\right)^3$$
$$= 2^4 \cdot 2^{2 \times 3}$$
$$= 2^4 \cdot 2^6$$
$$= 2^{4+6}$$
$$= 2^{10}$$

Special Exponents

An exponent may be equal to 0 or be equal to a negative integer.

- Any nonzero number with an exponent of 0 is equal to 1. For example, $3^0 = 1$ and $y^0 = 1$ provided that $y \neq 0$. The expression 0^0 is not defined.
- Any nonzero number or variable with a negative exponent is equal to the reciprocal of that same term with the sign of the exponent changed from negative to positive. For example:

$$5^{-2} = \frac{1}{5^2} = \frac{1}{25} \quad \text{and} \quad \frac{a^2}{a^5} = a^{2-5} = a^{-3} = \frac{1}{a^3} \text{ (provided that } a \neq 0)$$

Example 3

If $y \neq 0$, evaluate: (a) $2y^0$ (b) $(2y)^0$

   Solution:

(a) **2**
$2y^0 = 2 \times y^0 = 2 \times 1 = 2$

(b) **1**
A nonzero quantity raised to the zero power is 1.

The rules for working with exponents are summarized in Table 1.1.

TABLE 1.1 LAWS OF INTEGER EXPONENTS

Exponent Law	Rule	Example
Multiplication	$a^x \times a^y = a^{x+y}$	$n^5 \times n^2 = n^7$
Quotient	$a^x \div a^y = a^{x-y}$	$n^5 \div n^2 = n^3$
Power of a power	$(a^x)^y = a^{xy}$	$(n^2)^5 = n^{10}$
Power of a product	$(ab)^x = a^x b^x$	$(mn^5)^2 = m^2 n^{10}$
Zero power	$a^0 = 1$ (provided that $a \neq 0$)	$4^0 = 1$
Negative power	$\frac{1}{a^{-x}} = a^x$ and $a^{-x} = \frac{1}{a^x}$ (provided that $a \neq 0$)	$3^{-2} = \frac{1}{3^2} = \frac{1}{9}$

Scientific Notation

Very large and very small numbers may be easier to read if they are expressed in **scientific notation**. When written in scientific notation, 32,000,000 becomes 3.2×10^7 and 0.00000408 becomes 4.08×10^{-6}. To express a positive number in scientific notation, write it as a decimal number between 1 and 10 times a power of 10.

- When the original number is greater than 10, the number of decimal places that the decimal point needs to be moved to the *left* becomes the power of 10. If necessary, insert the decimal point. For example:

$$32,000,000 = 3\ 2\ 0\ 0\ 0\ 0\ 0 . = 3.2 \times 10^7$$

7 places

- When the original number is between 0 and 1, the number of decimal places that the decimal point needs to be moved to the *right* becomes the power of 10, with a negative sign in front of it. For example:

$$0 . 0\ 0\ 0\ 0\ 6\ 0\ 8 = 6.08 \times 10^{-5}$$

5 places

Check Your Understanding of Section 1.3

A. Multiple Choice

1. If $0.0000603 = 6.03 \times 10^n$, what is the value of n?
   (1) −5   (2) 5   (3) −4   (4) 4

2. The distance from the Sun to the planet Neptune is about 2,790,000,000 miles. Expressed in scientific notation, this distance, in miles, is
   (1) 2.79×10^9 (3) 27.9×10^7
   (2) 2.79×10^{-9} (4) 27.9×10^{-7}

3. Expressed in scientific notation, 0.003146 is equivalent to
   (1) 31.46×10^4 (3) 31.46×10^3
   (2) 3.146×10^{-3} (4) 3.146×10^{-2}

4. The distance from Earth to the Sun is approximately 93 million miles. A scientist would write that number of miles as
   (1) 9.3×10^6 (2) 9.3×10^7 (3) 93×10^7 (4) 93×10^{10}

5. Which expression is equivalent to 4.08×10^{20}?
   (1) 0.0408×10^{18} (3) 408×10^{18}
   (2) 40.8×10^{18} (4) 4080×10^{18}

6. If $10^k = \dfrac{1}{2}$, what is the value of 10^{k+3}?
   (1) 125 (2) 250 (3) 500 (4) 1000

7. Which expression is equivalent to 8.12×10^{-9}?
   (1) 812×10^{-7} (3) 8.12×10^{-11}
   (2) 0.0812×10^{-7} (4) 0.0812×10^{-11}

8. The expression $3^4 \cdot 9^3$ is equivalent to
   (1) 3^{10} (2) 3^{24} (3) 27^7 (4) 27^{12}

B. *In each case, show how you arrived at your answer by clearly indicating all of the necessary steps, formula substitutions, diagrams, graphs, charts, etc.*

9. If $3^y = 9$ and $2^{x+y} = 32$, what is the value of x?

10. Write $(-2y^3)^5$ as the product of a number and a base with a single exponent.

11. If $y = 3 \times 8^0 - 2^{-2}$, what is the value of y?

12. Express $(0.0005)^3$ in scientific notation.

13. (a) State whether $2^4 \times 4^2$ is or is *not* equal to 8^6, and explain your answer.
    (b) For what value of x does $2^4 \times 4^2 = 2^x$?

14. Charles claims that $x^2 > y^2$ whenever $x > y$. Explain why you agree or disagree.

1.4 ORDER OF OPERATIONS

KEY IDEAS

Arithmetic operations, as in $9 + 21 \div 3$, are not necessarily performed in the order in which they appear. Instead, they are performed according to an accepted set of rules called the **order of operations**. Since multiplication and division are performed before addition and subtraction, $9 + 21 \div 3 = 9 + 7 = 16$.

Rules for Order of Operations

When evaluating an arithmetic or algebraic expression involving more than one operation:

- Do the operations inside *P*arentheses, if any, first.
- Do any *E*xponents.
- *M*ultiply or *D*ivide in order from left to right.
- *A*dd or *S*ubtract in order from left to right.

As indicated above, the sentence "*P*lease *E*xcuse *M*y *D*ear *A*unt *S*ally" can help you remember the rules for order of operations. For example, to evaluate $6^2 \div (1 + 2) - 3$:

- Work inside the parentheses first: $\quad 6^2 \div (1+2) - 3 = 6^2 \div 3 - 3$
- Do the exponents: $\qquad\qquad\qquad\qquad = 36 \div 3 - 3$
- Divide: $\qquad\qquad\qquad\qquad\qquad\quad = 12 \quad -3$
- Subtract: $\qquad\qquad\qquad\qquad\qquad\; = 9$

Evaluating Algebraic Expressions

If $y = 3x^2 + 5x - 7$, then y represents some unknown number until x is assigned a specific value. For example, if $x = -2$:

$$
\begin{aligned}
y &= 3x^2 &&+ 5x &&- 7 \\
&= 3(-2)^2 + 5(-2) - 7 \\
&= 3(4) &&- 10 &&- 7 \\
&= 12 &&- 17 \\
&= -5
\end{aligned}
$$

Example 1

Find the value of $\frac{1}{2}ab^3$ when $a = 5$ and $b = -2$.

Solution: **−20**

$$
\begin{aligned}
\frac{1}{2}ab^3 &= \frac{1}{2}(5)(-2)^3 \\
&= \frac{1}{2}(5)(-8) \\
&= \frac{1}{2}(-40) \\
&= -20
\end{aligned}
$$

Example 2

Find the value of $-x^2 + 4xy$ when $x = -3$ and $y = -2$.

Solution: **15**

$$-x^2 + 4xy = -(-3)^2 + 4(-3)(-2)$$
$$= -9 \quad + 4(+6)$$
$$= -9 \quad + 24$$
$$= 15$$

Example 3

If $y = 3x^2 + 5x - 7$, find the value of y when $x = -2$.

Solution: **−5**

$$y = 3x^2 \quad +5x \quad -7$$
$$= 3(-2)^2 + 5(-2) - 7$$
$$= 3(4) \quad -10 \quad -7$$
$$= 12 \quad -17$$
$$= -5$$

Check Your Understanding of Section 1.4

A. Multiple Choice

1. When $x = 2$ and $y = 0.5$, which expression has the greatest value?

   (1) $x + y^2$ (2) $\dfrac{y}{x}$ (3) $\dfrac{x}{y} - 1$ (4) $xy + 1$

2. What is the value of $5xy^2$ if $x = -2$ and $y = -3$?
   (1) −90   (2) 90   (3) −180   (4) 180

3. If $a = 2$ and $b = -1$, the expression $3ab^2$ is equal to
   (1) 6   (2) 12   (3) 36   (4) −12

4. What is the value of $-3a^2b$ if $a = -2$ and $b = -4$?
   (1) −48   (2) 48   (3) −144   (4) 144

5. If $y = -\dfrac{1}{4}$ and $z = 8$, what is the value of $\frac{1}{2}yz^2$?

(1) 8 (2) 2 (3) −8 (4) −4

6. If $x = -2$, $y = 4$, and $z = -2$, what is the value of $\dfrac{x^3 y}{z}$?

(1) −16 (2) 16 (3) −12 (4) 12

B. *In each case, show how you arrived at your answer by clearly indicating all of the necessary steps, formula substitutions, diagrams, graphs, charts, etc.*

7–9. If $x = -2$ and $y = 5$, find the value of each expression.

7. $(xy)^2$ **8.** $(y + x)(y - x)$ **9.** $\dfrac{x^2 - y^2}{x + y}$

10–12. If $r = 24$ and $s = -2$, find the value of each expression.

10. $-\dfrac{1}{3}rs^2$ **11.** $r \div 3s^2$ **12.** $(r \div 3)s^2$

13. If $y = -x^2 + 3$, what is the largest possible value of y? Explain your answer.

14. If $y = (4 - x)^2$, what is the smallest possible value of y? Explain your answer.

1.5 TRANSLATING BETWEEN ENGLISH AND ALGEBRA

KEY IDEAS

Since algebra is used to represent or "model" problem situations, it is important to be able to translate commonly encountered English phrases and sentences into an *algebraic equation* or *inequality*.

An **equation** is a sentence that uses the equal symbol to indicate that two mathematical expressions have the same value. An **inequality** is a sentence that uses an inequality symbol to compare two mathematical expressions.

Interpreting Algebraic Expressions

When translating an algebraic expression into words, pay attention to the arithmetic operation that connects the variable to one or more numbers. For example:

- $n + 5$ means 5 *added* to n, or 5 more than n.
- $n - 3$ means 3 *subtracted* from n, 3 less than n, or n diminished by 3.
- $\frac{n}{2}$ means n *divided* by 2, or one-half of n.
- $4n + 1$ means n *multiplied* by 4 and then *increased* by 1, or 1 more than the product of 4 and n.

Representing One Quantity in Terms of Another

Sometimes it is necessary to compare two quantities by representing one in terms of the other. For example, suppose John is y years old. Table 1.2 shows how to represent different situations in terms of y.

TABLE 1.2 REPRESENTING JOHN'S AGE

Condition	Algebraic Expression
John's age 5 years ago	$y - 5$
John's age 9 years from now	$y + 9$
John's age when he is three times as old as he is now	$3y$
Twice John's age 8 years ago	$2(y - 8)$

Here are a few more examples:

- Tim's weight exceeds Sue's weight by 13 pounds. If Sue weighs x pounds, then Tim weighs $x + 13$ pounds.
- The number of dimes exceeds three times the number of pennies by 2. If there are x pennies, then there are $3x + 2$ dimes.
- Bill has 7 fewer dollars than twice the number Kim has. If Kim has x dollars, then Bill has $2x - 7$ dollars.
- The sum of two numbers is 25. If x represents one of these numbers, then $25 - x$ represents the other number.

Translating an English Sentence into an Equation or an Inequality

To translate an English sentence into an equation or an inequality, identify key phrases that can be translated directly into mathematical terms. For example:

• $\underbrace{\text{Two times}}_{2\times}\ \underbrace{\text{a number } n}_{n}\ \underbrace{\text{increased by 5}}_{+5}\ \underbrace{\text{is}}_{=}\ \underbrace{17.}_{17}$

The equation is $2n + 5 = 17$.

• $\underbrace{\text{Five times a number } n}_{5n}\ \underbrace{\text{exceeds}}_{=}\ \underbrace{2\text{ times that number}}_{2n}\ \underbrace{\text{by}}_{+}\ \underbrace{21.}_{21}$

The equation is $5n = 2n + 21$.

• $\underbrace{\text{When 2 is subtracted from a number } n,}_{n-2}\ \underbrace{\text{the result is } \textit{at most}}_{\leq}\ \underbrace{5.}_{5}$

The inequality is $n - 2 \leq 5$.

• $\underbrace{\text{When twice a number } x,}_{2x}\ \underbrace{\text{is increased by 5,}}_{+5}\ \underbrace{\text{the result is } \textit{at least}}_{\geq}\ \underbrace{27.}_{27}$

The inequality is $2x + 5 \geq 27$.

Check Your Understanding of Section 1.5

A. *Multiple Choice*

1. Dawn is 3 years older than her sister Sara. If Dawn's age is represented by x, which expression represents Sara's age?

   (1) $3x$ (2) $x + 3$ (3) $\dfrac{1}{3}x$ (4) $x - 3$

2. This year Juanita is twice as old as Aisha. If x represents Aisha's age now, which expression represents Juanita's age 1 year ago?

   (1) $2x$ (2) $\dfrac{1}{2}x - 1$ (3) $\dfrac{1}{2}x + 1$ (4) $2x - 1$

B. *3–9. In each case, represent the given information as an algebraic expression, equation, or inequality using the variable x.*

3. Five less than twice an unknown number.

4. Twice an unknown number diminished by 4 is less than 11.

5. Six more than three times an unknown number is at most 21.

6. Twice the sum of a number and 3 is at least 12.

7. When 2 is subtracted from a number, the difference exceeds one-half of the original number by 1.

8. If 5 less than a number is multiplied by 2, the product is at most 24.

9. In 7 years Vicky will be 3 years less than two times her current age, x.

10–12. In each case, represent the underlined sentence as an equation using the variable x.

10. The sum of two numbers is 13, and the smaller number is x. If three times the larger number is added to two times the smaller number, the sum is 33.

11. A number is five times as great as a positive number, x. When the smaller of the two numbers is subtracted from the larger number, the difference is 8 more than two times the smaller number.

12. One number exceeds another number by 7, and the smaller of the two numbers is x. The product of the two numbers exceeds five times the sum of the two numbers by 19.

13. A telephone call costs c cents for the first 3 minutes and m cents for each additional minute. What is the cost, in cents, of a 6-minute call?

14. Ari has a cellular phone that costs $12.95 per month plus $0.25 per minute for each call after the first 1 hour of calls. If Ari makes x hours and y minutes of calls in a month, where $x > 1$ and $y < 60$, how much money will Ari be charged?

CHAPTER 2

LINEAR EQUATIONS AND INEQUALITIES

2.1 SOLVING ONE-STEP EQUATIONS

KEY IDEAS

An equation such as $2x + 1 = 7$ is a **linear** or **first-degree equation** since the greatest exponent of the variable is 1. The **solution set** of an equation is the set of numbers that, when substituted for the variable, makes the equation a true statement. To find the solution set of an equation, isolate the variable by itself on one side of the equation.

Open Sentences

The equation $2x + 1 = 7$ is an *open sentence* since it can be either true or false depending on the number that is substituted for x. A number that makes an open sentence or equation a true statement is a **solution** or **root** of that equation. If x is replaced by 3, the equation $2x + 1 = 7$ becomes a true statement since $2(3) + 1$ evaluates to 7. Thus $x = 3$ is a *solution* or *root* of the equation.

Equivalent Equations

The three equations $2x + 1 = 7$, $x - 1 = 2$, and $2x = 6$ are *equivalent equations* since the same number, 3, is the solution of each equation. **Equivalent equations** are equations that have the same solution set. *Solving a linear equation* means isolating the variable, say x, until you are left with an equivalent equation that has the form $x =$ number.

Solving Linear Equations Using Inverse Operations

When the variable in a linear equation is involved in only one arithmetic operation, as in $x + 8 = 3$, isolate the variable in one step by performing the *inverse* of that arithmetic operation on both sides of the equation. Addition and subtraction are inverse operations, as are multiplication and division.

Example 1

Solve for x: $x + 8 = 3$.

Solution: **–5**

Since 8 is linked to x by addition, undo the addition by *subtracting* 8 from each side of the equation:

$$x + 8 = 3$$
$$\underline{-8 = -8}$$
$$x + 0 = -5, \quad \text{so } x = -5$$

Example 2

Solve for b: $b - 2.3 = 4.8$.

Solution: **7.1**

Isolate b by adding 2.3 to each side of the equation:

$$b - 2.3 = 4.8$$
$$\underline{+2.3 = +2.3}$$
$$b + 0 = 7.1, \quad \text{so } b = 7.1$$

Example 3

Solve for p: $\dfrac{p}{4} = -1.5$.

Solution: **–6.0**

Isolate p by multiplying each side of the equation by 4:

$$\frac{p}{4} = -1.5$$
$$4 \times \frac{p}{4} = 4 \times (-1.5)$$
$$p = -6.0$$

Example 4

Solve for t: $\dfrac{3}{2}t = 21$.

Solution: **14**

The product of a fraction and its reciprocal (multiplicative inverse) is 1. Hence, isolate t by multiplying each side of the equation by $\frac{2}{3}$, the reciprocal of the fractional coefficient of t:

$$\frac{3}{2}t = 21$$

$$\frac{2}{3} \times \frac{3}{2}t = \frac{2}{3} \times 21$$

$$t = 2 \times 7$$

$$= 14$$

Checking Roots

To make sure you have solved an equation correctly, you need to check that your solution works in the *original* equation. For example, to check whether $x = 3$ is the solution for the equation $2x - 1 = 5$, substitute 3 for x in the equation and use the order of operations to determine whether the left side of the equation equals the number on the opposite side:

- Write the original equation:

$$2x - 1 = 5$$

- Plug in 3 for x:

$$2(3) - 1 \overset{?}{=} 5$$

- Evaluate the left side of the equation:

$$6 - 1 \overset{?}{=} 5$$

- Compare the two sides of the equation:

$$\overset{\checkmark}{5 = 5}$$

Check Your Understanding of Section 2.1

1–21. In each case, solve for the variable and check.

1. $x + 5 = -9$

2. $-3x = 12$

3. $-7 = x - 1$

4. $-0.5w = -35$

5. $-4 = x + 2$

6. $8 + n = -7$

7. $-7.8 = 1.4 + x$

8. $1.2 = m + 3.5$

9. $t + 2 = \frac{4}{3}$

10. $\dfrac{2x}{3} = 16$ **14.** $\dfrac{r}{3} = -8.4$ **18.** $0.7t = 4.9$

11. $-21 = \dfrac{7x}{8}$ **15.** $-10 = \dfrac{-5a}{3}$ **19.** $\dfrac{a}{-3} = 1.6$

12. $\dfrac{x}{5} = -1.7$ **16.** $3x = 1\dfrac{1}{8}$ **20.** $b + \dfrac{1}{3} = 2\dfrac{1}{6}$

13. $y - \dfrac{1}{3} = -2$ **17.** $\dfrac{x}{4} = 2\dfrac{1}{3}$ **21.** $x - 1.9 = -7.6$

22. A number diminished by 7 is –3. What is the number?

23. The quotient of a number divided by –4 is 5. What is the number?

24. If –14 is 8 less than a number, what is the number?

25. Two-thirds of a number is 12. What is the number?

2.2 SOLVING MULTISTEP EQUATIONS

KEY IDEAS

Solving a linear equation in which the variable is involved in more than one arithmetic operation, as in $3x - 5 = 22$, requires more than one step. When isolating the variable in a multistep equation, follow the order of operations in reverse. In other words, undo any addition or subtraction before undoing any multiplication or division.

Solving an Equation Using Two Operations

To isolate a variable in an equation, it may be necessary to perform more than one inverse operation on both sides of the equation.

Example 1

Solve for x: $3x - 5 = 22$.

Solution: **9**

In the given equation, $3x - 5 = 22$, the variable is involved in two operations: multiplication, since $3x$ means three times x, and subtraction ($3x$ minus 5). Undo the subtraction before undoing the multiplication:

$$3x - 5 = 22$$
$$\underline{+5 = +5}$$
$$3x + 0 = 27$$
$$\frac{3x}{3} = \frac{27}{3}$$
$$x = 9$$

You can also show the addition or subtraction of the same number on each side of the equation horizontally:

Add 5 to each side.

$$3x - 5 + \boxed{5} = 22 + \boxed{5}$$
$$3x + 0 = 27$$
$$\frac{3x}{3} = \frac{27}{3}$$
$$x = 9$$

Example 2

Solve and check: $\frac{x}{3} - 2 = 13$.

Solution: **45**

$$\frac{x}{3} - 2 + (2) = 13 + (2)$$

$$\frac{x}{3} + 0 = 15$$

$$\cancel{3}\left(\frac{x}{\cancel{3}}\right) = 3(15)$$

$$x = 45$$

Check: $\frac{x}{3} - 2 = 13$

$$\frac{45}{3} - 2 \stackrel{?}{=} 13$$

$$15 - 2 \stackrel{?}{=} 13$$

$$13 \stackrel{\checkmark}{=} 13$$

Example 3

Solve for *x*: $0.03x - 0.7 = 0.8$.

Solution: **50**

When combining decimal numbers, make sure you align the decimal points. If you need to divide decimal numbers, you can use your calculator.

$$0.03x - 0.7 = \quad 0.8$$
$$\underline{+0.7 = +0.7}$$
$$0.03x \quad\quad = \quad 1.5$$
$$\frac{0.03x}{0.03} = \frac{1.5}{0.03}$$
$$x = 1.5 \div 0.03 = 50$$

Solving an Equation with Parentheses

If an equation has parentheses, remove them by multiplying each term inside the parentheses by the number in front of the parentheses. The distributive property of multiplication over addition guarantees that this procedure will produce an equivalent equation.

Example 4

Solve for x and check: $3(1 - 2x) = -15$.

Solution: **3**

- Write the given equation: $\quad 3(1 - 2x) = -15$
- Remove parentheses: $\quad 3(1) - 3(2x) = -15$
- Simplify: $\quad 3 - 6x = -15$
- Isolate x: $\quad (-3) + 3 - 6x = -15 + (-3)$

$$-6x = -18$$
$$\frac{-6x}{-6} = \frac{-18}{-6}$$
$$x = 3$$

Check: $\quad 3(1 - 2x) \overset{?}{=} -15$

$$3(1 - 2 \cdot 3) \overset{?}{=} -15$$

$$3(1 - 6) \overset{?}{=} -15$$

$$3(-5) \overset{?}{=} -15$$

$$-15 \overset{\checkmark}{=} -15$$

Solving Problems Involving Averages

The average (or mean) of a set of n numbers is the sum of the numbers divided by n.

Example 5

If 9 and y have the same average as 3, 6, and 27, what is the value of y?

    Solution: **15**

- The average of 9 and y is the sum of the two terms divided by 2, or $\frac{y+9}{2}$.
- The average of 3, 6, and 27 is the sum of the three numbers divided by 3, or $\frac{3+6+27}{3} = \frac{36}{3} = 12$.

Thus:

$$\frac{y+9}{2} = 12$$

$$\cancel{2}\left(\frac{y+9}{\cancel{2}}\right) = 2(12)$$

$$y+9 = 24$$

$$y = 24 - 9 = 15$$

Example 6

Bonnie receives grades of 79, 83, and 86 on her first three Math A exams. What grade must she receive on her next test in order to have an average grade of exactly 85 for the four exams?

    Solution: **92**

Let x represent Bonnie's next test grade. Then:

$$\frac{x+79+83+86}{4} = 85$$

$$\frac{x+248}{4} = 85$$

$$\cancel{4}\left(\frac{x+248}{\cancel{4}}\right) = 4(85)$$

$$x+248 = 340$$

$$x = 340 - 248 = 92$$

Weighted Averages

When finding an average, the scores may be weighted differently. Suppose Sally received test grades of 76, 88, and 92 on the final exam. If the final exam has a weighting factor of 2, then Sally's test average is

$$\frac{76+88+2(92)}{4} = \frac{348}{4} = 87$$

Example 7

In a Math A class, 12 students who attended after-school tutoring had an average midterm grade of 92. The remaining 18 students in the class had an average midterm grade of 76. What was the average midterm grade for the entire class?

Solution: **82.4**

Find the average of 92 and 76 by weighting 92 twelve times and weighting 76 eighteen times:

$$\text{Average} = \frac{(92 \times 12)+(76 \times 18)}{12+18}$$

$$= \frac{1104 \quad +1368}{30}$$

$$= 82.4$$

Check Your Understanding of Section 2.2

A. Multiple Choice

1. Rick's recorded times in four 1-mile runs are 4.8 minutes, 5.3 minutes, 4.7 minutes, and 5.4 minutes. For Rick's next run, which time will give him a mean of 5.0 minutes?
   (1) 4.8 min   (2) 5.3 min   (3) 5.7 min   (4) 6.0 min

2. The average grade on the first two algebra tests that Melita took was 88. On the third test she received a grade of 94. What was her average for the three tests?
   (1) 90   (2) 91   (3) 88   (4) 92

B. In each case, show how you arrived at your answer by clearly indicating all of the necessary steps, formula substitutions, diagrams, graphs, charts, etc.

3–22. In each case, solve for the variable and check.

3. $0.4y + 5 = 1$

4. $3x - 1 = -16$

5. $-2x + 7 = -13$

6. $32 = 3w + 5$

7. $\dfrac{x}{5} + 8 = 10$

8. $0.2x + 0.3 = 8.1$

9. $3(2x - 1) = 7$

10. $-17 = 2(5 + m)$

11. $19 - (1 - x) = 18$

12. $7 - 3(x - 1) = -17$

13. $-3(9 - 7p) = -12$

14. $\dfrac{3t}{2} + 5 = 17$

15. $0 = 18 - 2(h + 1)$

16. $1.04x + 8 = 60$

17. $2(3x - 5) + 12 = 0$

18. $-(8x - 3) = 19$

19. $0.25(3x - 5) = 2.5$

20. $6\left(\dfrac{x}{2} + 1\right) = -9$

21. $13 - \dfrac{3x}{4} = -8$

22. $2\left(7 - \dfrac{y}{4}\right) = 3$

23. Tamika's mathematics teacher computes each student's test average by weighting the final exam three times as much as a class test. If Tamika received grades of 87, 86, 88, 85, and 89 on her five class tests, what grade must Tamika receive on the final exam so that her test average will be 90?

24. When 2 is subtracted from one-third of a number, the result is 3. What is the number?

25. If 37 exceeds three times the sum of a number and 5 by 1, what is the number?

26. If 45 is 9 greater than the difference obtained by subtracting 7 from a number, what is the number?

27. If $3x + 7 = -17$ and $0.25y + 2.1 = 1.6$, what is the product of x and y?

28. If 11 and m have the same average as 5, 8, 17, and 32, what is the value of m?

29. Carol's average driving speed for a 4-hour trip was 48 miles per hour. During the first 3 hours her average rate was 50 miles per hour. What was her average rate for the last hour of the trip?

2.3 SOLVING EQUATIONS WITH LIKE TERMS

Terms such as $4x$ and $5x$ are called **like terms** because they differ only in their numerical coefficients. Sometimes it is necessary to simplify an equation by combining like terms before performing the inverse arithmetic operations needed to isolate the variable. To combine like terms, combine their numerical coefficients. For example:

- If $5x + 3x = 24$, then $8x = 24$, so $x = \dfrac{24}{8} = 3$.

- If $7x - x = -12$, then $6x = -12$, so $x = \dfrac{-12}{6} = -2$.

Variable Terms on the Same Side of the Equation

If the same variable appears in different terms of an equation, simplify the equation and combine like terms.

Example 1

Solve for y: $3(2y + 5) - 8y = 1$.

Solution: **7**

Simplify the equation by removing the parentheses. Then collect and combine like terms.

$$3(2y + 5) - 8y = 1$$
$$3(2y) + 3(5) - 8y = 1$$
$$6y + 15 - 8y = 1$$
$$(6y - 8y) + 15 = 1$$
$$-2y + 15 = 1$$
$$-2y + 15 - (15) = 1 - (15)$$
$$-2y = -14$$
$$\frac{-2y}{-2} = \frac{-14}{-2}$$
$$y = 7$$

Expressing Quantities Using the Same Variable

When a word problem involves two or more different quantities, set the variable equal to the quantity to which the others are being compared.

Example 2

Mike calculated that, of his day's intake of 2156 calories, four times as many calories were from carbohydrates as from protein, and twice as many calories were from fat as from protein. How many calories were from carbohydrates?

> *Solution*: **1232**
>
> Let x = number of calories from protein.
> Then $4x$ = number of calories from carbohydrates,
> and $2x$ = number of calories from fat.
>
> Since the sum of the calories is 2156:
>
> $$x + 4x + 2x = 2156$$
> $$7x = 2156$$
> $$x = \frac{2156}{7} = 308$$
>
> Number of calories from carbohydrates = $4x = 4(308) = 1232$

When a word problem involves the sum of two quantities, set the variable equal to either quantity. Then represent the remaining quantity as the difference between the sum and that variable. For example, if the sum of two numbers is 20 and x represents one of these numbers, then $20 - x$ represents the other number.

Example 3

A soda machine contains 20 coins. Some of the coins are nickels, and the rest are quarters. If the value of the coins is $4.40, find the number of coins of each kind.

> *Solution*: **3 nickels, 17 quarters**
>
> - Represent the number of nickels and the number of quarters by the same variable.
>
>   Let x = number of nickels.
>   Then $20 - x$ = number of quarters.
>
> - Translate the conditions of the problem into an equation:
>
> $$\underbrace{\text{Value of nickels}}_{0.05x} + \underbrace{\text{Value of quarters}}_{0.25(20 - x)} = \underset{= \ 4.40}{\$4.40}$$

- Solve the equation, and check. Clear the equation of decimals by multiplying each member of both sides of the equation by 100.

$$100(0.05x) + 100[0.25(20 - x)] = 100(4.40)$$
$$5x + 25(20 - x) = 440$$
$$5x + 500 - 25x = 440$$
$$-20x + 500 = 440$$
$$-20x = 440 - 500$$
$$x = \frac{-60}{-20}$$
$$= 3 \text{ nickels}$$
$$\text{and } 20 - x = 20 - 3$$
$$= 17 \text{ quarters}$$

Algebraic Modeling Versus Intelligent Guessing

The solution to Example 3 uses algebraic language to represent or *model* a problem situation. Sometimes an alternative problem-solving strategy such as intelligent *guessing and checking* can also be used to arrive at a correct answer. You could solve the problem in Example 3 by reasoning as follows:

- If all 20 coins were nickels, the value of the coins in the soda machine would be 20 × $0.05 = $1.00.
- If all 20 coins were quarters, the value of the coins in the soda machine would be 20 × $0.25 = $5.00.
- Since $5.00 is closer than $1.00 to the desired amount, $4.40, guess that the soda machine contains 19 quarters and 1 nickel. Then check the guess. If that guess doesn't work, reduce the number of quarters by 1 and check. Continue this process until you reach the correct answer. To help keep track of your guesses, organize your work in a table.

Guess		Total Value of Coins
Quarters	Nickels	
19	20 − 19 = 1	(19 × $0.25) + (1 × $0.05) = $4.80 ← Too high
18	20 − 18 = 2	(18 × $0.25) + (2 × $0.05) = $4.60 ← Too high
17	20 − 17 = 3	(17 × $0.25) + (3 × $0.05) = $4.40 ← Correct total!

Variable Terms on Different Sides of the Equation

If variable terms appear on opposite sides of an equation, collect like variable terms on the same side of the equation and collect number terms on the other side.

Example 4

Solve for x: $2x - 9 = 5x + 6$.

   Solution: **−5**

Collect terms with x on the left side of the equation, and collect numbers on the right side.

- Collect number terms by adding 9 to each side of the given equation:

$$\begin{array}{rcl} 2x - 9 &=& 5x + 6 \\ +9 = && +9 \\ \hline 2x + 0 &=& 5x + 15 \end{array}$$

- Collect variable terms by subtracting $5x$ from each side of the equation obtained in step 1:

$$\begin{array}{rcl} 2x &=& 5x + 15 \\ -5x &=& -5x \\ \hline -3x &=& 0 \ +15 \end{array}$$

- Solve for x by dividing each side of the equation obtained in step 2 by −3:

$$\frac{-3x}{-3} = \frac{15}{-3}$$
$$x = -5$$

Example 5

In 7 years Maria will be twice as old as she was 3 years ago. What is Maria's present age?

   Solution: **13**

Use an algebraic model.

- Represent Maria's past, present, and future ages, using the same variable.

Let x = Maria's present age.
Then $x + 7$ = Maria's age 7 years from now,
and $x - 3$ = Maria's age 3 years ago.

36

- Translate the conditions of the problem into an equation:

$$\underbrace{\text{Age in 7 years}}_{x+7} \underbrace{\text{will be}}_{=} \underbrace{\text{two times}}_{2\times} \underbrace{\text{age 3 years ago}}_{(x-3)}$$

- Solve the equation.

$$x+7 = 2x-6$$
$$x+7-(7) = 2x-6-(7)$$
$$x = 2x-13$$
$$x-2x = 2x-(2x)-13$$
$$-x = -13$$
$$\frac{-x}{-1} = \frac{-13}{-1}$$
$$x = 13$$

- Check the solution in the statement of the problem. If Maria is 13 years old now, 3 years ago she was 10 years old and in 7 years she will be 20 years old. Hence, it is true that in 7 years Maria will be twice as old as she was 3 years ago.

Consecutive Integer Problems

When a list of consecutive integers is arranged in increasing order, each integer after the first is 1 more than the integer that comes before it, as in $-2, -1,$ $0, 1, 2, 3,$ and 4. When a list of consecutive *even* integers is arranged in increasing order, each integer after the first is 2 more than the integer that comes before it, as in $-2, 0, 2, 4, 6,$ and 8. Similarly, if a list of consecutive *odd* integers is arranged in increasing order, each integer after the first is 2 more than the integer that comes before it, as in $-3, -1, 3, 5,$ and 7.

In general:

- If n is an integer, a set of consecutive integers that begins with n is

$$n+1, n+2, n+3, \ldots$$

- If n is an even integer, a set of consecutive even integers that begins with n is

$$n, n+2, n+4, \ldots$$

- If n is an odd integer, a set of consecutive odd integers that begins with n is

$$n, n+2, n+4, \ldots$$

Example 6

Find three consecutive odd integers such that twice the sum of the second and the third is 43 more than three times the first.

 Solution: **31, 33, 35**

$$\text{Let } x = \text{first odd integer.}$$
$$\text{Then } x + 2 = \text{second consecutive odd integer,}$$
$$\text{and } x + 4 = \text{third consecutive odd integer.}$$

Twice the sum of the second and third is 43 more than 3 times the first integer.

$$2[(x+2)+(x+4)] \qquad = 43 \qquad + 3x$$
$$2(2x+6) = 43+3x$$
$$4x+12 = 43+3x$$
$$4x+12-3x = 43+3x-3x$$
$$x+12 = 43$$
$$x+12-12 = 43-12$$
$$x = 31$$
$$\text{Then } x+2 = 33$$
$$x+4 = 35$$

Check Your Understanding of Section 2.3

A. Multiple Choice

1. If $15x = 3(x + 7)$, then x equals

   (1) $\dfrac{4}{7}$ (2) $\dfrac{7}{12}$ (3) 1.75   (4) 2.75

2. If $4(2x + 1) = 22 + 3(2x - 5)$, then x equals

   (1) $\dfrac{2}{3}$ (2) $\dfrac{3}{2}$ (3) $-\dfrac{1}{2}$ (4) 8

3. Which equation can be used to solve the problem below?

   If four times a number is increased by 15, the result is 3 less than six times the number. What is the number?

   (1) $4(x + 15) = 6x - 3$ (3) $4x + 15 = 6(x - 3)$
   (2) $4x + 15 = 6x - 3$ (4) $4(x + 15) = 3 - 6x$

B. In each case, show how you arrived at your answer by clearly indicating all of the necessary steps, formula substitutions, diagrams, graphs, charts, etc.

4–15. *Solve for the variable and check.*

4. $3x + 4x = -28$

5. $2x - 5x = -27$

6. $0.8x + 0.1x = -3.6$

7. $7t = t - 42$

8. $0.54 - 0.07y = 0.2y$

9. $3s + 2(8 - s) = 3$

10. $9b = 2b - 3(8 - b)$

11. $z + 1.5z = 35 - z$

12. $3(5 - 2n) = 2n - 9$

13. $-(8 - 3x) = 7x + 12$

14. $5(6 - q) = -3(q + 2)$

15. $3(4b - 7) = 2(3b + 11) + 5$

16. A postal clerk sold 50 postage stamps for $15.84. Some were 37-cent stamps, and the rest were 23-cent stamps. Find the number of 37-cent stamps that were sold.

17. Samantha, Lauren, and Jerry own shares of the same stock with a total value of $750. If Samantha owns 5 shares, Lauren owns 3 shares, and Jerry owns 2 shares, what is the value, in dollars, of Lauren's stock?

18. If one-half of a number is 8 less than two-thirds of the number, what is the number?

19. How old is Douglas if his age 6 years from now will be twice his age 7 years ago?

20. Tickets for a concert cost $4.00 for a balcony seat and $7.50 for an orchestra seat. If ticket sales totaled $1585 for 300 tickets sold, how many more tickets for balcony seats were sold than tickets for orchestra seats?

21. Find the largest of four consecutive integers whose sum is 15 less than five times the first.

22. A geologist collected 13 rocks that have exactly the same weight. If 9 of these rocks with an additional 5-ounce weight at one end of a balance-scale can balance the remaining rocks and a 23-ounce weight at the other end of the scale, what is the number of ounces in the weight of one of these rocks?

23. Allan has nickels, dimes, and quarters in his pocket. The number of nickels is 1 more than twice the number of quarters. The number of dimes is 1 less than the number of quarters. If the value of the change in his pocket is 85 cents, how many of each coin does Allan have?

24. Tickets to a concert that were purchased in advance cost $4.50 each, and tickets purchased at the box office on the day of the concert cost $8.00 each. The total amount of money collected in ticket sales was the same as if every ticket purchased had cost $6.00. If 180 tickets were purchased in advance, what was the total number of tickets purchased for the concert?

25. Carlos earns $9 per hour on weekdays and twice as much per hour on weekends. Last week he earned a total of $378, including the weekend. Of the 35 hours Carlos worked for that week, how many hours did he work during the weekend?

26. On Saturday, Jennifer's Bakery sold half as many apple pies as cheesecakes. The price of an apple pie is $6, and the price of a cheesecake is $8.50. If the total amount of the sales for these pastries was $391, what was the total number of each kind that was sold?

2.4 SOLVING PERCENT PROBLEMS

KEY IDEAS

Problems involving percent can be solved by writing algebraic equations.

Types of Percent Problems

There are three basic types of percent problems that you should know how to solve.

- **Type I.**   Finding a percent of a given number.

    Example: What is 15% of 80?

    $$n = 0.15 \times 80$$
    $$= 12$$

- **Type II.**   Finding a number when a percent is given.

    Example: 30% of what number is 12?

    $$0.30 \times n = 12$$
    $$0.30n = 12$$
    $$10(0.3n) = 10(12)$$
    $$3n = 120$$
    $$n = \frac{120}{3}$$
    $$= 40$$

- **Type III.**   Finding what percent one number is of another.

    Example: What percent of 30 is 9?

    $$\frac{p}{100} \times 30 = 9$$

$$\frac{p}{100} \times 30 = 9$$

$$\frac{\overset{3}{\cancel{30}}p}{\underset{10}{\cancel{100}}} = 9$$

$$p = \frac{10 \times 9}{3}$$
$$= \mathbf{30}$$

Example 1

If the rate of sales tax is 7.5%, what is the amount of sales tax on a purchase of $40.00?

Solution: **$3.00**

$$\begin{aligned} \text{Amount of tax} &= 7.5\% \text{ of } \$40 \\ &= 0.075 \times 40 \\ &= 3 \end{aligned}$$

Example 2

A pair of sneakers that regularly sells for $35 is on sale for $28. What is the percent of the discount?

Solution: **20**

The dollar amount of the discount is $35 − $28 −$7. Think: "What percent of 35 is 7?"

$$\frac{p}{100} \times 35 = 7$$

$$\frac{35p}{100} = 7$$

$$p = \frac{(100)\overset{1}{\cancel{7}}}{\underset{5}{\cancel{35}}} = 20$$

Example 3

The sales tax on the purchase of a new computer system is $68.25. If the sales tax rate is 7%, what is the cost of the computer system?

Solution: **$975**

Let c = cost of the computer system. Then:

$$\underbrace{7\%}_{0.07}\ \underbrace{\text{of}}_{\times}\ \underbrace{\text{cost of computer}}_{c}\ \underbrace{\text{is \$68.25}}_{= \$68.25}$$

$$0.07c = \$68.25$$

$$c = \frac{\$68.25}{0.07} = \$975$$

Example 4

Craig weighs exactly 148.5 pounds. When Craig weighs himself on a defective scale, his weight is shown as 155.0 pounds.

(a) What is the percent of error in measurement of the defective scale to the *nearest tenth*?
(b) If Robin weighs 110 pounds on the same scale, what is Robin's actual weight, correct to the *nearest tenth* of a pound, assuming the same percent of error in measurement?

Solution: (a) **4.4**

$$\text{Percent of error in measurement} = \frac{\text{amount of error}}{\text{original weight}} \times 100\%$$

$$= \frac{155.0 - 148.5}{148.5} \times 100\%$$

$$= \frac{4.5}{148.5} \times 100\%$$

$$\approx 4.4\%$$

(b) **105.4**

If x represents Robin's actual weight, then:

$$x + 0.044x = 110$$

$$1.044x = 100$$

$$x = \frac{110}{1.044} \approx 105.4\,\text{lb}$$

Circle Graphs

Fractions and percents compare the parts to the whole. A circle graph or pie chart shows visually how the parts that make up a whole are related to the whole and to each other.

The sectors ("slices") of a circle graph always add up to 100% of the whole. The circle graph in Figure 2.1 shows the enrollment by grade level at Central High School. If 1200 students attend Central High School, how many freshmen attend the school?

- Since all of the sectors of the circle graph must add up to 100%:

$$x\% + 25\% + 20\% + 20\% = 100\%$$
$$x\% = 100\% - 65\% = 35\%$$

- If 1200 students attend Central High School, the number of freshmen is

$$35\% \text{ of } 1200 = 0.35 \times 1200 = 420$$

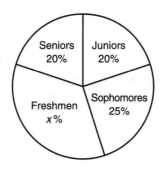

Figure 2.1   Enrollment at Central High School

The central angles of the sectors of a circle graph always add up to 360°. The circle graph in Figure 2.2 represents the results of a poll in which 210 seniors were asked whether they would be attending a college in New York State or in a different state. How many students were undecided?

- Since $360° - 140° - 100° = 120°$, the sector for "Undecided" represents $\dfrac{120°}{360°} = \dfrac{1}{3}$ of the whole.

- Since 210 seniors were polled, $\dfrac{1}{3} \times 210 = 70$ seniors were undecided.

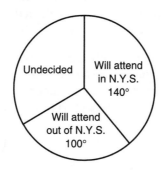

Figure 2.2   Poll of Seniors at Central High

Check Your Understanding of Section 2.4

A. Multiple Choice

1. John, Gary, and Melissa play a computer game for a total of 3 hours. If John plays the game for 28% of the total time, and Gary plays the game for 52% of the total time, how many *minutes* does Melissa play the game?
   (1) 30   (2) 36   (3) 42   (4) 48

2. Marcia paid $36 for a dress that was on sale for 25% of the original price. What was the original price of the dress?
   (1) $48   (2) $60   (3) $108   (4) $144

3. Carla paid $24 for slacks that were on sale. If the original price of the slacks was $32, what was the percent of discount?
   (1) 20   (2) 25   (3) 75   (4) 80

4. Vincent paid $48 for a computer game that was on sale for 60% off the original price. What was the original price of the computer game?
   (1) $76.80   (2) $80   (3) $120   (4) $128

5. After a 20% increase, the new price of a share of a computer stock is $78.00. What was the original price of a share of this stock?
   (1) $60   (2) $62.40   (3) $65   (4) $97.50

6. One year ago the average price of a home computer system was $1500. Today, the average price is $1200. By what percent has the average cost of a home computer system decreased?
   (1) 20%   (2) 25%   (3) 75%   (4) 80%

7. In a factory that manufactures light bulbs, 0.04% of the bulbs manufactured are defective. What is the minimum number of bulbs manufactured in which you would expect to find one defective bulb?
   (1) 250   (2) 1250   (3) 2500   (4) 4000

8. The accompanying circle graph shows how a total of $4800 has been invested in four different stocks. In which of the four stocks is the investment closest in value to $1200?
   (1) Stock A
   (2) Stock B
   (3) Stock C
   (4) Stock D

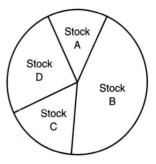

44

9. If 25% of a number is $6x$, the number is
(1) $2.4x$ (2) $24x$ (3) $1.5x$ (4) $15x$

B. *In each case, show how you arrived at your answer by clearly indicating all of the necessary steps, formula substitutions, diagrams, graphs, charts, etc.*

10. In a recent election poll, 1350 people were asked whether they would vote yes or no for a change in the local tax law. The results are summarized in the accompanying circle graph. How many people were undecided?

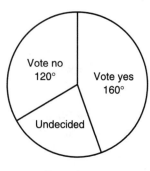

      Exercise 10   **Exercise 11**

11. The accompanying circle graph summarizes the results of a survey of 1400 students who named their favorite sports activities. If each student in the survey named exactly one activity, what is the total number of students who named either swimming or biking as their favorite sports activity?

12. Amy bought a $72 dress that was on sale for 25% off. If the sales tax rate is 7%, what did Amy pay for the dress?

13. A basketball player made 9 foul shots, which is 75% of the number he attempted. The coach tells this player to keep making foul shots until the total number made is 90% of the total number attempted. What is the minimum number of additional foul shots the basketball player needs to make?

14. In a school chorus of 35 boys and 28 girls, 80% of the boys and 25% of the girls have been members for more than 2 years. What fraction of the school chorus have been members for more than 2 years?

15. During a week when the 7% charge for sales tax on clothes was waived, Mary saved $13.65 in sales tax on the purchase of a new coat. What was the cost of the new coat?

16. Brittany bought a belt for 50% off the original price. The store charges 8% sales tax, and her final cost was $24.30. What was the original price of the belt?

17. A class consists of 14 boys and 19 girls. On a certain day all the boys are present but some girls are absent. If the girls represent only 30% of the class attendance on that day, how many girls are absent?

18. In an opinion poll of 50 men and 40 women, 70% of the men and 25% of the women said that they preferred fiction to nonfiction books. What percent of the number of people polled preferred to read fiction?

19. Mr. Perez owns a sneaker store. He bought 350 pairs of basketball sneakers and 150 pairs of soccer sneakers from the manufacturers for $62,500. He sold all the sneakers and made a 25% profit. If he sold the soccer sneakers for $130 per pair, how much did he charge for one pair of basketball sneakers?

20. A watch that sells regularly for $125 is on sale for 20% off. The sales tax on the watch is 7%. Matt and Tim decide to buy the same watch. When Matt pays for the watch, the salesperson deducts the discount *before* adding the sales tax. When Tim pays for his watch, the salesperson takes the discount *after* first adding the sales tax to the price of the watch. Was the total purchase price the same for Matt and Tim? Explain how you arrived at your answer.

21. A scarf is on sale at 20% off the list price. The following week the price of the scarf is reduced 25% of the sale price. Carol called a friend and told her that the price of the scarf is now 45% off the original price. Do you agree with Carol? If not, what is the actual percent of the discount off the original price?

2.5 WORKING WITH FORMULAS

KEY IDEAS

A **formula** is an equation that explains how one quantity depends on one or more other quantities. For example, the perimeter, P, of a rectangle depends on the length, L, and the width, W, of the rectangle. The formula $P = 2L + 2W$ shows exactly how P depends on L on W.

Solving for One Letter in Terms of Other Letters

When an equation has more than one variable, it is sometimes helpful to solve that equation for a particular variable in order to see how that variable depends on the other members of the equation. This variable is solved for *in terms of* the other variables of the equation by treating these other letters as constants (numbers) and isolating the variable of interest in the usual way.

Example 1

The formula to convert from F degrees Fahrenheit to C degrees Celsius is $C = \frac{5}{9}(F - 32)$.

(a) Solve the formula for F in terms of C.
(b) How many degrees Fahrenheit is equivalent to 20 degrees Celsius?

Solution: (a) $\mathbf{F = \dfrac{9}{5}C + 32}$

- Write the given formula:

$$C = \frac{5}{9}(F - 32)$$

- Multiply both sides of the equation by $\frac{9}{5}$ (the reciprocal of $\frac{5}{9}$):

$$\frac{9}{5}C = \frac{9}{5} \times \frac{5}{9}(F - 32)$$

- Simplify:

$$\frac{9}{5}C = F - 32$$

- Add 32 to each side of the equation:

$$\frac{9}{5}C + 32 = F - 32 + 32$$

- Write $\frac{9}{5}$ as 1.8:

$$1.8C + 32 = F$$

or

$$F = 1.8C + 32$$

(b) **68**

To find the number of degrees Fahrenheit that is equivalent to 20 degrees Celsius, evaluate the formula $F = \frac{9}{5}C + 32$ by substituting 20 for C and 1.8 for $\frac{9}{5}$:

$$F = 1.8 \times 20 + 32$$
$$= \quad 36 \quad + 32$$
$$= \quad 68$$

Example 2

Solve for y: $ay + 2x = b + x$ ($a \neq 0$).

Solution: $y = \dfrac{b-x}{a}$

- Write the given equation:
- Subtract $2x$ from each side:

$$ay + 2x = b + x$$
$$ay = b + x - 2x$$
$$ay = b - x$$

- Divide each side of the equation by a:

$$y = \dfrac{b-x}{a}$$

Perimeter and Area Formulas

Perimeter is the distance around a figure. **Area** represents the number of 1-by-1-unit squares that a figure can enclose.

You should remember the following facts related to squares and rectangles:

- Opposite sides of a rectangle have the same length. The area A of a rectangle is Length (base) × Width (height). Thus, $A = L \times W$. The perimeter P of a rectangle is the sum of the lengths of its four sides. Hence, $P = 2L + 2W$.

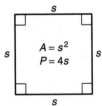

- A square is a rectangle in which all four sides have the same length. The area A of a square is side × side. Thus, $A = s \times s = s^2$. The perimeter P of a square is the sum of the lengths of its four sides, so $P = 4s$.

Example 3

Carpet costs $22 per square yard. Assuming no waste, what is the cost of carpeting a rectangular room that is 12 feet wide and 18 feet long?

Solution: **$528**

- The area of the room is $12\,\text{ft} \times 18\,\text{ft} = 216\,\text{ft}^2$.
- Since $1\,\text{yd} = 3\,\text{ft}$, $1\,\text{yd}^2 = 3 \times 3 = 9\,\text{ft}^2$.
- An area of $216\,\text{ft}^2$ is equivalent to $\dfrac{216}{9} = 24\,\text{yd}^2$.
- The cost of $24\,\text{yd}^2 = 24\,\text{yd}^2 \times \dfrac{\$22}{\text{yd}^2} = \$528$.

Formulas for the Volume of a Rectangular Solid

Volume is a measure of capacity and corresponds to the number of 1-by-1-by-1-unit cubes a solid figure can hold, as illustrated below.

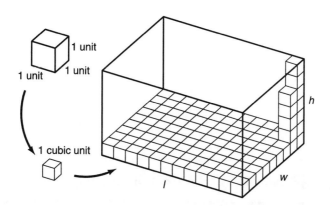

The volume, V, of a rectangular solid is the product of its length (l), width (w), and height (h), so $V = l \times w \times h$. When the six sides (faces) of a rectangular solid are equal squares, the solid is called a **cube**. If the length of an edge of a cube is e, then its volume is $V = e \times e \times e = e^3$.

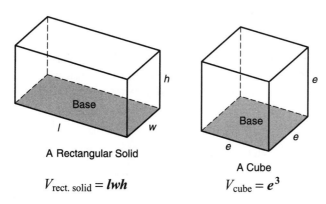

A Rectangular Solid	A Cube
$V_{\text{rect. solid}} = \boldsymbol{lwh}$	$V_{\text{cube}} = \boldsymbol{e^3}$

Example 4

What is the edge length of a cube that has the same volume as a rectangular box with dimensions 32 inches × 2 inches × 1 inch?

Solution: **4 in**

The volume of the rectangular box is $32\,\text{in} \times 2\,\text{in} \times 1\,\text{in} = 64\,\text{in}^3$. It is given that the box and cube have the same volume, so $V_{\text{cube}} = e^3 = 64$.
Since $4^3 = 4 \times 4 \times 4 = 64$, $e = 4\,\text{in}$.

Example 5

The dimensions of a brick, in inches, are 4 by 6 by 8. How many such bricks are needed to have a total volume of exactly 1 cubic foot?

Solution: **9**

- The volume of the brick is $4 \text{ in} \times 6 \text{ in} \times 8 \text{ in} = 192 \text{ in}^3$.
- A volume of 1 cubic foot is equivalent to a volume of $12 \text{ in} \times 12 \text{ in} \times 12 \text{ in} = 1728 \text{ in}^3$.
- Since $\dfrac{1728 \text{ in}^3}{192 \text{ in}^3} = 9$, 9 bricks are needed to have a total volume of exactly 1 ft^3.

Check Your Understanding of Section 2.5

A. *Multiple Choice*

1. The formula to convert from F degrees Fahrenheit to C degrees Celsius is $C = \dfrac{5}{9}(F - 32)$. What Celsius temperature is equivalent to 50 degrees Fahrenheit?
   (1) 5 (2) 10 (3) 18 (4) 27

2. Assume body weight and height are related by the formula $W = 2H + 13$, where W represents weight in pounds and H represents height in inches. What is the weight, in pounds, of a person whose height is 5 feet 6 inches?
   (1) 128 (2) 135 (3) 145 (4) 148

3. If $a(x + b) = c$, what is x in terms of a, b, and c?
   (1) $\dfrac{c - b}{a}$ (2) $\dfrac{c - ab}{a}$ (3) $\dfrac{b + c}{a}$ (4) $\dfrac{ac - b}{a}$

4. If the width of a rectangle is represented by w and the perimeter of the rectangle is represented by k, then the length of the rectangle can be represented by
   (1) $k - 2w$ (2) $k - \dfrac{w}{2}$ (3) $\dfrac{k - 2w}{2}$ (4) $2k - w$

5. If $4x + y = H$, what is x in terms of y and H?
   (1) $\dfrac{H}{4} - y$ (2) $\dfrac{H}{4} + y$ (3) $\dfrac{H + y}{4}$ (4) $\dfrac{H - y}{4}$

6. In terms of c, y, and a, what is the value of x in the equation $2ax + 2y = c$?

   (1) $\dfrac{c-y}{a}$ (2) $\dfrac{c-2y}{2a}$ (3) $c-2y-2a$ (4) $\dfrac{c+2y}{2a}$

7. A cube whose edge has a length of 6 has the same volume as a rectangular box whose length is 12 and whose width is 9. The height of the rectangular box is
   (1) 6   (2) 2   (3) 3   (4) 4

8. The length of a rectangular box is doubled, and its length is tripled. In order for the volume of the box to remain the same, the original height of the box must be
   (1) multiplied by 6   (3) divided by 5
   (2) divided by 6   (4) decreased by 6

B. In each case, show how you arrived at your answer by clearly indicating all of the necessary steps, formula substitutions, diagrams, graphs, charts, etc.

9. The width of a rectangle is 3 less than its length. If the length is multiplied by 2 and the width is increased by 4, the perimeter of the new rectangle is 50. Find the area of the new rectangle.

10. When Raymond was on a diet, his weight, w, in pounds after x days on the diet could be approximated by the formula $w = 240 - 0.6x$.
    (a) After how many days on the diet did Raymond's weight drop to 210 pounds?
    (b) At the end of 30 days on the diet, Raymond's weight was what percent of his original weight?

11. What is the least number of squares each of side length 2 inches needed to cover completely, without overlap, a larger square with a side length of 8 inches?

12. The length and width of a rectangle are consecutive odd integers. If the perimeter of the rectangle is 48, find the dimensions of the rectangle.

13. The lengths of the sides of a triangle are consecutive even integers. The perimeter of the triangle is the same as the perimeter of a square whose side is 5 less than the shortest side of the triangle. Find the length of the longest side of the triangle.

14. From each of the four corners of a rectangular piece of cardboard 8 inches wide and 12 inches long, a 2-inch square is cut off, as shown in the accompanying diagram.
    (a) What is the area of the new figure?
    (b) If the four sides of the new figure are folded straight up to form a box that is open at the top, what is the volume of this box?

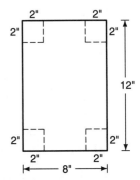

15. Brand X paint costs $14 per gallon. One gallon of brand X paint provides coverage of an area of at most 150 square feet. What is the minimum cost of the amount of brand X paint needed to cover the four walls of a rectangular room that is 12 feet wide, 16 feet long, and 8 feet high?

16. What is the least number of square inches of wrapping paper needed to cover a cube whose volume is 27 cubic inches?

17. The width of a rectangular box is $\frac{4}{9}$, and its height is $\frac{3}{8}$. If the volume of this box is 2, what is the length of the box?

18. A rectangular solid has the same volume as a cube whose edge length is 6 inches. What is the height of the rectangular solid if its length is 9 inches and its width is 5 inches?

19. A 7-inch by 5-inch rectangular picture is surrounded by a rectangular mat having a uniform width of 2 inches. If the matting costs $2.25 per square inch, what is the total cost of the matting?

20. A square is cut from a rectangular board to make the figure shown in the accompanying diagram. The rectangular board measured 9 inches by 13 inches before the cut, and the square cutout measures 3 inches on each side.
    (a) What is the perimeter of the new figure?
    (b) What is the area of the new figure?

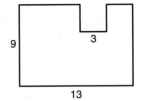

21. A cube-shaped container with an edge length of 16 inches is to be filled with water, using a rectangular cup with dimensions of 3 by 6 by 10 inches. What is the maximum number of full cups of water that can be poured into the container without the water overflowing the container?

22. In the accompanying diagram, the dimensions of the rectangular box are integers greater than 1. If the area of face *ABCD* is 12 and the area of face *CDEF* is 21, what is the volume of the box? Explain how you arrived at your answer.

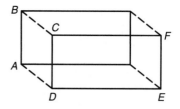

2.6 RATIO AND PROPORTION

$\equiv$ **KEY IDEAS** $\equiv$

If *x* and *y* are two quantities that have the same unit of measurement, then a comparison of *x* to *y* using division is called the **ratio** of *x* to *y* provided $y \neq 0$. The ratio of *x* and *y* can be written in any one of three ways:

$$\frac{x}{y}, \quad x:y, \quad \text{or} \quad x \text{ to } y$$

An equation that states that two ratios are equal is called a **proportion**.

Ratio Versus Rate

A *ratio* and a *rate* each compare two numbers.

- A **ratio** compares two numbers that have the *same* unit of measurement. For example, if John is 24 years old and Glen is 8 years old, then the ratio of John's age to Glen's age is 3 to 1 since

$$\frac{\text{John's age}}{\text{Glen's age}} = \frac{24 \text{ years}}{8 \text{ years}} = \frac{3}{1} \text{ or 3 to 1}$$

Notice that a ratio has no units of measurement.
- A **rate** compares two numbers that have *different* units of measurement. For example, if Jack earns $300 for working 16 hours, then his average rate of pay per hour is

$$\frac{\overset{18.75}{\cancel{300} \text{ dollars}}}{\underset{1}{\cancel{16} \text{ hours}}} = \frac{18.75 \text{ dollars}}{1 \text{ hour}} = 18.75 \text{ dollars per hour}$$

Rates are useful in making comparisons. If Jill earns \$243 for working 12 hours, then Jill earns more dollars per hour than Jack since her average rate of pay per hour is

$$\frac{\overset{20.25}{\cancel{243} \text{ dollars}}}{\underset{1}{\cancel{12} \text{ hours}}} = \frac{20.25 \text{ dollars}}{1 \text{ hour}} = 20.25 \text{ dollars per hour}$$

Solving Problems with Ratios

If two whole numbers are in the ratio of 2 to 1, the pair of numbers could be 2 and 1, 4 and 2, 6 and 3, and so forth. Since in each of these possibilities 2 and 1 are multiplied by the same nonzero number, say x, we can represent the set of all numbers whose ratio is 2 to 1 by $2x$ and x.

Example 1

The perimeter of a rectangle is 56 centimeters. If the length and width of the rectangle are in the ratio of 3 to 1, as shown in the accompanying diagram, what are the dimensions of the rectangle?

Solution: **width = 7 cm, length = 21 cm**

Let x = width of the rectangle. Then $3x$ = length of the rectangle.

Since the perimeter of a rectangle is the sum of the lengths of the four sides:

$$x + 3x + x + 3x = 56 \text{cm}$$
$$8x = 56 \text{cm}$$
$$x = \frac{56}{8} = 7 \text{cm}$$
$$\text{and } 3x = 21 \text{cm}$$

Example 2

Three business partners divide profits of \$174,789 in the ratio of 1 to 3 to 5. What is the number of dollars in the greatest share?

Solution: **97,105**

Let x, $3x$, and $5x$ represent the dollar amounts of the three shares. Then:

$$x + 3x + 5x = \$174{,}789$$
$$9x = \$174{,}789$$
$$x = \frac{\$174{,}789}{9} = \$19{,}421$$
$$\text{and } 5x = 5(\$19{,}421) = \$97{,}105$$

Example 3

An appliance store sells only two brands of refrigerators. The ratio of the number of brand X refrigerators sold at this store to the number of brand Y refrigerators sold is about 4 to 11. Of the next 100 refrigerators the store sells, approximately how many will be brand Y?

Solution: **Approximately 74**

The ratio of 4 to 11 means that for every 4 brand X refrigerators that are sold, 11 brand Y refrigerators are sold. Thus, if 4×2 or 8 brand X refrigerators are sold, then 11×2 or 22 brand Y refrigerators are sold. If 4×3 or 12 brand X refrigerators are sold, then 11×3 or 33 brand Y refrigerators are sold. In general, if $4 \times n$ brand X refrigerators are sold, $11 \times n$ brand Y refrigerators are sold. Hence:

$$4n + 11n = 100$$
$$15n = 100$$
$$n = \frac{100}{15}$$
$$\approx 6.7$$
$$\text{brand Y refrigerators} = 11n = 11(6.7) \approx 74$$

Solving Proportions

In a proportion, the cross-products are equal. For example:

$$\frac{2}{6} = \frac{4}{12}$$
$$6 \times 4 = 2 \times 12$$
$$24 = 24$$

| **MATH FACTS** |

EQUAL CROSS-PRODUCTS RULE

$$\text{If } \frac{a}{b} = \frac{c}{d}, \text{ then } b \times c = a \times d,$$

provided that denominators b and d are not 0. Terms b and c of the proportion are called the *means*, and terms a and d are the *extremes*. Thus, **in a proportion the product of the means is equal to the product of the extremes**.

Example 4

The rate of currency exchange between U.S. dollars and Canadian dollars is 1.4909 Canadian dollars per U.S. dollar. How many U.S. dollars are equivalent to 210 Canadian dollars?

Solution: **140.85**

Since the ratio of Canadian dollars to U.S. dollars is fixed at 1.4909 to 1:

$$\frac{\text{Canadian dollars}}{\text{U.S. dollars}} = \frac{1.4909}{1} = \frac{210}{x}$$
$$1.4909x = 210$$
$$x = \frac{210}{1.4909}$$
$$= 140.85$$

Example 5

In a certain Math A class, the ratio of the number of girls to the number of boys is $3:5$. If there is a total of 32 students in this class, how many are girls and how many are boys?

Solution: **12 girls, 20 boys**

Method 1	Method 2
Let x = number of girls.	Let $3x$ = number of girls.
Then $32 - x$ = number of boys.	Then $5x$ = number of boys.
$\dfrac{\text{Girls}}{\text{Boys}} = \dfrac{3}{5} = \dfrac{x}{32 - x}$	$3x + 5x = 32$
	$8x = 32$

Cross-multiply:

$$5x = 3(32 - x)$$
$$5x = 96 - 3x$$
$$5x + 3x = 96 - 3x + 3x$$
$$\frac{8x}{8} = \frac{96}{8}$$
$$x = 12 \text{ girls}$$
$$32 - x = 32 - 12 = 20 \text{ boys}$$

$$\frac{8x}{8} = \frac{32}{8}$$
$$x = 4$$

Therefore
$$3x = 3(4) = 12 \text{ girls}$$
and $5x = 5(4) = 20$ boys

Using Proportions to Solve Percent Problems

When comparing 4 to 8, we know that 4 is 50% of 8, which can be represented by the following proportion:

$$\frac{4 \text{ (number being compared to base)}}{8 \text{ (base)}} = \frac{50 \text{ (amount of percent)}}{100}$$

In comparing 4 to 8, 8 is the *base* and 4 is the number that is being compared to that base. In general, the statement "*a* is *p* percent of *b*" can be translated into the percent equation

$$\frac{a}{b} = \frac{p}{100}$$

where *a* is the number that is being compared to the base *b* and *p* is the percent that *a* is of *b*.

Example 6

On a 40-question multiple-choice test, how many questions would you need to answer correctly in order to get 80% of the questions correct?

Solution: **32**

Let *x* represent the number of questions that must be answered correctly. Since *x* is being compared to a base of 40:

$$\frac{x}{40} = \frac{80}{100}$$
$$100x = 3200$$
$$x = \frac{3200}{100} = 32$$

Check Your Understanding of Section 2.6

A. *Multiple Choice*

1. On a certain map, $\frac{3}{8}$ inch represents 120 miles. How many miles do $1\frac{3}{4}$ inches represent?
   (1) 400   (2) 480   (3) 520   (4) 560

2. The population of the bacteria culture doubles in number every 12 minutes. The ratio of the number of bacteria at the end of 1 hour to the number of bacteria at the beginning of that hour is
   (1) $60:1$ (2) $32:1$ (3) $16:1$ (4) $8:1$

3. If four pairs of socks cost $10.00, how many pairs of socks can be purchased for $22.50?
   (1) 7   (2) 8   (3) 9   (4) 11

4. In a set of five consecutive positive even integers, the ratio of the greatest integer to the least integer is 2 to 1. If these integers are arranged from lowest to highest, which is the middle integer in the list?
   (1) 10   (2) 12   (3) 14   (4) 16

5. If the ratio of p to q is 3 to 2, what is the ratio of $2p$ to q?
   (1) $1:3$ (2) $2:3$ (3) $3:1$ (4) $3:4$

B. *Show how you arrived at your answer by clearly indicating all of the necessary steps, formula substitutions, diagrams, graphs, charts, etc.*

6–11. Solve each proportion for the variable.

6. $\dfrac{2}{6} = \dfrac{8}{x}$

8. $\dfrac{2}{3} = \dfrac{2-x}{12}$

10. $\dfrac{4}{11} = \dfrac{x+6}{2x}$

7. $\dfrac{1}{x+1} = \dfrac{10}{5}$

9. $\dfrac{2x-5}{3} = \dfrac{9}{4}$

11. $\dfrac{10-x}{5} = \dfrac{7-x}{2}$

12. Three relatives will share an inheritance of $196,800 in the ratio of 3 to 7 to 2. How many dollars of the inheritance will each of the three relatives receive?

13. At a certain college the ratio of the number of students who are liberal arts majors to the number of students who are science majors is 8 to 3. If 480 students are science majors, how many students are liberal arts majors?

14. If four compact discs cost $27, at the same rate what is the cost of seven compact discs?

15. During an hour of prime-time television programming, the ratio of the number of minutes of television programs to the number of minutes of commercial interruptions is about 11 to 2. What is the approximate number of minutes of commercial interruptions in $1\frac{1}{2}$ hours of prime-time television programming?

16. In Jill's purse the ratio of the number of quarters to the number of dimes to the number of nickels is 3 to 4 to 2. If the value of these coins is $8.75, how many of each coin does Jill have?

17. A blueprint of a house is drawn to scale so that 1 inch on the blueprint represents 2.5 feet. If a bedroom on the blueprint is 4.8 inches in width and 7.2 inches in length, what is the number of square feet in the area of the actual bedroom?

18. Three numbers are in the ratio of $2:3:5$. If the smallest number is multiplied by 8, the result is 32 more than the sum of the second and third numbers. What is the smallest of the three numbers?

19. The denominator of a fraction is 4 less than twice the numerator. If 3 is added to both the numerator and the denominator, the new fraction is equal to $\frac{2}{3}$. What is the original fraction?

20. A team has won $\frac{3}{5}$ of the 25 games it has played. What is the minimum number of additional games the team must win in order to have won $\frac{3}{4}$ of the games it has played?

2.7 SOLVING INEQUALITIES

 KEY IDEAS

An inequality is solved in much the same way as an equation *except* that, when both sides of an inequality are multiplied or divided by the same *negative* number, you must reverse the sign of the inequality.

Graphing Inequalities

The set of values that satisfies an inequality of the form $x \geq a$ can be graphed by darkening a and all points to the right of a on the number line. All points to the left of a satisfy the inequality $x < a$. For example, the graph of $x \geq 2$ is

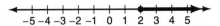

The darkened circle around 2 indicates that 2 is included in the set of values that satisfies $x \geq 2$.
The graph of $x < -1$ is

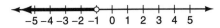

The open circle around -1 indicates that the set of values that satisfies $x < -1$ does *not* include -1.

Solving Inequalities Algebraically

When solving an inequality algebraically, isolate the variable in the usual way.

- An equivalent inequality results when the same quantity is added (or subtracted) on both sides of an inequality or when both sides of an inequality are multiplied (or divided) by the same *positive* quantity.
- If both sides of the inequality must be multiplied or divided by the same *negative* number, the direction of the inequality must be changed. For example, if $-2x < 6$, then

$$\frac{-2x}{-2} \boxed{>} \frac{6}{-2} \text{ so } x > -3$$

- Restrictions on the variable may limit the values that are contained in the solution set. For example, if x represents a positive integer, then the solution set of $x < 4$ is 1, 2, 3, which is represented graphically as

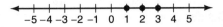

- If the replacement set is not indicated, you can assume that it is the entire set of real numbers. If $x < 2$ and the replacement set is not indicated, the solution set is represented as

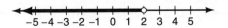

- In a word problem the phrase "is at least" is translated as ≥, and the phrase "is at most" becomes ≤. If the average grade x on an exam is *at least* 87 and *at most* 100, then $x \geq 87$ and $x \leq 100$.

Example 1

Find the solution set of $3x - 4 \leq 11$ if the replacement set (domain) of x is the set of positive odd integers.

> *Solution*: **{1, 3, 5}**

Since $3x - 4 \leq 11$, adding 4 to each side of the inequality gives $3x \leq 15$. Then, dividing each side of $3x \leq 15$ by $+3$ gives $x \leq 5$. Since x is limited to the set of positive odd integers, the solution set consists of the positive odd integers that are less than or equal to 5; that is, the solution set is 1, 3, and 5.

Example 2

Solve for x: $4 > 1 - x$.

> *Solution*: $x > -3$

- To isolate x in $4 > 1 - x$, add x on each side of the inequality:

$$4 > 1 - x$$
$$\underline{+x = +x}$$
$$x + 4 > 1$$

- Then subtract 4 from each side of the inequality:

$$x + 4 > \quad 1$$
$$\underline{-4 = -4}$$
$$x \quad > -3$$

Example 3

One positive integer is four times another, and their sum is *at most* 18. Find the largest possible values for the two numbers.

> *Solution*: **3, 12**

Method 1: Use an algebraic model.
Let x = the smaller of the two integers. Then $4x$ = the larger of the two integers.

$$x + 4x \leq 18$$
$$5x \leq 18$$
$$x \leq \frac{18}{5} \text{ or, equivalently, } x \leq 3\frac{3}{5}$$

Since x must be an integer, the greatest integer value of x is 3. Therefore, the greatest value of $4x$, the larger of the two integers, is 4 times 3 or 12.

Method 2: Guess, check, and revise.
Start by assuming the smaller of the positive integers is 1. Organize your guesses in a table.

Guess			
Smaller Number	Larger Number	Sum	Conclusion
1	$4 \times 1 = 4$	$1 + 4 = 5$	Make numbers larger.
2	$4 \times 2 = 8$	$2 + 8 = 10$	Make numbers larger.
3	$4 \times 3 = 12$	$3 + 12 = 15$	Make numbers larger.
4	$4 \times 4 = 16$	$4 + 16 = 20$	Sum is too high, so answer must be preceding set of numbers.

Example 4

Graph the solution set of $1 - 2x \le x + 13$.

Solution: See the accompanying graph.

The solution set is $x \ge -4$.

$$1 - 2x \le x + 13$$

- Subtract x from each side: $\quad 1 - 2x - x \le x + 13 - x$
- Combine like terms: $\quad 1 - 3x \le 13$
- Subtract 1 from each side: $\quad 1 - 3x - 1 \le 13 - 1$

- Divide each side by -3: $\quad \dfrac{-3x}{-3} \le \dfrac{12}{-3}$

- Reverse the inequality sign: $\quad x \ge -4$

The graph of $x \ge -4$ is

Check Your Understanding of Section 2.7

A. *Multiple Choice*

1. The inequality $5 \le x - 2$ is equivalent to
   (1) $x \le 7$ (2) $x \ge 7$ (3) $x \ge 3$ (4) $x \le -3$

2. Which element is in the solution set for the inequality $8 > 5x - 2$?

(1) 0 (2) 2 (3) 3 (4) 5

3. Which number is *not* a member of the solution set of $-4x \geq 17$?

(1) −4.2 (2) −4.3 (3) −4.4 (4) −4.5

4. Which is the least integer that makes the inequality $2x + 3 > 5$ true?

(1) 1 (2) 2 (3) 5 (4) −4

5. Which graph represents the solution set of the inequality $7 < 2x - 5$?

(1) 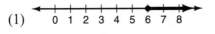 (3)

(2) 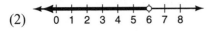 (4)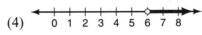

6. Which inequality is equivalent to $11 > \frac{2}{3}x - 5$?

(1) $x < 6$ (2) $x < 9$ (3) $x < 16$ (4) $x < 24$

7. What is the greatest integer value of p that satisfies the inequality $4 + 3p < p + 1$?

(1) −2 (2) −1 (3) 0 (4) 4

8. If the sum of a number and the original number increased by 5 is at least 12, which could be the original number?

(1) −5 (2) −1 (3) 3 (4) 4

9. If $a + c < b + d$ and $b + d < 2a$, which inequality *must* be true?

(1) $a < b$ (2) $a > b$ (3) $a < c$ (4) $a > c$

10. If $ab < 0$ and $c > 0$, which statement *must* be true?

(1) $abc > 0$ (2) $c - ab > 0$ (3) $\frac{ab}{c} > 0$ (4) $\frac{a}{b} > \frac{c}{b}$

11. If $a + b < c + d$ and $d + e < a + b$, which inequality *must* be true?

(1) $e < c$ (2) $e > c$ (3) $e < d$ (4) $e > d$

12. If $a + b > c + d$ and $c - e > b - d$, which statement *must* be true?

(1) $a < c$ (2) $a > c$ (3) $a < e$ (4) $a > e$

B. *In each case, show how you arrived at your answer by clearly indicating all of the necessary steps, formula substitutions, diagrams, graphs, charts, etc.*

13–18. Find and graph the solution set.

13. $9 - 2x > 1$ **16.** $7 \leq 3x - 2$

14. $3(x + 1) > -6$ **17.** $1 - \dfrac{x}{2} \geq 4$

15. $2x + 1 > 5x - 8$ **18.** $0.1x - 0.02x \leq 0.24$

19. If the replacement set is the set of integers, what is the greatest value of x that makes the inequality $7x - 6 < 8$ a true statement?

20. If the replacement set is $\{-2, -1, 0, 1, 2\}$, what is the solution set of $8 - 3(3 - x) \leq 1$?

21. Brian has $89 and wants to purchase compacts discs through a music club. Each CD costs $9.50. The music club will add a total postage and handling charge of $3.50 to his order. What is the greatest number of CDs Brian can purchase?

22. Steve is 5 years older than Peter. If the sum of their ages is *at most* 37 years, what is the oldest that Peter can be?

23. An architect wants to design a rectangular room so that its length is 8 meters more than its width, and its perimeter is greater than 56 meters. If each dimension of the room must be a whole number of meters, what are the smallest possible dimensions, in meters, of the room?

24. Susan has seven more nickels than dimes. If the total value of these coins is at least $3.00, what is the least number of nickels that she has?

25. At a movie theater, a cashier sold 250 more adult admission tickets than children's tickets. The adult's tickets were $6.50 each, and the children's tickets were $3.50 each. What is the *least* number of children's admission tickets that the cashier had to sell for the total receipts to be *at least* $2750?

26. George is twice as old as Edward, and Edward's age exceeds Robert's age by 4 years. If the sum of their ages is at least 56 years, what is the minimum age of each person?

27. A portion of a wire 60 inches in length is bent to form a rectangle whose length exceeds twice its width by 1 inch. What are the smallest possible dimensions of the rectangle if, at most, 4 inches of the wire is unused?

CHAPTER 3

PROBLEM SOLVING AND LOGICAL REASONING

3.1 PROBLEM-SOLVING STRATEGIES

KEY IDEAS

Knowing different *strategies* can help you solve a problem. Strategies do not tell you the specific steps to follow when solving a problem. They merely suggest approaches that you can try when figuring out the answer to an unfamiliar type of problem.

Strategy 1: Draw a Diagram.

Drawing a diagram can help you visualize a problem situation and organize the important facts.

Example 1

Amy goes shopping and spends one-third of her money on a new dress. She then goes to another store and spends one-half of the money she has left on shoes. If Amy has $56 left after these two purchases, how much money did she have when she started shopping?

Solution: **$168**

- Use a rectangle to represent the amount of money Amy had when she started shopping. Since she spends $\frac{1}{3}$ of her money on a dress, divide the rectangle into three equal parts, and write "Dress" in the first part:

Dress		

- Two parts of the rectangle remain. Amy spends $\frac{1}{2}$ of the *remaining* money on shoes. Hence, the second part represents the amount of money spent on shoes. Since $56 remains, enter this amount in the third part:

65

Dress	Shoes	$56

- Since the three parts of the rectangle are equal, Amy started shopping with $3 \times \$56 = \$168$.

Strategy 2: Look at a Specific Case.

If a problem does not give specific numbers, try finding the answer by using easy numbers.

Example 2

The perimeter of a rectangle is 10 times as great as the width of the rectangle. The length of the rectangle is how many times as great as the width?

    Solution: **4**

Draw a diagram and consider a specific rectangle.

- Consider a specific rectangle whose width is 1. If the perimeter of the rectangle is 10 times as great as the width, the perimeter is 10.
- Because $(2 \times \text{length}) + 2 \times 1 = 10$, $2 \times \text{length} = 8$, so the length of the rectangle is 4.
- Since length $= 4$ and width $= 1$, the length is 4 times as great as the width.

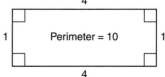

Strategy 3: Find a Pattern.

Examining a series of specific cases may help you to discover an equation or pattern that solves a problem.

Example 3

A store rents video tapes at the following rate: $3 for the first day and $2.50 for each additional day the tape is not returned. If C represents the cost, in dollars, of renting a tape for d days, write a formula in which C is expressed in terms of d.

    Solution: $C = 3 + (d - 1) \times 2.50$

- Write the cost, C, for renting the tape for different values of d in table form until you notice a pattern and can make a generalization:

d	C
1	$3
2	$3 + $2.50
3	$3 + 2 × $2.50
4	$3 + 3 × $2.50
.	.
.	.
.	.
d	$3 + (d − 1) × $2.50

The general formula is $C = 3 + (d - 1) \times 2.50$.

Example 4

People enter a room one at a time and are given a name tag in one of four possible colors. The colors are given out in this order: red, blue, white, and green. What is the color of the name tag that is given to the 93rd person who enters the room?

Solution: **Red**

Determine the colors of the name tags that the first 8 people who enter the room receive:

person 1: red	person 5: red
person 2: blue	person 6: blue
person 3: white	person 7: white
person 4: green	person 8: green

The pattern of colors repeats every four people. Therefore, if the number of a person entering the room is divisible by 4, he or she will receive a green tag. For example, since 8 is divisible by 4, the 8th person will be given a green tag. If the number of the person is not divisible by 4, the remainder tells where the person fits in the cycle of four colors. For example, since $6 \div 4 = 1$ remainder 2, the 6th person will receive a blue tag since blue is the second color in the repeating pattern. Since $93 \div 4 = 23$ remainder 1, the 93rd person will receive a red tag, which is the first color in the repeating cycle of colors.

Strategy 4: Work Backwards.

When you know only the end result of a computation and want to find the beginning value, reverse the steps that led to that final result.

Example 5

Sara's telephone service costs $21 per month plus $0.25 for each local call; long-distance calls are extra. Last month, Sara's bill was $36.64 and included $6.14 in long-distance charges. How many calls did she make?

Solution: **38**

Work back from the fact that Sara's final bill was $36.64.

- Since the final bill included $6.14 in long-distance charges, the part of the bill that did not include these long-distance charges was $36.64 − $6.14 = $30.50.
- The final bill included a fixed monthly charge of $21. Thus, the remaining part of the bill, for local calls only, was $30.50 − $21 = $9.50.
- Since each local call cost $0.25, the number of local calls was $\dfrac{\$9.50}{\$0.25} = 38$.

Strategy 5: Organize Data.

Shopping in a supermarket for a large number of groceries becomes easier and more efficient if you first prepare an organized list of the items you need to buy. Similarly, organizing the facts of a problem in a convenient form can help you find the answer.

Example 6

Gary and Sue are thinking of the same number. Gary says, "The number is a positive odd integer that is *at most* 25." Sue says, "The number is a prime number." If Gary's statement is true and Sue's statement is false, what are all the possible numbers?

Solution: **1, 9, 15, 21, 25**

Summarize the facts of the problem by making a list of the possible numbers.
- If Gary's statement is *true*, the number may be one of the following:

$$1, 3, 5, 7, 9, 11, 13, 15, 17, 19, 21, 23, 25$$

- A prime number is a whole number greater than 1 that is divisible only by itself and 1. If Sue's statement is *false*, the number is not a prime number, so eight numbers can be eliminated from Gary's list:

$$1, \cancel{3}, \cancel{5}, \cancel{7}, 9, \cancel{11}, \cancel{13}, 15, \cancel{17}, \cancel{19}, 21, \cancel{23}, 25$$

The remaining numbers in Gary's list are 1, 9, 15, 21, and 25.

Example 7

Rene, Vanessa, and Jose each hold one of the following offices in the student government: president, treasurer, and secretary. At a meeting of the student government, Rene and the secretary arrived together. The president spoke before Jose. Vanessa left the meeting with Jose and the treasurer. Which of the three students is the president of the student government?

Solution: **Vanessa**

Organize the clues and their logical relationships by preparing a table with three columns and three rows. Label each column with a different student government office, and label each row with a different student name.

- Since Rene and the secretary arrived together, Rene is not the secretary, so put an ✗ in the last column of the table, as shown below. Also, since the president spoke before Jose, put an ✗ in the first column to show that Jose is not the president.

	President	Treasurer	Secretary
Vanessa			
Rene			✗
Jose	✗		

- Vanessa left the meeting with Jose and the treasurer. Put an ✗ in the second column of the table to show that Vanessa is not the treasurer and another ✗ in the second column to show that Jose is not the treasurer.

	President	Treasurer	Secretary
Vanessa		✗	
Rene			✗
Jose	✗	✗	

- Since one of the three students must be the treasurer, Rene is the only possibility. Put a check mark in the second column to show that Rene is the treasurer, and an ✗ in the first column to show that she cannot, therefore, be the president. Jose is not the president and not the treasurer, so he is the secretary. Show this with a check mark in the third column.

	President	Treasurer	Secretary
Vanessa	✓	✗	✗
Rene	✗	✓	✗
Jose	✗	✗	✓

- Since neither Rene nor Jose is the president, Vanessa must hold this office.

Strategy 6: Guess, Check, and Revise.

When you can't figure out an algebraic solution, you may be able to make a reasonable guess and then use it to come up with a better guess. You can then repeat this process until you reach the correct answer.

Example 8

Beth plans to swim *at least* 100 laps during a 6-day period. During this period, she will increase the number of laps completed each day by one lap. What is the *least* number of laps Beth must complete the first day?

Solution: **15**

Make a reasonable guess. Then organize the results of that guess in a table.

- Since $\dfrac{100}{6} \approx 16.6$, use 16 as your first guess for the number of laps completed on the first day.

Guess = 16

Day	Number of Laps	
1	16	
2	17	
3	18	
4	19	
5	20	
6	21	
Total	111	← Too high

- Since 111 is greater than 100 by more than 6, refine your guess. Try 14 for the number of laps completed on the first day.

Guess = 14

Day	Number of Laps	
1	14	
2	15	
3	16	
4	17	
5	18	
6	19	
Total	99	← Too high

- Since 14 laps for the first day gives a total less than 100 by only 1, and 16 laps for the first day gives a total that is too high, the least number of laps Beth must swim on the first day must be between 14 and 16, or 15 laps.

If you use this strategy and your first guess works, check that it is the correct answer by trying at least one more guess above and below the value that works.

Strategy 7: Use an Algebraic Equation.

Sometimes the most efficient way of solving a problem is to use an algebraic equation. Problems that compare quantities lend themselves to algebraic solutions. In this type of problem, first identify the "base" quantity to which one or more other quantities are being compared. After assigning a variable to the base quantity, represent the other quantities in terms of the same variable. Then translate the conditions of the problem that tell how these quantities are related into an equation.

Example 9

Ticket sales for a music concert totaled $2160. Three times as many tickets were sold for the Saturday night concert as were sold for the Sunday afternoon concert. Two times as many tickets were sold for the Friday night concert as were sold for the Sunday afternoon concert. Tickets for all three concerts sold for $2.00 each. Find the number of tickets sold for the Saturday night concert.

Solution: **540**

- Identify the quantity to which the others are being compared. The numbers of tickets sold for Friday and Saturday night concerts are being compared to the number of tickets sold for the Sunday afternoon concert.
- Assign variables. Since the base quantity is the number of tickets sold for the Sunday afternoon concert:

  Let x = number of tickets sold for the Sunday afternoon concert.
  Then $3x$ = number of tickets sold for the Saturday night concert,
  and $2x$ = number of tickets sold for the Friday night concert.

- Identify the condition that relates the ticket sales for the three concerts, and write and solve an equation. Since each ticket costs $2 and the total amount of sales was $2160:

$$\overbrace{x \cdot \$2 + 3x \cdot \$2 + 2x \cdot \$2}^{Total\ receipts} = \$2160$$
$$\$2x + \quad \$6x + \$4x \quad = \$2160$$
$$\$12x = \$2160$$
$$x = \frac{\$2160}{\$12} = 180$$

- Answer the question that is asked. The number of tickets sold for the Saturday night concert = $3x = 3(180) = 540$.

Strategy 8: Solve a Simpler Analogous Problem.

You may be able to solve an unfamiliar problem by simplifying it so that it becomes an equivalent problem that is easier to solve. Look back at Example 8 on page 70. If you remove the context of this problem, you are left with an equivalent problem: "The sum of six consecutive integers is *at least* 100. What is the *least* possible value of the first of the six integers?"
   Let x represent the first of six consecutive integers, then

$$x + (x+1) + (x+2) + (x+3) + (x+4) + (x+5) \geq 100$$
$$6x + 15 \geq 100$$
$$6x \geq 85$$
$$x \geq \frac{85}{6}$$
$$\geq 14.166\ldots$$

Since x is an integer and $x \geq 14.166\ldots$, the *least* possible value of x is 15.

Example 10

The length and width of a rectangular box are each doubled, and the height is tripled. The volume of the new box is how many times as great as the volume of the original box?

Solution: **12**

Simplify the problem by considering a cube that measure 1 inch by 1 inch by 1 inch.

- The volume of the original box is 1 cubic inch.
- If the length and width of this box are each doubled and the height is tripled, the dimensions of the new box are 2 inches by 2 inches by 3 inches. Hence, the volume of the new box is 12 cubic inches.
- The volume of the new box is 12 times the volume of the original box.

Example 11

An ordered list of numbers begins with −3. Every number after −3 is 5 more than the number that precedes it:

$$-3, 2, 7 \ 12, 17, \ldots$$

What is the 102nd number in the list?

Solution: **502**

Simplify the problem by changing the first number in the list from −3 to 5. Then the second number in the list becomes 10 ($= 2 \times 5$), the third number is 15 ($= 3 \times 5$), and so forth. The 102nd number in the new list is $102 \times 5 = 510$. Because the starting value in the original list was increased by 8 from −3 to 5, each number in the new list is greater by 8 than the corresponding value in the original list. To find the 102nd number in the original list, *subtract* 8 from 510 to obtain 502.

Strategy 9: Adopt a Different Point of View.

A word problem typically directs your attention to a particular unknown quantity. Changing your point of view by considering a different but related quantity may help you to discover a simple way of solving the problem.

Example 12

A jar contains 110 marbles of which 50 are red and 60 are green. The probability of picking a red marble from the jar without looking is, therefore,

$\frac{50}{110}$. How many red marbles must be added to the jar so that the probability of picking a red marble will be $\frac{2}{3}$?

Solution: **70**

If x represents the number of red marbles that need to be added to the jar, then $\frac{50+x}{110+x} = \frac{2}{3}$. Rather than solving this equation, you could adopt a different point of view by considering the probability of picking a *green* rather than a red marble:

- If the probability of picking a red marble is $\frac{2}{3}$, the probability of picking a green marble is $\frac{1}{3}$ since the sum of the probabilities of the two possible outcomes (red marble vs. green marble) must add up to 1.
- The number of green marbles stays fixed at 60. The probability of picking a green marble will be $\frac{1}{3}$ only when the total number of marbles in the jar is three times 60 or 180, because $\frac{60}{180} = \frac{1}{3}$.
- Since $180 - 110 = 70$, 70 red marbles must be added to the jar.

Strategy 10: Account for All Possible Cases.

Solving a problem may depend on breaking it down so that all possible cases are considered.

Example 13

In a certain homeroom class, 21 students are enrolled in Math A, 17 students are enrolled in biology, 9 are enrolled in both Math A and biology, and 3 students are not enrolled in either course. How many students are in the homeroom class?

Solution: **32**

Account for all possible nonoverlapping sets of students that comprise the homeroom class. Students can be enrolled in both courses, in exactly one of the two courses, or in neither of the two courses:

- Students in both courses $= 9$
- Students in Math A but not biology $= 21 - 9 = 12$
- Students in biology but not Math A $= 17 - 9 = 8$
- Students not in either course $= 3$

Total students in homeroom class $= 32$

By accounting for all possible cases, you know that the total number of students in the homeroom class is $9 + 12 + 8 + 3 = 32$.

Ten Problem-Solving Strategies

Different people may solve the same problem using different strategies or a combination of strategies. You should use whatever strategy works for you, even if it is not one of the ten problem-solving strategies summarized in the accompanying table.

TEN PROBLEM-SOLVING STRATEGIES

1. **Draw a diagram.** Visualize a problem situation by drawing a diagram that may help you find the clues needed to solve the problem.

2. **Look at a specific case.** When no numbers are given, solve the problem using easy numbers.

3. **Find a pattern.** Look at a series of specific cases until you discover a pattern that solves the problem.

4. **Work backwards.** When the final answer but not the starting value is given, reverse the steps needed to get that answer.

5. **Organize data.** Arrange the data in an organized list or table so that any patterns or logical relationships become easier to see.

6. **Guess, check, and revise.** Make a reasonable guess, and then use it to obtain a better guess. Repeat this process until you find the correct answer.

7. **Use an algebraic equation.** After setting a variable equal to an unknown quantity, translate the conditions of the problem into an algebraic equation that can be used to solve the problem.

8. **Solve a simpler analogous problem.** Make a problem more manageable by working with a simpler but equivalent version of it.

9. **Adopt a different point of view.** Change your point of view by focusing on a different but related quantity so that the problem becomes easier to solve.

10. **Account for all possible cases.** Break down a problem situation into all possible cases. Consider all possible values of a variable.

Check Your Understanding of Section 3.1

A. Multiple Choice

1. If n represents an odd number, which computation results in an answer that is an even number?
   (1) $2 \times n + 1$ (2) $2 \times n - 1$ (3) $3 \times n - 2$ (4) $3 \times n + 1$

2. The perimeter of a square is four times as great as the perimeter of a smaller square. What is the ratio of the area of the smaller square to the area of the larger square?
   (1) 1 to 2   (2) 1 to 4   (3) 1 to 8   (4) 1 to 16

3. The length of each side of a cube is multiplied by 3. What is the ratio of the volume of the larger cube to the volume of the smaller cube?
   (1) 3 to 1   (2) 9 to 1   (3) 27 to 1   (4) 81 to 1

4. If $\dfrac{b-2}{5}$ represents a positive integer, what is the remainder when b is divided by 5?
   (1) 1   (2) 2   (3) 3   (4) 4

5. If n pencils cost c cents, what is the cost, in cents, of p pencils?
   (1) $\dfrac{pc}{n}$ (2) $\dfrac{nc}{p}$ (3) $\dfrac{c}{np}$ (4) $c - \dfrac{p}{n}c$

6. If the length of each side of a square is increased by 10%, by what percent does the area of the square increase?
   (1) 20   (2) 21   (3) 40   (4) 100

7. At the same school, 61 girls tried out for the soccer team, 35 girls tried out for the basketball team, and 16 girls tried out for both teams. What percent of the girls who tried out for at least one of the two teams tried out for both teams?
   (1) 12.5   (2) 20   (3) 25   (4) 87.5

8. Frank, George, and Hernando each have exactly one of the following jobs: plumber, cabinet maker, and electrician. Frank is not able to install a new electric line. Hernando is not able to make a new cabinet. George hired one of the other two men to do electrical work. Which statement *must* be true?
   (1) Hernando is an electrician. (3) Frank is a plumber.
   (2) George is a cabinet maker.   (4) Frank is an electrician.

B. Show how you arrived at your answer by clearly indicating all of the necessary steps, formula substitutions, diagrams, graphs, charts, etc.

9. Sue bought a skirt on sale for 50% off the original price. The store charged $3.40 sales tax, and the final cost was $22. What was the original price of the skirt?

10. After Carl unloaded $\frac{1}{8}$ of the cartons in a truck and Tom unloaded $\frac{2}{7}$ of the remaining cartons, 45 cartons were left. How many cartons were originally in the truck?

11. Cindy makes a list of numbers in which each number after the first is 3 more than twice the number that precedes it. If the third number in Cindy's list is 19, what is the first number?

12. Bob and Ray are describing the same number. Bob says, "The number is a positive even integer less than or equal to 20." Ray says, "The number is divisible by 4." If Bob's statement is true and Ray's statement is false, what are all the possible numbers?

13. The number of bacteria doubles every 20 minutes. The bacteria population was 16 million at 1:00 P.M. How many bacteria, expressed in thousands, were present at 11:00 A.M. on the same day?

14. Margarita spends one-fifth of her salary on a belt and one-fourth of the remaining money on a shirt. If she has $72 left after making these two purchases, what is her salary?

15. Give a possible value of x, if any, that makes each statement true.

   (a) $x > x^2$ (b) $y(x+2) = y$ (c) $3x^2 + 5x^4 = 0$ (d) $\dfrac{1}{x} < -1$

16. Dalia and Miguel are describing the same positive number. Dalia says, "If the number is increased by 4, it is a perfect square less than 100." Miguel adds, "If 3 is subtracted from the number, the result is a prime number greater than 2." What is the number that Dalia and Miguel are describing?

17. The EZ-Car Rental Agency charges $35 for the first day and $22 for each additional day. If John's car rental bill was $167, for how many days did John rent the car?

18. Rick, Mark, Vanessa, Sandy, and Ariela solved the same math problem but required different amounts of time. These times, in increasing order, were 5 minutes, 6 minutes, 9 minutes, 10 minutes, and 14 minutes. Vanessa needed 4 minutes longer than Sandy. Only one person took more time than Rick. Ariela finished before Vanessa. How many minutes did each person need to solve the problem?

19. Three boys agree to divide a package of baseball cards in the following way: Jose takes one-half of the cards, Shawn takes one-third of the remaining cards, and Yin takes the rest. If Yin takes 12 cards, how many cards does Jose take?

20. Assume the cost of a telephone call from Wilson, N.Y., to East Meadow, N.Y., is $0.60 for the first 3 minutes plus $0.17 for each additional minute. What is the greatest number of whole minutes of a telephone call if the cost cannot exceed $2.50?

21. For a concert, 100 more balcony tickets than main-floor tickets were sold. The balcony tickets cost $4, and the main-floor tickets cost $12. If the total amount of sales for both types of tickets was $3056, find the number of balcony tickets that were sold.

22. There are six teams in a high school bowling league. If each team will play against every other team two times, how many matches are scheduled?

23. The accompanying figure shows a correctly worked out addition problem in which each letter represents a different digit from 1 to 9. What digit does D represent?

$$83A$$
$$+DBB$$
$$\overline{CAC2}$$

24. A hotel charges $20 for the use of its dining room and $2.50 a plate for each breakfast. An association gives a breakfast and charges $3 a plate but invites four nonpaying guests. If each person has one plate, how many paying persons must attend for the association to collect the exact amount needed to pay the hotel?

25. A movie theater charges $7 for an adult's ticket and $4 for a child's ticket. On a recent night, the sale of children's tickets was three times the sale of adults' tickets. If the total amount collected for ticket sales was not more than $2000, what is the greatest number of adults who could have purchased tickets?

26. The pages of a book are numbered using the digits 0, 1, 2, 3, . . . , 9. For example, page 31 contains the two digits 3 and 1. If 897 digits are needed to number the pages of a book, how many numbered pages does the book contain?

27. If 3^{35} is multiplied out, what is the units (ones) digit of the product?

28. The average of four numbers is exactly 32. The largest of the four numbers is 5 more than one of the other numbers, and exceeds twice the smaller number by 10. The remaining number is one-half as great as the largest of the four numbers. What is the largest of the four numbers?

29. In his will, a man leaves one-half of his money to his wife, one-half of what is then left to his older child, and one-half of the remainder to his younger child. His two cousins divide the rest equally, each receiving $2000. What was the total amount of money bequeathed in the man's will?

30. Mindy, Arturo, Renee, and Joan each belong to a different one of the following school clubs: Chorus, Tennis, Chess, and Drama. Joan and Renee were in the audience at a performance of the Drama Club. Arturo wished the tennis player good luck in her match. Mindy and Joan were members of the Chess and Drama clubs, but each dropped out of one of these clubs. Who belongs to each club?

31. Allan, Barry, and Craig have dates with Stephanie, Vickie, and Debbie. Barry said that his date knows Vickie. Allan got the phone number of his date from Debbie. Craig wants to date either Stephanie or Debbie. Barry does not want to date Stephanie. Match each boy with his date.

32. Dennis, Joni, Elvis, and Taisha each play in exactly one of the following types of music groups: jazz, rock, classical, and country. The four music groups appeared at the same concert. Dennis, Taisha, and the classical musician arrived at the concert together. The country group performed after Elvis and Joni performed. The jazz group performed before Taisha and Elvis performed. Taisha left the concert with the country musician. Who plays in each group?

3.2   SOLVING MOTION PROBLEMS

=== KEY IDEAS ===

Motion problems depend on this relationship: rate × time = distance. For example, a train traveling at an average rate of 40 miles per hour for 3 hours travels a distance of

$$40\,\frac{\text{miles}}{\text{hour}} \times 3\,\text{hours} = 120\,\text{miles}.$$

Using Rate × Time = Distance

To solve a motion problem, you may need to use one or more problem-solving strategies.

Example 1

Two trains leave the same station at the same time and travel in opposite directions. One train travels at 80 miles per hour and the other at 100 miles per hour. In how many hours will the trains be 900 miles apart?

Solution: **5**

Method I: Use a diagram and an equation.

- Draw a diagram to show the distance traveled by each train in h hrs:

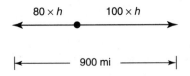

- Hence, $80h + 100h = 900$, so $180h = 900$ and $h = \dfrac{900}{180} = 5$.
- The two trains will be 900 mi apart in 5 hrs.

Method II: Organize the data in a list.

In 1 hr the two trains are $80 + 100 = 180$ mi apart. Hence:

- After 2 hrs, the trains are $180\,\text{mi} \times 2 = 360\,\text{mi}$ apart.
- After 3 hrs, the trains are $180\,\text{mi} \times 3 = 540\,\text{mi}$ apart.
- After 4 hrs, the trains are $180\,\text{mi} \times 4 = 720\,\text{mi}$ apart.
- After 5 hrs, the trains are $180\,\text{mi} \times 5 = 900\,\text{mi}$ apart.

You could also have reasoned arithmetically that, since the two trains are $80 + 100 = 180\,\text{mi}$ apart after 1 hr, they will be 900 mi apart in $\dfrac{900}{180} = 5\,\text{hrs}$.

Example 2

A truck traveling at a constant rate of 45 miles per hour leaves Albany at 9:00 A.M. One hour later a car traveling at a constant rate of 60 miles per hour leaves from the same place, traveling in the same direction, on the same highway. At what time will the car overtake the truck, if both vehicles continue in the same direction on the highway?

Solution: **1:00 P.M.**

- Let x represent the number of hours the car has traveled when it overtakes the truck. Then, since the truck started 1 hr before the car, the number of hours the truck has traveled when it is overtaken by the car is $x + 1$.
- Organize the given facts in a table.

	Rate	Time	Distance
Truck	45 mph	$x + 1$	$45(x + 1)$
Car	60 mph	x	$60x$

- Write an equation. When the car overtakes the truck, the two vehicles have traveled the same distance. Thus:

$$60x = 45(x+1)$$
$$60x = 45x + 45$$
$$15x = 45$$
$$x = \frac{45}{15} = 3$$

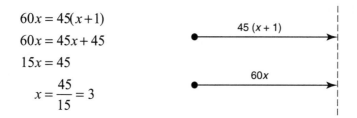

- The car starts at 10:00 A.M. since it leaves 1 hr after the truck. Hence, the car overtakes the truck 3 hrs after the car starts, that is, at 1:00 P.M.

Check Your Understanding of Section 3.2

A. Multiple Choice

1. A bicyclist travels 6 miles in 20 minutes. Which expression does *not* represent the average rate of speed of the bicyclist?

   (1) $\dfrac{3}{10}$ mile per minute

   (2) $3\dfrac{1}{3}$ minutes per mile

   (3) 18 miles per hour

   (4) 20 miles per hour

2. A car travels 110 miles in 2 hours. At the same rate of speed, how far will the car travel in *h* hours?

   (1) $55h$ (2) $220h$ (3) $\dfrac{h}{55}$ (4) $\dfrac{h}{220}$

3. If Jamar can run $\frac{3}{5}$ of a mile in 2 minutes 30 seconds, what is his average rate, in miles per minute?

   (1) $\dfrac{4}{5}$ (2) $\dfrac{6}{25}$ (3) $3\dfrac{1}{10}$ (4) $4\dfrac{1}{6}$

4. Two cars leave from the same location and travel in opposite directions. One car travels at a rate of *m* miles per hour. The rate of the other car exceeds the rate of the first car by 15 miles per hour. Which expression represents the number of miles that the two cars will be apart after 4 hours?

   (1) $4m + 15$ (2) $8m^2 + 15m$ (3) $8m + 60$ (4) $4m^2 + 60$

5. The distance from Earth to the imaginary planet Zota is 1.74×10^7 miles. If a spaceship is capable of traveling at an average rate of 1450 miles per hour, what is the approximate number of days the spaceship will take to reach Zota?

   (1) 50 (2) 1200 (3) 500 (4) 8333

B. In each case, show how you arrived at your answer by clearly indicating all of the necessary steps, formula substitutions, diagrams, graphs, charts, etc.

6. A train leaves a station at 9:00 A.M. traveling at a constant rate of speed. Two hours later a second train, traveling at an average rate of 25 miles per hour more than the first train, leaves the same station, going in the same direction, on a parallel track. If the faster train overtakes the first train at 3:00 P.M. on the same day, what is the average rate of speed of the first train?

7. A freight train and a passenger train start toward each other at the same time from two towns that are 500 miles apart. After 3 hours the trains are still 80 miles apart. If the speed of the passenger train is 20 miles per hour greater than the speed of the freight train, find the speed of each train.

8. A truck left a rest stop on the highway at 12 noon going north at the rate of 50 miles per hour. One hour later a car left the same rest stop going south at the rate of 60 miles per hour. At what time were the trains 325 miles apart?

9. Ms. Ruiz drove from her home to her job at an average speed of 60 miles per hour and returned home along the same route at an average speed of 40 miles per hour. If her total driving time for the trip was 2 hours, how many *minutes* did she take to drive from her job to her home?

10. Gregory hiked from the base of a mountain to its top at an average rate of 2 miles per hour and hiked down the mountain, following the same path, at an average rate of 3 miles per hour. If Gregory hiked, excluding rest stops, for a total of 8 hours, how many miles was it from the base to the top of the mountain?

11. John rode his bicycle from his house to Vincent's house at the rate of 15 miles per hour. He left the bicycle with Vincent, and walked home along the same route at the rate of 3 miles per hour. If the entire trip took 3 hours, how long did John take to walk back to his house?

12. Brian is walking along a trail at an average rate of 3 miles per hour. Carol is bicycling along the same trail in the same direction at an average rate of 9 miles per hour. At 12:00 noon Brian is 5 miles ahead of Carol. If Brian and Carol maintain these rates, at what time will Carol overtake Brian?

13. A boy riding a bicycle at the rate of 10 miles per hour and a car going 40 miles per hour are traveling in the same direction on the same road. If the car passes a road sign 30 minutes after the boy does, in how many minutes will the car overtake the bicycle?

3.3 SOLVING PROBLEMS IN DIFFERENT WAYS

KEY IDEAS

Mathematics problems can be solved algebraically, graphically, or numerically. If you get stuck trying to solve a problem algebraically, you may be able to solve it by making a graph or a table.

The Cable TV Problem

For example, suppose you are signing up for a cable television service plan. Plan A costs $11 per month plus $7 for each premium channel. Plan B costs $27 per month plus $3 for each premium channel. For what number of premium channels will the two plans cost the same? This problem can be represented and solved in at least three different ways.

Algebraic Representation

Algebraic language allows you to generalize from a pattern that becomes evident when you look at a few specific cases.

- Write the cost of each TV service plan for a few specific cases until you can find a pattern that allows you to write the cost of each plan when x premium channels are ordered.

	Plan A	Plan B
1 premium channel:	$11 + 7 \times 1$	$27 + 3 \times 1$
2 premium channels:	$11 + 7 \times 2$	$27 + 3 \times 2$
	. . . . .	. . . . .
Generalize for x channels: $\rightarrow$	$11 + 7 \times \underline{x}$	$27 + 3 \times \underline{x}$

- Write and solve an equation.

$$\text{cost of plan A} = \text{cost of plan B}$$
$$11 + 7x = 27 + 3x$$
$$7x - 3x = 27 - 11$$
$$4x = 16$$
$$x = \frac{16}{4} = 4$$

- Check that the solution $x = 4$ works:

Cost of plan A $= \$11 + \$7x$
$$= \$11 + \$7(4)$$
$$= \$11 + \$28$$
$$\overset{\checkmark}{=} \$39$$

Cost of plan B $= \$27 + \$3x$
$$= \$27 + \$3(4)$$
$$= \$27 + \$12$$
$$\overset{\checkmark}{=} \$39$$

Table Representation

You can make a table by hand or by using a graphing calculator. If you have a TI-83 graphing calculator:

- First store the equation for each TV service plan by pressing the $\boxed{Y=}$ key. Set $Y_1 = 11 + 7x$ by pressing:

$$\boxed{1} \quad \boxed{1} \quad \boxed{+} \quad \boxed{7} \quad \boxed{X, T, \theta, n} \quad \boxed{\text{ENTER}}$$

Then set $Y_2 = 27 + 3x$.
- Enter the TABLE SETUP mode by pressing $\boxed{2nd}$ followed by $\boxed{\text{WINDOW}}$. Set TblStart = 1, as shown in Figure 3.1. This makes 1 the starting value of X when you view the table. The consecutive values of X in the table will increase by 1 when ΔTbl = 1.

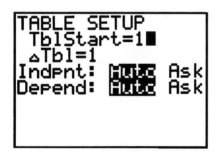

Figure 3.1 Table Setup

Figure 3.2 Representation of the Two Plans in a Table

- Display the table by pressing $\boxed{2nd}$ followed by $\boxed{\text{GRAPH}}$. The plans will have the same cost for the value of X that makes $Y_1 = Y_2$. According to the table in Figure 3.2, when $X = 4$ premium channels, $Y_1 = Y_2 = \$39$.

Graphical Representation

Instead of using the table feature of your graphing calculator, represent the cost of each TV service plan by a graph. The plans will have the same cost at the point where the two graphs intersect.

- Set $Y_1 = 11 + 7x$ and $Y_2 = 27 + 3x$ in the $Y =$ editor.
- Adjust the size of the viewing rectangle by pressing WINDOW and then choosing appropriate values for the screen variables so that the graph fits within the viewing rectangle. After setting $X_{min} = 0$, $X_{max} = 10$, $Y_{min} = 0$, and $Y_{max} = 60$, press GRAPH to obtain the display in Figure 3.3.
- Use the cursor keys to move the cursor along one of the graphs until the cursor appears to be at the point of intersection of the two lines. Select the **intersect** feature from the CALC menu to find the x-coordinate of the point where the two lines meet. To get the display shown in Figure 3.4, press:

Figure 3.4 Finding Where the Lines Meet

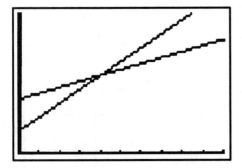

Figure 3.3 Graphs of Plan A and Plan B   **Figure 3.4** Finding Where the
                                                          Lines Meet

Why Represent Problems in Different Ways?

As you become more comfortable in approaching problems in different ways, you will become better at solving new and unfamiliar problems. Furthermore:

- One way of solving a particular problem may prove to be easier than another.
- A table or graph may provide additional information that is not available when solving a problem algebraically with an equation.

- When a problem is solved in two different ways, one method can serve as a check on the other method. For example, answers arrived at algebraically can often be confirmed graphically with the help of a graphing calculator.

Check Your Understanding of Section 3.3

A. Multiple Choice

1. A cellular telephone company has two plans. Plan A charges $11 a month and $0.21 per minute. Plan B charges $20 a month and $0.10 per minute. After how much time, to the *nearest minute*, will the cost of plan A be equal to the cost of plan B?
   (1) 1 hr 22 min   (2) 1 hr 36 min   (3) 81 hr 8 min   (4) 81 hr 48 min

2. A store advertises that during its Labor Day sale $15 will be deducted from every purchase over $100. Furthermore, after the deduction is taken, the store offers an early-bird discount of 20% to any person who makes a purchase before 10 A.M. If Hakeem makes a purchase of x dollars, where $x > 100$, at 8 A.M., what is the cost of his purchase?
   (1) $0.20x - 15$ (2) $0.20x - 3$ (3) $0.85x - 20$ (4) $0.80x - 12$

B. Show how you arrived at your answer by clearly indicating all of the necessary steps, formula substitutions, diagrams, graphs, tables, etc.

3. Carlos has $365 in savings and saves $20 each week. His brother has $590 in savings and spends $25 of his savings each week. After how many weeks will Carlos and his brother have the same amount in savings?

4. A plumbing bill includes the costs of parts and labor. The bill for repairing the plumbing in a house was $458. The cost of parts was $271. If the cost of labor was $68 per hour, find the number of hours of labor.

5. At Central High School, 434 students are enrolled in Spanish and 271 students are enrolled in French. The number of students enrolling in Spanish has been increasing at a rate of about 21 students per year, while the number of students enrolling in French has been decreasing at a rate of about 3 students per year. If these rates continue, in how many years will there be two times as many students taking Spanish as taking French?

6. Juan has a cellular phone that costs $12.95 per month plus 25 cents per minute for each call. Tiffany has a cellular phone that costs $14.95 per month plus 15 cents per minute for each call. For what number of minutes do the two plans cost the same?

7. The formula $F = 1.8C + 32$ gives the temperature in degrees Fahrenheit when the Celsius temperature is known. John claims that it is not possible for the measured temperature of the same object to read the same on both the Fahrenheit and Celsius scales. Do you agree or disagree? Explain your reasoning.

8. The Excel Cable Company has a monthly fee of $32.00 and an additional charge of $8.00 for each premium channel. The Best Cable Company has a monthly fee of $26.00 and an additional charge of $10.00 for each premium channel.
   (a) For what number of premium channels will the total monthly subscription fee for the two cable companies be the same?
   (b) If a family subscribes to two premium channels for a period of 1 year with one of these two companies, how much money will the family save by choosing the less expensive cable company?

9. Mr. Day and Ms. Knight gave the same Math A test consisting of 16 questions. Mr. Day gave each student 5 points for each correct answer and then added 20 points to the total. Ms. Knight gave each student 7 points for each correct answer and then subtracted 6 points from the total. A student in Mr. Day's class answered the same number of questions correctly as a student in Ms. Knight's class and received the same grade. What grade did each student receive on the test?

10. An empty tub has a 40-gallon capacity. Kristin turns on the hot-water faucet, which releases water into the tub at a constant rate of 0.8 gallon per minute. Exactly 3 minutes later, Kristin turns on the cold-water faucet, which releases water into the tub at a constant rate of 1.3 gallons per minute. Kristin turns both faucets off when the tub has received the same number of gallons of hot water as cold water.
   (a) How many minutes after Kristin turned on the hot-water faucet does she turn both faucets off?
   (b) When both faucets are turned off, what percent of the tub is filled with water?

11. The market supply and demand of a certain type of computer memory chip are related to the price of the chip according to these equations:

Supply: $Y = 4x + 108$

Demand: $Y + 3x = 150$

where Y is the price, in dollars, and x is the number of chips.
(a) Use an algebraic method to find the value of x for which the supply price equals the demand price.
(b) Use a graphing calculator to confirm your answer by creating either a graph or a table.

12. Over a 12-month period, the price of stock A increased according to the equation $y = 7.5x + 180$, and the price of stock B fell according to the equation $y = 328 - 11x$. In each equation, y is the average price of the stock, in dollars, during month x, where $x = 1$ represents January.
(a) Use an algebraic method to find the month during which the average monthly prices of the two stocks were the same. What was the average price of each stock during this month?
(b) Use a graphing calculator to confirm your answer by creating either a graph or a table.

3.4   VENN DIAGRAMS AND COUNTING

KEY IDEAS

Some counting problems can be solved with the help of **Venn diagrams**, which use circles to represent two or more sets of people or objects that have members in common.

Venn Diagrams

The Venn diagram in Figure 3.5 refers to a group of 50 students who attend Pleasantville High School. Circle A indicates that 20 of these 50 students are members of the soccer team, and circle B indicates that 28 of the 50 students are members of the track team.

50 Students at Pleasantville High School

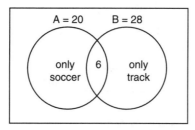

Figure 3.5   Venn Diagram

Based on the numbers in Figure 3.5, you can conclude that:

- 6 students belong to *both* teams since the region in which both circles overlap contains the number 6.
- 20 – 6 = 14 students belong to the soccer team but not the track team.
- 28 – 6 = 22 students belong to the track team but not the soccer team.
- 50 – (6 + 14 + 22) = 8 of the 50 students do not belong to either team.

Solving Counting Problems

A Venn diagram and logical reasoning can help you to solve counting problems involving overlapping sets of people or objects.

Example

In a class of 30 students, 19 are studying French, 12 are studying Spanish, and 7 are studying both French and Spanish. If no other foreign language is being studied, what percent of the students in the class are *not* studying any foreign language?

Solution: **20**

- Draw a Venn diagram to help organize the facts.

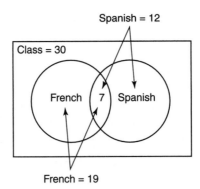

- List the number of students in each of the possible nonoverlapping categories of students.

Category	Number of Students
• Studying both languages	7
• Studying *only* French	19 – 7 = 12
• Studying *only* Spanish	12 – 7 = 5
• *Not* studying any language	x

- By accounting for all possible cases, the total number of students in the class is $7 + 12 + 5 + x$, where x represents the number of students who are not studying any foreign language. Since it is given that there are 30 students in the class:

$$7 + 12 + 5 + x = 30$$
$$24 + x = 30$$
$$x = 6$$

- Because $\dfrac{6}{30} = 0.20 = 20\%$, 20% of the students in the class are not studying any foreign language.

Check Your Understanding of Section 3.4

A. *Multiple Choice*

1. In the accompanying figure, circle X represents the set of all integers that are divisible by 2, circle Y represents the set of all integers that are divisible by 3, and circle Z represents the set of all integers whose tens digit is greater than the units digit. Which integer could be a member of the set represented by the shaded region?

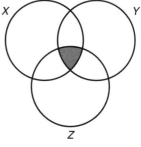

   (1) 21   (2) 24   (3) 42   (4) 48

2. The students in a physical education class all belong to either the basketball team or to the tennis team, or are not on either team. If 15 students are on the basketball team, 18 students are on the tennis team, 11 students are on both teams, and 14 students are not on either of the two teams, how many students are in the class?

   (1) 28   (2) 32   (3) 36   (4) 40

B. *In each case, show how you arrived at your answer by clearly indicating all of the necessary steps, formula substitutions, diagrams, graphs, charts, etc.*

3. Of 25 city council members, 12 voted for proposition A, 19 voted for proposition B, and 10 voted for both propositions. If every council member voted on both propositions, what percent of the council members voted *against* both propositions?

4. In a school of 500 students, 65 students are in the school chorus, 255 students are on sports teams, and 40 students participate in both activities. What percent of the number of students in the school participate in either the school chorus or on sports teams but not on both teams?

5. The accompanying Venn diagram shows the number of students who take mathematics (circle A), science (circle B), and technology (circle C).

   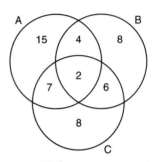

   Find the percentage of students who take:
   (a) mathematics *and* science
   (b) mathematics *or* technology

6. In a survey of 200 seniors at a high school it was found that 75 students play soccer but not softball, 45 students play softball but not soccer, and 28 students do not participate in either sport. Of the seniors surveyed, what percent play both soccer and softball?

7. Of the 210 juniors attending a high school, 120 are enrolled in chemistry and 30 are enrolled in art. If a total of 100 juniors are enrolled in only one of the two courses, find the number of juniors who are not enrolled in either course.

8. When report cards were distributed in a certain homeroom class, it was found that 13 students received at least one grade of A, 15 students received at least one grade of B, and 6 students received both A's and B's. If one-third of the students in the homeroom class received no A's and no B's, how many students are in the homeroom class?

3.5   CONNECTING TWO STATEMENTS

KEY IDEAS

In mathematics, a sentence that can be judged to be either true or false is called a **statement**. Two or more statements can be connected using the logical connectives *and*, *or*, and *if . . . then*.

Connecting Statements by Using *And* or *Or*

Two different statements may be combined using the word *and* or *or* to form a new statement.

- A **conjunction** uses *and* to connect two statements. The conjunction "It is spring *and* the birds are chirping" is true only when the two statements joined by the word *and* are both true. Each of the statements that make up a conjunction is called a **conjunct**.
- A **disjunction** uses *or* to connect two statements. The disjunction "I will study *or* I will go to the movies" is true only when at least one of the statements joined by the word *or* is true. Each of the statements that make up a disjunction is called a **disjunct**.

Forming the Negation of a Statement

The negation of a statement is formed by inserting the word *not* so that the resulting statement and the original statement have opposite truth values. The statement "Tuesday follows Monday" is true, but its negation "Tuesday does *not* follow Monday," is false.

Drawing Inferences

Based on the truth values of a set of statements, you may be able to draw a conclusion about the truth value of a related statement.

- When the disjunction <u>statement 1</u> **or** <u>statement 2</u> is *false*, you can conclude that the truth value of statement 1 is false and the truth value of statement 2 is also false.
- When the disjunction <u>statement 1</u> **or** <u>statement 2</u> is *true*, then at least one of the two statements (disjuncts) is true.

Example 1

Given the *false* statement, "Ben has a driver's license or Ben is not 18 years old," what is the truth value of the statement, "Ben is 18 years old"?

Solution: **True**

- Because the given disjunction is false, both disjuncts must be false.
- Hence, the truth value of the statement "Ben is not 18 years old" is false.
- Since a statement and its negation have opposite truth values, the truth value of the statement "Ben is 18 years old" is true.

Example 2

Given the true statement "I will buy a new suit or I will not go to the dance" and the false statement "I will buy a new suit," what is the truth value of the statement, "I will go the dance"?

>   *Solution*: **False**

- Because the given disjunction is true, at least one of the two disjuncts must also be true.
- It is given that the disjunct "I will buy a new suit" is false. This means that the disjunct "I will not go to the dance" is true.
- Since a statement and its negation have opposite truth values, the truth value of the statement "I will go to the dance" is false.

The Conditional Statement

In the conditional statement "If <u>statement 1</u>, then <u>statement 2</u>," statement 1 is called the **hypothesis** (or **antecedent**) and statement 2 is the **conclusion** (or **consequent**). A conditional is always true *except* in the single situation in which the hypothesis (statement 1) is true and, at the same time, the conclusion (statement 2) is false. For example, the truth value of the open sentence

$$\overbrace{\text{If } x \text{ is divisible by 4}}^{p=\text{hypothesis}}, \text{ then } \overbrace{x \text{ is divisible by 3}}^{q=\text{conclusion}}.$$

depends on the replacement value of x:
- Suppose $x = 12$:

$$\overbrace{\text{If } 12 \text{ is divisible by 4}}^{p \text{ is true}}, \text{ then } \overbrace{12 \text{ is divisible by 3}}^{q \text{ is true}}.$$

  Since the hypothesis (p) and the conclusion (q) are both true, the conditional statement is true.
- Suppose $x = 8$:

$$\overbrace{\text{If } 8 \text{ is divisible by 4}}^{p \text{ is true}}, \text{ then } \overbrace{8 \text{ is divisible by 3}}^{q \text{ is false}}.$$

  Since the hypothesis (p) is true and the conclusion (q) is false, the conditional statement is false.
- Suppose $x = 9$:

$$\overbrace{\text{If } 9 \text{ is divisible by 4}}^{p \text{ is false}}, \text{ then } \overbrace{9 \text{ is divisible by 3}}^{q \text{ is true}}.$$

Since the hypothesis (p) is false and the conclusion (q) is true, the conditional statement is true.

- Suppose $x = 7$:

$$\underbrace{\text{If 7 is divisible by 4,}}_{p \text{ is false}} \text{ then } \underbrace{\text{7 is divisible by 3.}}_{q \text{ is false}}$$

Since the hypothesis (p) and the conclusion (q) are both false, the conditional statement is true.

Converse, Inverse, and Contrapositive

To form the **converse** of a conditional statement, switch the hypothesis with its conclusion. The **inverse** of a conditional statement is formed by negating the hypothesis and the conclusion. To form the **contrapositive** of a conditional statement, switch and negate the hypothesis and the conclusion.

Statement	General Form
Original	If statement 1, then statement 2.
Converse	If statement 2, then statement 1.
Inverse	If *not* statement 1, then *not* statement 2.
Contrapositive	If *not* statement 2, then *not* statement 1.

The converse and inverse do not necessarily have the same truth value as the original conditional. For example:

Conditional:	If it is January, then it is winter.	(True)
Converse:	If it is winter, then it is January.	(False)
Inverse:	If it is not January, then it is not winter.	(False)

The contrapositive always has the same truth value as the original conditional.

- When a conditional statement is true, its contrapositive is also true. For example:

Conditional:	If it is January, then it is winter.	(True)
Contrapositive:	If it is not winter, then it is not January.	(True)

- When a conditional statement is false, its contrapositive is also false. For example:

$$\text{Conditional:} \quad \text{If } 2 + 2 = 4, \text{ then } 1 + 1 = 3. \quad \text{(False)}$$
$$\text{Contrapositive:} \quad \text{If } 1 + 1 \neq 3, \text{ then } 2 + 2 \neq 4. \quad \text{(False)}$$

MATH FACTS

Different statements that always have the same truth value are **logically equivalent**. Since a conditional statement and its contrapositive always have the same truth value, they are logically equivalent statements. To form a statement that is logically equivalent to a given conditional, write its contrapositive.

Example

Multiple Choice:
Which statement is logically equivalent to the statement "If we recycle, then the amount of trash in landfills is reduced"?

(1) If we do not recycle, then the amount of trash in landfills is not reduced.
(2) If the amount of trash in landfills is not reduced, then we did not recycle.
(3) If the amount of trash in landfills is reduced, then we recycled.
(4) If we do not recycle, then the amount of trash in landfills is reduced.

Solution: **(2)**

The given statement and its contrapositive are logically equivalent. To form the contrapositive of a conditional, switch and then negate the hypothesis and conclusion:

Given conditional:
"If we recycle, then the amount of trash in landfills is reduced.
Contrapositive:
"If the amount of trash in landfills is *not* reduced, then we did *not* recycle.

Check Your Understanding of Section 3.5

A. Multiple Choice

1. Let p represent the statement "x is divisible by 3," and q represent the statement "x is divisible by 4." The statement p *and* q is true when
   (1) $x = 6$ (2) $x = 8$ (3) $x = 12$ (4) $x = 14$

2. Let p represent the statement "x is divisible by 6," and q represent the statement "x is a prime number." The statement p *or* q is false when
(1) $x = 2$ (2) $x = 6$ (3) $x = 9$ (4) $x = 30$

3. For which value of x is the sentence "If x is divisible by 6, then it is divisible by 8" a false statement?
(1) 8   (2) 15   (3) 24   (4) 30

4. What is the converse of the statement "If you used too much milk, then your cereal is soggy"?
(1) If your cereal is soggy, then you did not use too much milk.
(2) If you did not use too much milk, then your cereal is not soggy.
(3) If your cereal is not soggy, then you did not use too much milk.
(4) If your cereal is soggy, then you used too much milk.

5. Which statement is logically equivalent to the statement "If a number is even, then it is a multiple of 4"?
(1) If a number is not even, then it is not a multiple of 4.
(2) If a number is a multiple of 4, then the number is even.
(3) If a number is not a multiple of 4, then the number is not even.
(4) If a number is not even, then it is not a multiple of 4.

6. Which statement is logically equivalent to the statement "If it is sunny, then it is hot"?
(1) If it is hot, then it is sunny.
(2) If it is not hot, then it is not sunny.
(3) If it is not sunny, then it is not hot.
(4) If it is not hot, then it is sunny.

7. Which statement is then converse of "If I pass this test, then I am happy"?
(1) If I am not happy, then I did not pass this test.
(2) If I am happy, then I pass this test.
(3) If I did not pass this test, then I am happy.
(4) If I did not pass this test, then I am not happy.

8. Which statement is logically equivalent to "If I do not eat, then I do not watch TV"?
(1) If I eat, then I which TV.
(2) If I watch TV, then I eat.
(3) If I do not eat, then I watch TV.
(4) If I watch TV, then I do not eat.

9. If the statement "If the World Series is finished, then it is not September" is true, which statement *must* also be true?
   (1) If the World Series is not finished, then it is not September.
   (2) If it is September, then the World Series is not finished.
   (3) If it is not September, then the World Series is not finished.
   (4) If the World Series is finished, then it is September.

10. What is the inverse of the statement "If it is spring, then flowers bloom"?
   (1) If it is not spring, then flowers do not bloom.
   (2) If it is not spring, then flowers bloom.
   (3) If flowers do not bloom, then it is not spring.
   (4) If flowers bloom, then it is spring.

11. If the statement "If I am happy, then it is the winter" is false, which statement *must* also be false?
   (1) If I am not happy, then it is the winter.
   (2) If it is the winter, then I am not happy.
   (3) If it is not the winter, then I am not happy.
   (4) If I am not happy, then it is not the winter.

B. In each case, show how you arrived at your answer by clearly indicating all of the necessary steps, formula substitutions, diagrams, graphs, charts, etc.

12. Given the true statement "John is not handsome" and the false statement "John is handsome or smart," what is the truth value of the statement "John is smart"?

13. Given the true statement, "It is January or there is no pollen in the air" and the true statement "There is pollen in the air," what is the truth value of the statement, "It is not January"?

14. Given the true statement "It is not humid" and the true statement "It is not hot or it is humid," what is the truth value of the statement, "It is hot"?

3.6 COMBINED INEQUALITIES

KEY IDEAS

Inequalities and equations can be combined using the logical **and** and logical **or**. For example, $x \geq 3$ represents the disjunction $x = 3$ or $x > 3$. The inequality $3 < x < 7$ represents the conjunction $x > 3$ and $x < 7$.

Graphing Combined Inequalities

The graph of the conjunction $x \geq 1$ *and* $x < 4$ shown in Figure 3.6 consists of the portion on the number line over which the graphs of the two simple inequalities $x \geq 1$ and $x < 4$ overlap.

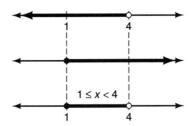

Figure 3.6   Graph of $1 \leq x < 4$

The solution interval includes all numbers from 1 to 4, including 1 but not 4. The compound inequality that represents this set of numbers is $1 \leq x < 4$, read as "1 is less than or equal to x *and* x is less than 4."

Hence:

$$x \geq 1 \text{ and } x < 4 \text{ means } 1 \leq x < 4$$

The graph of the solution set of $(x \leq -1)$ or $(x \geq 3)$ (see Figure 3.7) consists of all values less than or equal to -1 *or* greater than or equal to 3. Hence the graph consists of two nonoverlapping intervals since all values *between* -1 and 3 do *not* satisfy this inequality condition.

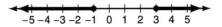

Figure 3.7   Graph of a Solution Set

Example 1

If x is an *integer*, list the numbers that satisfy $-4 < x < 1$.

   Solution: $\{-3, -2, -1, 0\}$

The solution set consists of the set of all integers between -4 and 1, but not including -4 and 1.

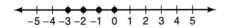

99

Example 2

Multiple Choice:
Which integer is a member of the solution of $-3 < x \le 1$?
(1) -3 (2) -4 (3) -2 (4) 2

    Solution: **(3)**

The solution set consists of the set of all integers between -3 and 1, including 1 and also -2.

Example 3

Multiple Choice:
Which inequality is represented by the following graph?

(1) $2 < x \le -3$ (2) $-3 \le x < 2$ (3) $-3 < x \le 2$ (4) $-3 \le x \le 2$

    Solution: **(2)**

The graph includes all values between -3 and 2, including -3.

Writing Combined Inequalities

A combined inequality tells the lowest and highest possible values for a variable quantity in a single inequality. For example, the measure of an acute angle lies between $0°$ and $90°$, as shown in the accompanying diagram. If x represents the number of degrees in the measure of an acute angle, then $0 < x$ *and* $x < 90$, which can be combined into the single inequality $0 < x < 90$. The inequality $0 < x < 90$ is 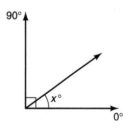 read as either "x is between 0 and 90" or "x is greater than 0 and less than 90."

    Suppose that on a summer day the temperature in New York City ranges from a low of $74°$ to a high of $89°$. If x represents the temperature at any given time during that day, then $74°$ (low) $\le x \le 89°$ (high). The inequality $73° < x < 90°$ means that the temperature, x, is *between* $73°$ and $90°$, so x is greater than $73°$ and, at the same time, less than $90°$.

Solving Combined Inequalities

To solve a combined inequality such as $-5 < 3x + 7 < 28$, isolate the variable by performing the same operation on each part of the inequality:

- Subtract 7 from each member of the inequality:

$$
\begin{array}{rcl}
-5 < & 3x + 7 & < 28 \\
-7 & -7 & -7 \\
\hline
-12 < 3x & & < 21
\end{array}
$$

- Divide each member of the inequality by 3:

$$\frac{-12}{3} < \frac{3x}{3} \qquad < \frac{21}{3}$$

- Simplify:

$$-4 < x \qquad < 7$$

Check Your Understanding of Section 3.6

A. *Multiple Choice*

1. A member of the solution set of $-1 \le x < 4$ is
   - (1) -1
   - (2) -2
   - (3) 5
   - (4) 4

2. If y is an integer, what is the solution set of $-3 \le y < 2$?
   - (1) $\{-3, -2, -1, 0, 1\}$
   - (2) $\{-2, -1, 0, 1, 2\}$
   - (3) $\{-3, -2, -1\}$
   - (4) $\{0, 1\}$

3. If x is an integer, what is the solution set of $5 < x \le 8$?
   - (1) $\{5, 6, 7, 8\}$
   - (2) $\{6, 7, 8\}$
   - (3) $\{5, 6, 7\}$
   - (4) $\{6, 7\}$

4. If n is an integer, how many different values of n make the inequality $-7 < 3n \le 15$ a true statement?
   - (1) 9
   - (2) 8
   - (3) 7
   - (4) 5

5. Which inequality is represented by the graph below?

   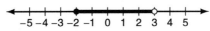

   - (1) $-2 \le x < 3$
   - (2) $-2 < x \le 3$
   - (3) $x > 3$ or $x \le 2$
   - (4) $x \ge 3$ or $x < -2$

6. Which inequality is represented by the graph below?

   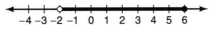

   - (1) $-2 < x \le 6$
   - (2) $-2 \le x < 6$
   - (3) $-2 < x < 6$
   - (4) $-2 \le x \le 6$

7. Which inequality is true if $x = \dfrac{2.54}{1.28}$, $y = \sqrt{3.61}$, and $z = \left(\dfrac{5}{4}\right)^3$?

(1) $x < z < y$ (2) $y < z < x$ (3) $z < y < x$ (4) $x < y < z$

8. Which graph represents the solution set of $-2 \le x < 1$?

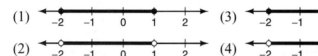

9. Vicki is shorter than Gail but taller than Toni. If v, g, and t, respectively, represent the numbers of inches in each girl's height, which of the following is true?

(1) $g < v < t$ (2) $v < g < t$ (3) $t < v < g$ (4) $t < g < v$

10. Let x, y, and z represent positive integers greater than 1. If $xy = 35$ and $yz = 21$, which inequality *must* be true?

(1) $x < y < z$ (2) $y < x < z$ (3) $z < y < x$ (4) $z < x < y$

11. Which statement is true when $-1 < x < 0$?

(1) $x < x^2 < x^3$ (2) $x^3 < x < x^2$ (3) $x < x^3 < x^2$ (4) $x^3 < x^2 < x$

12. On the same day that the temperature in New York City ranges from 80° to 89°, the temperature in Niagara Falls ranges from 60° to 69°. At any given time on that day, the difference in the temperatures of these two cities must be

(1) between 10° and 20°   (3) between 10° and 30°
(2) between 20° and 30°   (4) 20°

B. *In each case, show how you arrived at your answer by clearly indicating all of the necessary steps, formula substitutions, diagrams, graphs, charts, etc.*

13. A stream has a normal level of acidity if the average of three pH measurements taken at different locations is between 7.0 and 7.8. The first two pH measurements for the stream are 7.6 and 8.0. What possible range of values for the third measurement will make the lake have a normal level of acidity?

14. It is estimated that 1 of 75 lightbulbs that are manufactured are defective and that the actual number of defective lightbulbs does not vary more than 5% above or below the estimate.
(a) In a production run of 30,000 lightbulbs, estimate the number of defective lightbulbs.
(b) If x represents the actual number of defective lightbulbs, which interval best describes the set of possible values for x?
(1) $360 \leq x \leq 400$ (3) $400 \leq x \leq 400$
(2) $380 \leq x \leq 420$ (4) $375 \leq x \leq 425$

15. If two times an integer, x, is increased by 5, the result is always greater than 16 and less than 29. What is the smallest value of x?

CHAPTER 4

ANGLES, LINES, AND POLYG

4.1 BUILDING A GEOMETRY VOCABULARY

KEY IDEAS

Cement and bricks are used to give a house a strong foundation. The building blocks of geometry take the form of *undefined terms, defined terms, postulates,* and *theorems.*

Undefined Terms

Some basic terms in geometry can be described but cannot be defined by using simpler terms. *Point, line,* and *plane* are undefined terms.

A **point** indicates location and has no size or dimensions. A point is represented by a dot and named by a capital letter (Figure 4.1).

A **line** is a set of continuous points that form a straight path that extends without ending in two opposite directions. A line has no width. A line is identified by naming two points on the line or by writing a lowercase letter after the line (Figure 4.2). The notation $\overleftrightarrow{AB}$ is read as "line *AB*" and refers to the line that contains points *A* and *B*.

A **plane** is a flat surface that has no thickness and extends without ending in all directions. A plane is represented by a "window pane" and named by writing a capital letter in one of its corners (Figure 4.3).

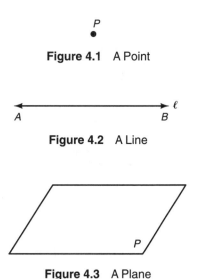

P
•

Figure 4.1 A Point

Figure 4.2 A Line

Figure 4.3 A Plane

ed Terms

Defined terms are terms whose distinguishing characteristics can be explained by using either undefined or previously defined terms. The accompanying table lists some basic geometric terms and their definitions.

Term	Definition	Illustration
Line segment	A **line segment** is a part of a line consisting of two points, called endpoints, and the set of all points between them.	 Notation: $\overline{AB}$
Ray	A **ray** is a part of a line consisting of a given point, called the endpoint, and the set of all points on one side of the end-point.	 Notation: $\overrightarrow{LM}$ A ray is always named by using two points, the first of which must be the endpoint. The arrow on top must always point to the right.
Opposite rays	**Opposite rays** are rays that have the same endpoint and that form a line.	 $\overrightarrow{KX}$ and $\overrightarrow{KB}$ are opposite rays.
Angle	An **angle** is the union of two rays having the same endpoint. The endpoint is called the vertex of the angle; the rays are called the sides of the angle.	 Vertex: K Sides: $\overrightarrow{KJ}$ and $\overrightarrow{KL}$
Collinear points	**Collinear points** are points that lie on the same line.	 Points A, B and C are collinear. Points B, C, and D are *not* collinear.

Naming Angles

An angle may be named in three different ways.

1. By using three letters, with the middle letter corresponding to the vertex of the angle and the other letters naming one point on each side of the angle. In Figure 4.4, the name of the angle may be ∠*RTB or* ∠*BTR.*

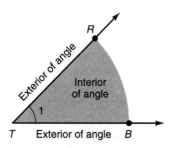

Figure 4.4

2. By placing a number at the vertex and in the interior of the angle. The angle may then be referred to by the number. In Figure 4.4, the name of the angle may be ∠1 *or* ∠*RTB or* ∠*BTR.*
3. By using a *single* capital letter that corresponds to the vertex, provided that this causes no confusion. In Figure 4.5 there is no question that ∠*A* is another name for ∠*BAD.* On the other hand, ∠*D* may not be used since it is not clear whether this name refers to ∠*ADB or* ∠*CDB or* ∠*ADC.*

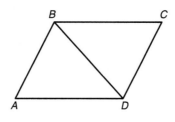

Figure 4.5

Postulates Versus Theorems

A **theorem** is a generalization that can be proved to be true, while a **postulate** is a statement that is *assumed* to be true without proof. Here is a familiar theorem: *The sum of the measures of the three angles of a triangle is 180 degrees.*

Not all statements can be proved since there must be some basic assumptions, that is, *postulates*, that are needed as a beginning. Here are two postulates:

1. Exactly one line segment can connect two different points.

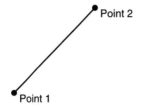

•Point 2

•Point 1

2. If two lines intersect, they meet at exactly one point.

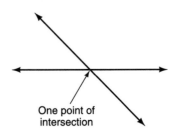

One point of
intersection

Measuring Segments and Angles

The **length of a line segment** is the distance between its endpoints. If, for example, the distance between points A and B is 2 inches, then the length or measure of $\overline{AB}$ is 2, which is abbreviated by writing $AB = 2$.

The **degree measure** of an angle is the amount of rotation from one side of the angle to the other side. A rotation of 1 degree (1°) is $\frac{1}{360}$ of one complete circular rotation, as shown in Figure 4.6. If the degree measure of $\angle AOX$ is 50, we write $m\angle AOX = 50$.

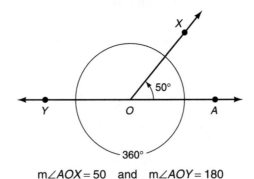

$$m\angle AOX = 50 \quad \text{and} \quad m\angle AOY = 180$$

Figure 4.6 Angle Rotation and Degree Measure

Classifying Angles

An angle whose sides form a straight line is called a **straight angle** and has a degree measure of 180. Angles whose degree measures fall between 0 and 180 can be classified as shown in Figure 4.7.

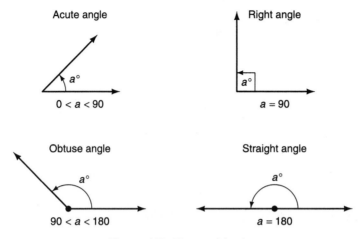

Acute angle

$a°$

$0 < a < 90$

Right angle

$a°$

$a = 90$

Obtuse angle

$a°$

$90 < a < 180$

Straight angle

$a°$

$a = 180$

Figure 4.7   Types of Angles

Perpendicular Lines

Perpendicular lines are lines that intersect to form a right angle. If line ℓ is perpendicular to line m, as shown in Figure 4.8, we write $\ell \perp m$, where the symbol $\perp$ is read as "is perpendicular to."

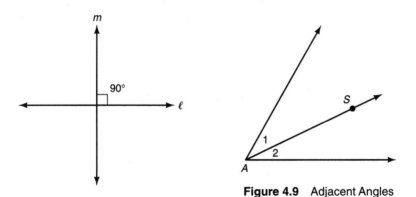

m

$90°$

ℓ

Figure 4.8   Perpendicular Lines

S

1

2

A

Figure 4.9   Adjacent Angles

Adjacent Angles

Two angles are **adjacent** if they have the same vertex, share one side, and do not overlap. In Figure 4.9, $\angle 1$ and $\angle 2$ are adjacent angles since they have the same vertex, A, share $\overrightarrow{AS}$ as their common side, and do not overlap. Two or more adjacent angles may form a straight angle, as shown in Figure 4.10, where m∠1 + m∠2 + m∠3 = 180.

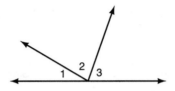

Figure 4.10 Straight Angle Formed by Angles 1, 2, and 3

Congruent Segments and Angles

If line segments AB and XY are each 2 inches in length, then line segment AB is *congruent to* line segment XY. This can be abbreviated by writing $\overline{AB} \cong \overline{XY}$, where the symbol $\cong$ means "is congruent to."

Angles that have the same degree measure are also congruent. If m∠A = 60 and m∠B = 60, then $\angle A \cong \angle B$. Since all right angles measure 90 degrees, *all right angles are congruent.* Thus, if $\angle A$ and $\angle X$ are right angles, then $\angle A \cong \angle X$.

Bisectors and Midpoints

The **midpoint** of a line segment divides the segment into two segments that have the same length. Any line or segment that passes through the midpoint of a line segment, as shown in Figure 4.11, *bisects* the line segment. As shown in Figure 4.12, a ray such as BD that divides an angle into two angles with equal measures *bisects* the angle and is called an **angle bisector**. A line segment has exactly one midpoint, and an angle has exactly one bisector.

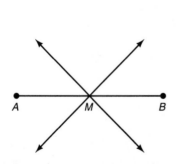

Figure 4.11 Bisectors Through Midpoint M, Where $AM = MB$

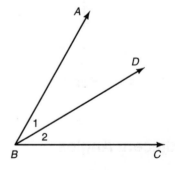

Figure 4.12 Angle Bisector $\overrightarrow{BD}$ Makes m∠1 = m∠2

Check Your Understanding of Section 4.1

A. *Multiple Choice*

1. If M is the midpoint of $\overline{AB}$, which statement is false?
   - (1) $\dfrac{AB}{2} = MB$
   - (3) $AM + AB = MB$
   - (2) $AB - MB = AM$
   - (4) $AM = MB$

2. If C is the midpoint of $\overline{AB}$ and D is the midpoint of $\overline{AC}$, which statement is true?
   - (1) $AC > BC$
   - (3) $DB = AC$
   - (2) $AD < CD$
   - (4) $DB = 3CD$

B. *In each case, show how you arrived at your answer by clearly indicating all of the necessary steps, formula substitutions, diagrams, graphs, charts, etc.*

3. Points P, I, and Z are collinear, and $IZ = 8$, $PI = 14$, and $PZ = 6$.
   (a) Which of the three points is between the other two?
   (b) If point M is the midpoint of $\overline{PI}$ and point N is the midpoint of $\overline{IZ}$, what is the length of $\overline{MN}$?

4. Ray PL lies between the sides of $\angle RPH$. If $m\angle RPL = x - 5$, $m\angle LPH = 2x + 18$, and $m\angle RPH = 58$, what is the degree measure of the smallest angle formed that has $\overrightarrow{PL}$ as one side?

5. If $\overrightarrow{PQ}$ bisects $\angle HPJ$, $m\angle QPJ = 2x - 9$, and $m\angle QPH = x + 29$, what is the degree measure of $\angle HPJ$?

6. If R is the midpoint of $\overline{XY}$, $XR = 3n + 1$, and $YR = 16 - 2n$, what is the length of $\overline{XY}$?

7 and 8. In each case, find the value of x.

7.

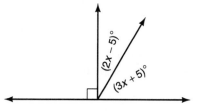

8.

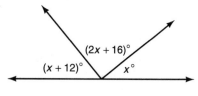

111

4.2 SPECIAL PAIRS OF ANGLES

=========== **KEY IDEAS** ===========

If two angles form a pair of *complementary, supplementary,* or *vertical* angles, a special relationship exists between the measures of the two angles.

Complementary and Supplementary Angles

Two angles are **complementary** if the sum of their degree measures is 90. If the degree measures of two angles add up to 180, the angles are **supplementary**.

Example 1

If the degree measures of two complementary angles are in the ratio of 2 to 13, what is the degree measure of the smaller angle?

Solution: **12**

> Let $2x$ = degree measure of the smaller angle.
> Then $13x$ = degree measure of the larger angle.

Since the two angles are complementary, the sum of their degree measures is 90. Thus:

$$2x + 13x = 90$$
$$15x = 90$$
$$x = \frac{90}{15} = 6$$
$$2x = 2(6) = 12$$

Example 2

If the degree measure of an angle exceeds twice the degree measure of its supplement by 30, what is the degree measure of the angle?

Solution: **130**

> Let x = degree measure of the angle.
> Then $180 - x$ = degree measure of the supplement of the angle.

$$x = 2(180 - x) + 30$$
$$= 360 - 2x + 30$$
$$x + 2x = 390$$
$$3x = 390$$
$$x = \frac{390}{3} = 130$$

Vertical Angles

In Figure 4.13, opposite angles 1 and 3 are called **vertical angles**, as are opposite angles 2 and 4. Suppose than m∠1 = 50. Then, since ∠1 and ∠2 are supplementary:

$$m\angle 2 = 180 - m\angle 1$$
$$= 180 - 50$$
$$= 130$$

Since ∠2 and ∠3 are also supplementary:

$$m\angle 3 = 180 - m\angle 2$$
$$= 180 - 130$$
$$= 50$$

Since m∠1 = 50 and m∠3 = 50, m∠1 = m∠3. Similarly, m∠4 = (180 − m∠3) = 130, so m∠2 = m∠4.

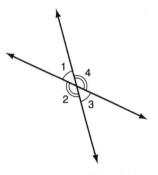

Figure 4.13   Vertical Angles

MATH FACTS

Opposite angles formed when two lines intersect are called **vertical angles**. Vertical angles are equal in measure and are, therefore, congruent.

Example 3

In the accompanying diagram, what is the value of y?

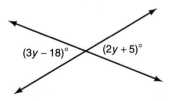

$(3y - 18)°$ $(2y + 5)°$

Solution: **23**

Since vertical angles are equal in measure:

$$3y - 18 = 2y + 5$$
$$3y = 2y + 5 + 18$$
$$3y - 2y = 23$$
$$y = 23$$

Check Your Understanding of Section 4.2

A. *Multiple Choice*

1. If the measures of two supplementary angles are in the ratio of $4:5$, the degree measure of the larger angle is
   (1) 20   (2) 80   (3) 100   (4) 120

2. If the measures of two complementary angles are in the ratio of $1:5$, the degree measure of the large angle is
   (1) 72   (2) 75   (3) 144   (4) 150

3. If the complement of $\angle A$ is greater than the supplement of $\angle B$, which statement must be true?
   (1) $m\angle A + m\angle B = 180$ (3) $m\angle A < m\angle B$
   (2) $m\angle A + m\angle B = 90$ (4) $m\angle A > m\angle B$

4. Two angles are complementary. The measure of one angle exceeds two times the measure of the other angle by 21. What is the degree measure of the smaller angle?
   (1) 23   (2) 53   (3) 67   (4) 127

5. In the accompanying figure, what is the value of y?

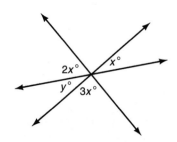

   (1) 20   (2) 30   (3) 45   (4) 60

6. In the accompanying figure, line *m* is perpendicular to line *p*. What is the value of *x*?

(1) 15   (3) 24
(2) 20   (4) 30

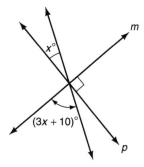

B. *In each case, show how you arrived at your answer by clearly indicating all of the necessary steps, formula substitutions, diagrams, graphs, charts, etc.*

7–9. In each case, find the value of x.

7.

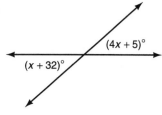

9.

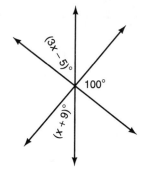

8.

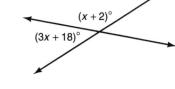

10. If the measures of two supplementary angles are in the ratio of 2 : 3, find the measure of the smaller angle.

11. The degree measure of an angle exceeds four times the degree measure of its complement by 6. Find the degree measure of the angle.

12. The measure of the supplement of ∠*x* is three times as great as the measure of the complement of ∠*x*. What is the measure of ∠*x*?

13. If the difference between the degree measures of an angle and its complement is 14, find the degree measure of the smaller angle.

14. If the difference between the degree measures of an angle and its supplement is 22, find the degree measure of the larger angle.

15. If $\overleftrightarrow{XY}$ and $\overleftrightarrow{AB}$ intersect at point *C*, m∠*XCB* = 4*y* − 9, and m∠*ACY* = 3*y* + 29, find m∠*XCB*.

115

4.3 PARALLEL LINES

◇ KEY IDEAS

Lines that lie in the same plane and never meet are called **parallel lines**. The notation $\overleftrightarrow{AB}\|\overleftrightarrow{CD}$ is read as "line AB is parallel to line CD." When two parallel lines are cut by a third line, called a **transversal**, any two of the eight angles that are formed either have equal measures or are supplementary.

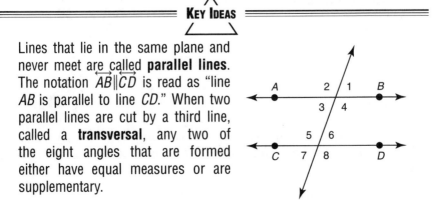

Recognizing Special Paris of Angles

If two lines are cut by a transversal, pairs of angles between the two lines and on opposite sides of the transversal are called **alternate interior angles**. A pair of **corresponding angles** consists of one interior angle and one exterior angle both of which lie on the same side of the transversal. When looking at a diagram, you will see that the sides of a pair of alternate interior angles form a Z-shape, while the sides of a pair of corresponding angles form an F-shape, as shown in Figure 4.14. Keep in mind that in different diagrams these letter shapes may appear reversed, sideways, upside down, or otherwise rotated.

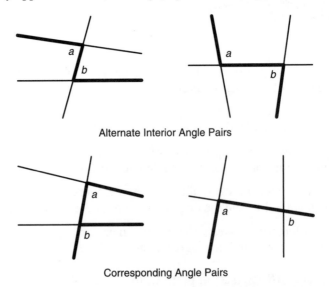

Alternate Interior Angle Pairs

Corresponding Angle Pairs

Figure 4.14 Letter Shapes Formed by Angle Pairs

116

Properties of Parallel Lines

If two parallel lines are cut by a transversal, pairs of acute angles are equal in measure, pairs of obtuse angles are equal in measure, and any acute angle is supplementary to any obtuse angle.

MATH FACTS

If $\ell \parallel m$, then:

- *Alternate interior angles* are equal in measure [$c = e$ and $d = f$].
- *Corresponding angles* are equal in measure [$a = e$, $c = g$, $b = f$, and $d = h$].

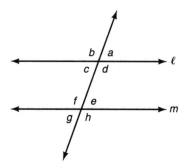

- The degree measures of two interior angles on the same side of the transversal add up to 180 [$d + e = 180$ and $c + f = 180$]. Furthermore, the sum of the degree measures of *any* acute angle and *any* obtuse angle formed is 180, as in $b + g = 180$.

Example 1

In the accompanying diagram, $\ell \parallel m$. What is the value of x?

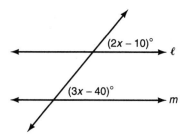

Solution: **30**

Since corresponding angles formed by parallel lines are equal in measure:

$$3x - 40 = 2x - 10$$
$$3x = 2x - 10 + 40$$
$$= 2x + 30$$
$$3x - 2x = 30$$
$$x = 30$$

Example 2

In the accompanying diagram, $\ell \parallel m$. What is the value of x?

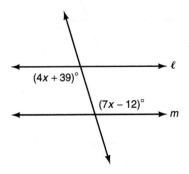

Solution: **17**

Since alternate interior angles formed by parallel lines are equal in measure:

$$7x - 12 = 4x + 39$$
$$7x = 4x + 39 + 12$$
$$= 4x + 51$$
$$7x - 4x = 51$$
$$3x = 51$$
$$x = \frac{51}{3} = 17$$

Example 3

In the accompanying diagram, $\overleftrightarrow{ALB} \parallel \overleftrightarrow{CJD}$ and $\overleftrightarrow{LJ}$ is a transversal. If $m\angle LJB = 6x - 7$ and $m\angle LJD = 7x + 5$, find the value of x.

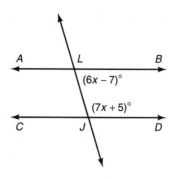

Solution: **14**

When two lines are parallel, the sum of the measures of a pair of interior angles on the same side of the transversal is 180°:

$$m\angle JLB + m\angle LJD = 180$$
$$(6x - 7) + (7x + 5) = 180$$
$$13x - 2 = 180$$
$$13x = 182$$
$$x = \frac{182}{13}$$
$$= 14$$

Example 4

In the accompanying diagram, line k is parallel to line n, and line ℓ is a transversal that intersects lines k and n. If $m\angle 1 = x + 25$ and $m\angle 2 = 5x - 25$, find x.

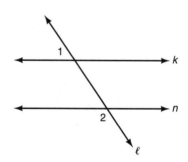

Solution: **30**

Since $\angle 1$ is an acute angle and $\angle 2$ is an obtuse angle formed by parallel lines, the sum of their degree measures must be 180:

$$m\angle 1 + m\angle 2 = 180$$
$$(x + 25) + (5x - 25) = 180$$
$$6x = 180$$
$$x = \frac{180}{6} = 30$$

Determining When Lines Are Parallel

Two lines are parallel if any one of the following statements is true:

- A pair of alternate interior angles have the same measure.
- A pair of corresponding angles have the same measure.
- A pair of interior angles on the same side of the transversal are supplementary.

Each of these statements is the *converse* of a statement that gives a property of parallel lines.

Example 5

Multiple Choice:
Which of the accompanying diagrams contains a pair of parallel lines?

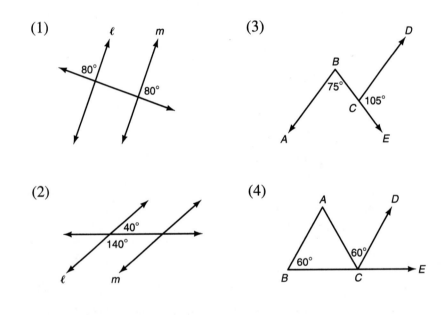

Solution: **(3)**

In (3), m∠*BCD* = 180 − 105 = 75. Since alternate interior angles are equal in measure, $\overleftrightarrow{BA} \parallel \overleftrightarrow{CD}$.

Check Your Understanding of Section 4.3

A. Multiple Choice

1. In the accompanying diagram, parallel lines ℓ and m are cut by transversal t. Which statement is true?
   (1) $m\angle 1 + m\angle 2 + m\angle 5 = 360$
   (2) $m\angle 1 + m\angle 2 + m\angle 3 = 180$
   (3) $m\angle 1 + m\angle 2 = m\angle 2 + m\angle 3$
   (4) $m\angle 1 + m\angle 3 = m\angle 4 + m\angle 5$

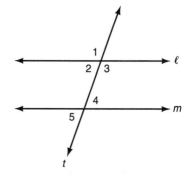

2. In the accompanying diagram, parallel lines $\overleftrightarrow{AB}$ and $\overleftrightarrow{CD}$ are cut by transversal $\overleftrightarrow{EF}$ at P and Q, respectively. Which statement *must always* be true?
   (1) $m\angle APE = m\angle CQF$
   (2) $m\angle APE + m\angle CQF = 90$
   (3) $m\angle APE < m\angle CQF$
   (4) $m\angle APE + m\angle CQF = 180$

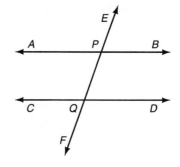

3. In the accompanying diagram parallel lines $\overleftrightarrow{AB}$ and $\overleftrightarrow{CD}$ are cut by transversal $\overleftrightarrow{EF}$. If $m\angle BEF = 2x + 60$ and $m\angle EFD = 3x + 20$, what is $m\angle BEF$?
   (1) 100   (3) 140
   (2) 20   (4) 40

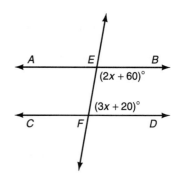

4. In the accompanying diagram, $\overrightarrow{AD}\|\overrightarrow{BC}$ and $\overrightarrow{AC}$ bisects $\angle BAD$. If $m\angle ABC = x$, what is $m\angle 1$ in terms of x?

(1) $90 - x$

(2) $\dfrac{90 - x}{2}$

(3) $90 - \dfrac{x}{2}$

(4) $\dfrac{90 + x}{2}$

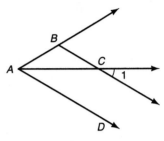

5. If, in the accompanying diagram, $\ell_1 \| \ell_2$ and $\ell_3 \| \ell_4$, then $\angle x$ is not always congruent to which other angle?

(1) a
(2) b
(3) c
(4) d

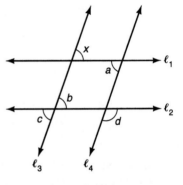

B. *In each case, show how you arrived at your answer by clearly indicating all of the necessary steps, formula substitutions, diagrams, graphs, charts, etc.*

6. Explain why two lines that are perpendicular to the same line are or are not *always* parallel to each other.

7. In the accompanying diagram, $\overleftrightarrow{WX}\|\overleftrightarrow{YZ}$; $\overleftrightarrow{AB}$ and $\overleftrightarrow{CD}$ intersect $\overleftrightarrow{WX}$ at E and $\overleftrightarrow{YZ}$ at F and G, respectively. If $m\angle CEW = m\angle BEX = 50$, find $m\angle EGF$.

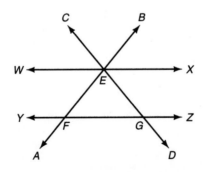

8. Two parallel lines are cut by a transversal so that the measures of a pair of interior angles on the same side of the transversal are in the ratio of $5:13$. Find the measure of the smaller of these angles.

9. In the accompanying diagram, parallel lines $\overleftrightarrow{HE}$ and $\overleftrightarrow{AD}$ are cut by transversal $\overleftrightarrow{BF}$ at points G and C, respectively. If $m\angle HGF = 5n$ and $m\angle BCD = 2n + 66$, find the value of n.

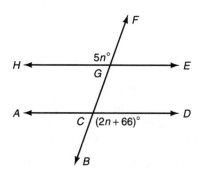

10. In the accompanying diagram, lines p and q are parallel.

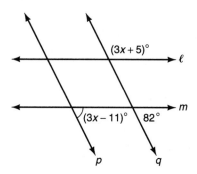

(a) Find the value of x.
(b) Explain why line ℓ is or is not parallel to line m.

4.4 CLASSIFYING POLYGONS AND TRIANGLES

KEY IDEAS

A **triangle** is a *polygon* with three sides. A **polygon** is a "closed" figure whose sides are line segments. Each corner of a polygon where two sides intersect is called a **vertex** of the polygon.

Classifying Polygons

A polygon may be identified by the number of sides. A polygon with 4 sides is called a **quadrilateral**, a polygon with 5 sides is a **pentagon**, and a polygon with 6 sides is a **hexagon**. An **octagon** has 8 sides, a **decagon** has 10 sides, and a polygon with 12 sides is a **dodecagon**.

- In an **equiangular polygon**, each angle has the same measure. A rectangle is an equiangular polygon since the degree measure of each of the four angles is 90.
- In an **equilateral polygon**, each side has the same length. A triangle in which the length of each side is 9 inches is equilateral.
- A **regular polygon** is both equiangular and equilateral. A square is a regular polygon since its four sides have the same length and its four angles have the same measure.

Classifying Triangles

A triangle may be classified according to the number of sides that have the same length, as shown in Figure 4.15.

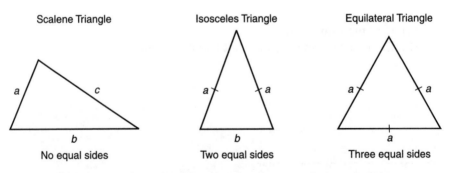

Figure 4.15 Triangles Classified by Number of Congruent Sides

A triangle may also be classified by the degree measure of its largest angle (Figure 4.16).

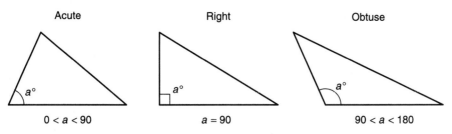

Acute	Right	Obtuse
$0 < a < 90$	$a = 90$	$90 < a < 180$

Figure 4.16 Triangles Classified by Largest Angle Measurement

Median and Altitude

In every triangle a median and an altitude can be drawn from any vertex to the side opposite that vertex.

- A **median** of a triangle is a segment drawn from a vertex of the triangle to the midpoint of the opposite side.

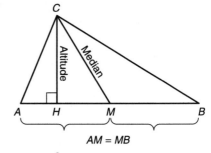

$AM = MB$

- An **altitude** of a triangle is a segment drawn from a vertex of the triangle perpendicular to the opposite side or, as in the accompanying diagram, to the opposite side extended.

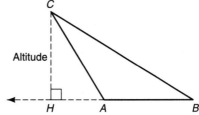

Distance

The term *distance* in geometry is always interpreted as the *shortest* path. The **distance between a point and a line** is the length of the perpendicular segment drawn from the point to the line.

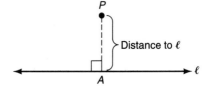

125

Check Your Understanding of Section 4.4

A. *Multiple Choice*

1. In $\triangle ABC$, $AB = x + 7$, $BC = 2x - 1$, and $AC = 3x - 9$. If the perimeter of $\triangle ABC$ is 45, which statement *must* be true?
   (1) $\triangle ABC$ is scalene.
   (2) $\triangle ABC$ is isosceles.
   (3) $\triangle ABC$ is equilateral.
   (4) $\triangle ABC$ contains a right angle.

2. In $\triangle ABC$, $AB = 2x$, $BC = 3x - 10$, and $AC = x + 9$. If the perimeter of $\triangle ABC$ is 59, which statement *must* be true?
   (1) $\triangle ABC$ is scalene.
   (2) $\triangle ABC$ is isosceles.
   (3) $\triangle ABC$ is equilateral.
   (4) $\triangle ABC$ contains a right angle.

3. In the accompanying diagram of equilateral triangle *ABC*, *DE* = 4, and $\overline{DE} \parallel \overline{AB}$. If the length of $\overline{DE}$ is one-third of the length of $\overline{AB}$, what is the perimeter of quadrilateral *ABED*?
   (1) 20   (3) 32
   (2) 24   (4) 36

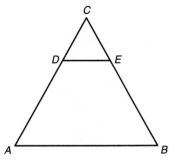

4. In $\triangle ABC$, $AB = x + 10$, $BC = 2x$, and $AC = 3x - 10$. If the perimeter of $\triangle ABC$ is 66, which statement must be true?
   (1) $\triangle ABC$ is scalene.
   (2) $\triangle ABC$ is isosceles.
   (3) $\triangle ABC$ is equilateral.
   (4) $\triangle ABC$ contains a right angle.

5. Assume that the Pentagon building in Washington, D.C., has the shape of a regular pentagon. If the length of one side of the pentagon is represented by $2x + 1$, then its perimeter is represented by
   (1) $10x + 1$ (2) $10x + 5$ (3) $2x + 5$ (4) $5x + 5$

6. Line ℓ contains points *P* and *Q*. The number of different points whose distance from point *P* is two times its (their) distance from point *Q* is
   (1) 1   (2) 2   (3) more than 4 (4) 4

B. *In each case, show how you arrived at your answer by clearly indicating all of the necessary steps, formula substitutions, diagrams, graphs, charts, etc.*

7. Express in terms of x the perimeter of a regular hexagon if the length of each side is represented by $3x + 2$.

8. If the perimeter of an equilateral polygon with seven sides is represented by $21x^2 - 35x$, express the length of one of its sides in terms of x.

9. The perimeter of a regular pentagon is equal to the perimeter of a certain regular hexagon. If the length of a side of the pentagon exceeds the length of a side of the hexagon by 2, find the perimeter of each polygon.

4.5 TRIANGLE ANGLE SUM

KEY IDEAS

The accompanying diagrams illustrate that, after $\angle 1$ and $\angle 3$ have been "torn off," their sides can be aligned with one of the sides of $\angle 2$ so that the exterior sides of the three angles form a straight line.

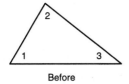
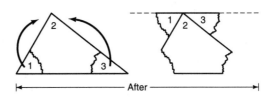

Before — |———————————— After ————————————|

Since the degree measure of a straight angle is 180, this experiment suggests that the sum of the degree measures of angles 1, 2, and 3 is 180 and, therefore, the sum of the degree measures of the angles of *any* triangle is 180.

Angles of a Triangle

To *prove* that the sum of the degree measures of the angles of a triangle is 180, evidence that does not depend on "tearing off" angles is needed. In Figure 4.17 line ℓ has been drawn parallel to side *AC* of the triangle and through one of its vertices (*B*).

127

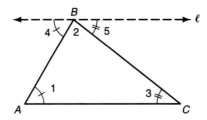

Figure 4.17 Angles of a Triangle

Since $\ell \parallel AC$, alternate interior angles are equal in measure:

$$m\angle 1 = m\angle 4,$$

and

$$m\angle 3 = m\angle 5.$$

The sum of the degree measures of the angles formed at vertex B is 180:

$$m\angle 4 + m\angle 2 + m\angle 5 = 180.$$

We may therefore substitute in the above equation $m\angle 1$ for $m\angle 4$ and $m\angle 3$ for $m\angle 5$:

$$\mathbf{m\angle 1 + m\angle 2 + m\angle 3 = 180.}$$

This analysis provides an "informal" proof of the following theorem:

MATH FACTS

TRIANGLE ANGLE SUM THEOREM

The sum of the degree measures of the angles of a triangle is 180.

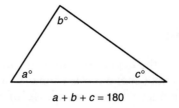

$$a + b + c = 180$$

Example 1

In the accompanying figure, $\overleftrightarrow{AB}$ is parallel to $\overleftrightarrow{CD}$, $\overline{AE} \perp \overline{BC}$, m$\angle BCD = 2x$, m$\angle BAE = 3x$. Find the value of x.

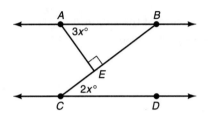

Solution: **18**

Since alternate interior angles formed by parallel lines have the same degree measure, m$\angle ABE = $ m$\angle BCD = 2x$. In right triangle *AEB*, angles *BAE* and *ABE* are the acute angles and, as a result, are complementary. Hence:

$$\text{m}\angle BAE + \text{m}\angle ABE = 90$$
$$3x + 2x = 90$$
$$5x = 90$$
$$x = \frac{90}{5}$$
$$= 18$$

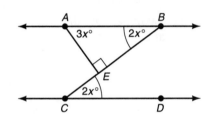

Example 2

If a triangle is equiangular, what is the degree measure of each of its angles?

Solution: **60**

In an equiangular triangle all of the angles have the same measure. Let x represent the degree measure of each angle. Then:

$$x + x + x = 180$$
$$3x = 180$$
$$x = \frac{180}{3} = 60$$

Example 3

The degree measures of the angles of a triangle are in the ratio of $2:3:5$. What is the measure of the smallest angle of the triangle?

Solution: **36**

Let $2x$ = degree measure of the smallest angle of the triangle. Then $3x$ and $5x$ = degree measures of the remaining angles.

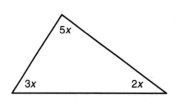

$$2x + 3x + 5x = 180$$
$$10x = 180$$
$$\frac{10x}{10} = \frac{180}{10}$$
$$x = 18$$
$$2x = 2(18) = 36$$

Example 4

In the accompanying diagram, $\overline{DE} \perp \overline{AEC}$. If m$\angle ADB = 80$ and m$\angle CDE = 60$, what is m$\angle DAE$?

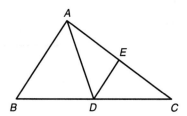

Solution: **50**

Since angles *ADB*, *ADE*, and *CDE* form a straight angle, the sum of their degree measures is 180. Hence:

$$80 + m\angle ADE + 60 = 80$$
$$m\angle ADE = 180 - 140 = 40$$

In $\triangle ADE$:

$$m\angle DAE + m\angle ADE + m\angle AED = 180$$
$$m\angle DAE + \quad 40 \quad + \quad 90 \quad = 180$$
$$m\angle DAE + \quad\quad 130 \quad\quad = 180$$
$$m\angle DAE = 180 - 130$$
$$= 50$$

Example 5

In the accompanying diagram, $\overline{AD} \parallel \overline{EC}$, $\overline{DF} \parallel \overline{CB}$, m∠$DAE$ = 34, and m∠DFE = 57. Find m∠ECB.

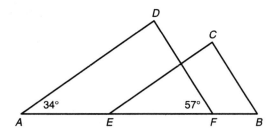

Solution: **89**

Since $\overline{AD} \parallel \overline{EC}$, transversal $\overline{AEFB}$ forms congruent and corresponding angles, so m∠CEB = m∠DAE = 34. Also, since $\overline{DF} \parallel \overline{CB}$, transversal $\overline{AEFB}$ forms congruent corresponding angles, so m∠CBE = m∠DFE = 57. In $\triangle CEB$:

$$\text{m}\angle ECB + \text{m}\angle CEB + \text{m}\angle CBE = 180$$
$$\text{m}\angle ECB + \quad 34 \quad + \quad 57 \quad = 180$$
$$\text{m}\angle ECB + \qquad 91 \qquad = 180$$
$$\text{m}\angle ECB = 180 - 91$$
$$= 89$$

Exterior Angles of a Triangle

At each vertex of a triangle an *exterior* angle of the triangle may be formed by extending one side of the triangle (Figure 4.18). Notice that:

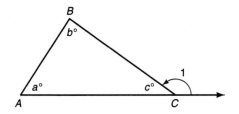

Figure 4.18 Exterior Angles of a Triangle

$$(a+b)+c = 180,$$
$$\text{m}\angle 1 + c = 180.$$

By comparing these two equations, we may conclude that:

$$m\angle 1 = a + b.$$

The angles whose measures are represented by a and b are the two interior angles of the triangle that are the most remote from $\angle 1$. With respect to $\angle 1$, these angles are nonadjacent. We may generalize as follows:

MATH FACTS

TRIANGLE EXTERIOR ANGLE THEOREM

The measure of an exterior angle of a triangle is equal to the sum of the measures of the two remote (nonadjacent) interior angles of the triangle.

Example 6

In each of the accompanying diagrams, find the value of x.

(a)

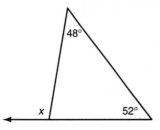

(b)

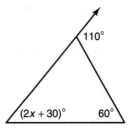

(c)

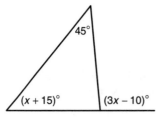

(d)

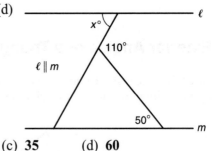

Solution: (a) **100**   (b) **10**   (c) **35**   (d) **60**

(a) $x = 48 + 52 = 100$

(b) $110 = 2x + 30 + 60$
$110 = 2x + 90$
$20 = 2x$
$x = 10$

(c) $3x - 10 = (x + 15) + 45$
$3x - 10 = x + 60$
$3x = x + 70$
$2x = 70$
$x = 35$

(d) $x = 60$

Check Your Understanding of Section 4.5

A. Multiple Choice

1. In the accompanying diagram, $\overline{RT}$ is extended to W, $\overrightarrow{RQ}$ and $\overrightarrow{TP}$ intersect at S to form $\triangle RST$, $m\angle PSQ = 40$, and the degree measure of exterior angle WRQ is 135. What is $m\angle STR$?
   (1) 85   (3) 105
   (2) 95   (4) 175

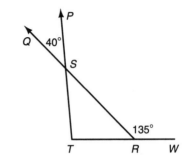

2. In the accompanying diagram of $\triangle ABC$, $\overline{AB}$ is extended to D, $m\angle ACB = 75$, and the degree measure of exterior angle CBD exceeds three times $m\angle CAB$ by 5. What is $m\angle CAB$?
   (1) 25   (3) 45
   (2) 35   (4) 60

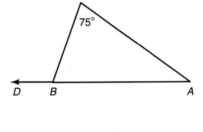

3. In the accompanying diagram, parallel lines $\overleftrightarrow{AB}$ and $\overleftrightarrow{CD}$ are cut by transversal $\overleftrightarrow{PQ}$ at points P and Q, $\overleftrightarrow{PX}$ bisects $\angle QPB$, and $\overleftrightarrow{QX}$ bisects $\angle PQD$. Which statement is *always* true?
   (1) $\angle PXQ$ is a right angle.
   (2) $m\angle PXQ = m\angle APQ$.
   (3) $m\angle PXQ = m\angle XPB$.
   (4) $m\angle PXQ + m\angle DQX = 180$.

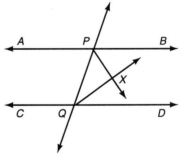

4. If, in $\triangle ABC$, $m\angle A = x + 24$, $m\angle B = 4x + 17$, and $m\angle C = 2x - 15$, what type of triangle is $\triangle ABC$?
   (1) right   (2) obtuse   (3) isosceles   (4) acute

5. One angle of a triangle measures 30°. If the measures of the other two angles are in the ratio 3 : 7, the measure of the largest angle of the triangle is
(1) 15°   (2) 45°   (3) 105°   (4) 126°

6. Two angles of a triangle measure 72° and 46°. Which could be the measure of an exterior angle of this triangle?
(1) 46°   (2) 62°   (3) 108°   (4) 144°

7. In the accompanying diagram of △ABC, side $\overline{BC}$ is extended to D, m∠B = 2y, m∠BCA = 6y, and m∠ACD = 3y. What is the degree measure of ∠A?
(1) 15   (3) 20
(2) 17   (4) 24

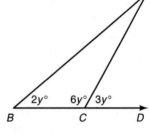

8. In the accompanying diagram, the bisectors of ∠A and ∠B in acute triangle ABC meet at D, and m∠ADB = 130. What is the degree measure of ∠C?
(1) 50   (3) 70
(2) 60   (4) 80

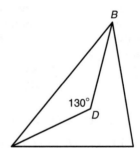

9. In the accompanying diagram of △ABC, $\overline{AC}$ is extended to D, $\overline{DEF}$ is drawn, m∠B = 50, m∠BFE = 105, and m∠ACB = 65. What is the degree measure of ∠D?
(1) 40   (3) 50
(2) 45   (4) 55

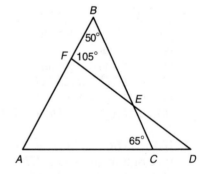

10. In the accompanying diagram, $\overleftrightarrow{AB} \parallel \overleftrightarrow{GCD}$, $\overleftrightarrow{AED}$ is a transversal, and $\overline{EC}$ is extended to F. If m$\angle CED$ = 60, m$\angle DAB$ = $2x$, and m$\angle FCG$ = $3x$, what is m$\angle GCE$?

(1) 36   (3) 108
(2) 72   (4) 144

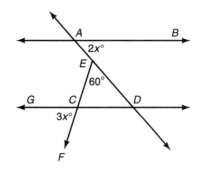

11. In the accompanying diagram, $\angle ACD$ is an exterior angle of $\triangle ABC$, m$\angle A$ = $3x$, m$\angle ACD$ = $5x$, and m$\angle B$ = 50.
What is the value of x?

(1) 15   (3) 25
(2) 20   (4) 32

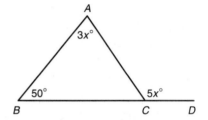

B. *In each case, show how you arrived at your answer by clearly indicating all of the necessary steps, formula substitutions, diagrams, graphs, charts, etc.*

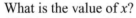

12. In the accompanying diagram, $\overline{ABC} \parallel \overline{DE}$, $\overline{BE}$ bisects $\angle DBC$, m$\angle FDE$ = 25, and m$\angle DFE$ = 105. What is m$\angle ABD$?

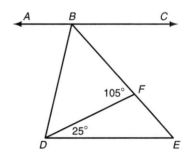

13. In the accompanying diagram of $\triangle ABC$, $\overline{AB}$ is extended to D. If m$\angle ACB$ = $x + 30$, m$\angle CAB$ = $2x + 10$, and m$\angle CBD$ = $4x + 30$, what is m$\angle ABC$?

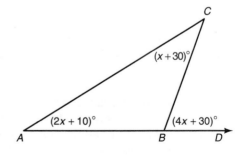

14. In the accompanying diagram, $\overline{DE} \parallel \overline{BC}$, m$\angle AED = 37$, m$\angle A = x$, and m$\angle B = 2x + 5$. What is the degree measure of $\angle ADE$?

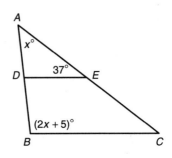

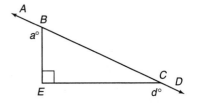

15. In the accompanying diagram, $\overleftrightarrow{ABCD}$ is a straight line, and $\angle E$ in triangle $\triangle BEC$ is a right angle. What does $a + d$ equal?

16. In $\triangle DEF$, m$\angle D$ is 1 more than twice m$\angle E$, and m$\angle F$ is 7 less than m$\angle D$. What is the number of degrees in the angle of the triangle with the greatest measure?

4.6 ANGLES OF A POLYGON

KEY IDEAS

Each point of a polygon at which two sides intersect is called a **vertex** of the polygon.

At each vertex there is an interior angle of the polygon. If the polygon has n sides, the sum, S, of the degree measures of these n interior angles is given by the formula

$$S = (n - 2) \times 180$$

Angles of a Quadrilateral and a Polygon

In Figure 4.19, $\overline{BD}$ is a diagonal. A **diagonal** of a polygon is a line segment joining two nonconsecutive vertices of the polygon. Notice that $\overline{BD}$ divides quadrilateral $ABCD$ into two triangles the sum of whose angle measures is the same as the sum of all the angle measures of the polygon. Thus, the sum of the degree measures of the four angles of the quadrilateral is equal to 2 × 180 or 360.

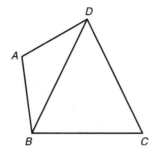

Figure 4.19 Diagonal of a Polygon

A polygon having n sides can be separated into $n - 2$ nonoverlapping triangles, so the sum, S, of the degree measures of the angles of a polygon having n sides is given by the formula

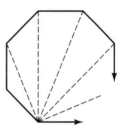

$$S = (n - 2) \times 180$$

Example 1

The degree measures of the angles of a quadrilateral are in the ratio of $1 : 2 : 3 : 4$. Find the degree measure of the largest angle of the quadrilateral.

Solution: **144**

Let x = degree measure of smallest angle of quadrilateral.

Then $2x$, $3x$, $4x$ = degree measures of the remaining angles of the quadrilateral.

$$x + 2x + 3x + 4x = 360$$
$$10x = 360$$
$$x = \frac{360}{10} = 36$$
$$4x = 4(36) = 144$$

Example 2

(a) Find the sum of the degree measures of the angles of a hexagon.
(b) If the hexagon is regular, find the degree measure of each interior angle.

137

Solution: (a)   **720**

A hexagon has six sides, so

$$S = (n-2) \cdot 180$$
$$= (6-2) \cdot 180$$
$$= 4 \cdot 180$$
$$= 720$$

(b)   **120**

A regular polygon is equiangular, so the degree measure of each angle can be found by dividing the sum of the angle measures by 6 (the number of interior angles of the regular hexagon).

$$\text{Degree measure of each angle} = \frac{\text{Sum of measures of angles}}{\text{Number of angles.}}$$

$$= \frac{720}{6} = 120$$

Example 3

If the sum of the degree measures of the angles of a polygon is 900, determine the number of the sides.

Solution: **7**

$$900 = 180(n-2)$$
$$\frac{900}{180} = n-2$$
$$5 = n-2$$
$$n = 7$$

Exterior Angles of a Polygon

At each vertex of a polygon, an *exterior* angle may be formed by extending one side of the polygon so that the interior and exterior angles at that vertex are supplementary. In Figure 4.20, angles 1, 2, 3, and 4 are exterior angles and the sum of their degree measures is 360.

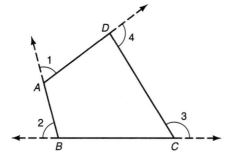

Figure 4.20   Exterior Angles of a Polygon

In general:

MATH FACTS

EXTERIOR ANGLES OF POLYGONS

- The sum of the degree measures of the exterior angles of a polygon having any number of sides (one exterior angle at each vertex) is 360.
- If a regular polygon has n sides, then the degree measure of each exterior angle is $\frac{360}{n}$.

Example 4

Find the degree measure of each interior angle and each exterior angle of a regular decagon.

Solution: **Interior angle = 144, exterior angle = 36**

A decagon has 10 sides.

<u>Method 1: Use the formula Sum = $180 \times (n - 2)$.</u>

- Since $n = 10$, sum = $180 \times 8 = 1440$.
- Because there are 10 interior angles each with the same measure, the measure of each interior angle is $\frac{1440}{10} = 144$.
- Since interior and exterior angles are supplementary at each vertex, the measure of each exterior angle is $180 - 144 = 36$.

Method 2: Adopt a different point of view.

- Instead of first finding the measure of an interior angle, find the measure of each exterior angle.
- The sum of the measures of the exterior angles of any polygon is 360. Since a regular decagon has 10 congruent exterior angles, one at each vertex, the measure of each exterior angle is $\dfrac{360}{10} = 36$.
- Since interior and exterior angles are supplementary at each vertex, the measure of each interior angle is $180 - 36 = 144$.

Example 5

If the measure of each interior angle of a regular polygon is 150, how many sides does the polygon have?

Solution: **12**

Since the measure of each interior angle is 150, the measure of an exterior angle drawn at any vertex is $180 - 150 = 30$. For a regular polygon:

$$\text{Measure of an exterior angle} = \frac{360}{n}$$
$$30 = \frac{360}{n}$$
$$30n = 360$$
$$n = \frac{360}{30} = 12 \text{ sides}$$

Check Your Understanding of Section 4.6

A. Multiple Choice

1. The number of sides of a regular polygon for which the measure of an interior angle is equal to the measure of an exterior angle is
   (1) 8   (2) 6   (3) 3   (4) 4

2. The number of sides of a regular polygon whose interior angle measures 108 is
   (1) 5   (2) 6   (3) 7   (4) 4

3. If the degree measures of four interior angles of a pentagon are 116, 138, 94, and 88, what is the degree measure of the other interior angle?
(1) 76   (2) 104   (3) 120   (4) 144

4. If the ratio of the measures of the interior angles of a quadrilateral is 2:3:4:6, what is the measure of the smallest angle of the quadrilateral?
(1) 12   (2) 24   (3) 36   (4) 48

5. If the measures of the angles of a triangle are in the ratio 3:4:5, the degree measure of an exterior angle of the triangle can *not* be
(1) 165   (2) 135   (3) 120   (4) 105

6. If the sum of the degree measures of the interior angles of a polygon is 1620, how many sides does the polygon have?
(1) 9   (2) 10   (3) 11   (4) 12

7. Which of the following could *not* represent the degree measure of an exterior angle of a regular polygon?
(1) 72   (2) 15   (3) 27   (4) 45

B. In each case, show how you arrived at your answer by clearly indicating all of the necessary steps, formula substitutions, diagrams, graphs, charts, etc.

8. What is the sum of the degree measures of the interior angles of a polygon having 13 sides?

9. Find the number of sides of a regular polygon in which the degree measure of an interior angle is three times the degree measure of an exterior angle.

10. Find the number of sides in a regular polygon in which the degree measure of an interior angle exceeds six times the degree measure of an exterior angle by 12.

CHAPTER 5

CONGRUENT TRIANGLES AND SPECIAL QUADRILATERALS

5.1 CONGRUENT TRIANGLES

KEY IDEAS

The size and shape of a triangle are determined by the measures of its six parts: the lengths of its *three* sides and the degree measures of its *three* angles. Triangles that have the same size and the same shape are **congruent**.

Matching Vertices of Two Congruent Triangles

In Figure 5.1, if it were possible to "slide" $\triangle ABC$ to the right, we could determine whether the two triangles can be made to exactly coincide. If the triangles can be made to coincide, then $\triangle ABC$ is congruent to $\triangle RST$.

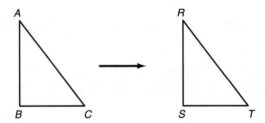

Figure 5.1　Testing Whether Triangles Are Congruent

Notice that, when "sliding" $\triangle ABC$ to the right, we would try to match the vertices of the angles that appear to be congruent. Angles that are paired off in this way are called **corresponding angles**. **Corresponding sides** of the two triangles lie opposite corresponding angles. In Figure 5.1:

Corresponding Angles	Corresponding Sides
$\angle A$ and $\angle R$	$\overline{BC}$ and $\overline{ST}$
$\angle B$ and $\angle S$	$\overline{AC}$ and $\overline{RT}$
$\angle C$ and $\angle T$	$\overline{AB}$ and $\overline{RS}$

Knowing When Two Triangles Are Congruent

Two triangles are congruent when all of their corresponding angles and all of their corresponding sides are congruent. To determine whether two triangles are congruent, you can use one of the short-cut methods described below. In diagrams, pairs of sides with the same number of tic marks are congruent, and pairs of angles with the same number of arcs are congruent.

- **Side-Side-Side (SSS ≅ SSS) Method**
  Two triangles are congruent when their corresponding sides are congruent.

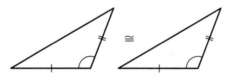

- **Side-Angle-Side (SAS ≅ SAS) Method**
  Two triangles are congruent when two sides and the included angle of one triangle are congruent to the corresponding parts of the second triangle.

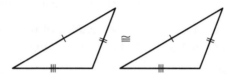

- **Angle-Side-Angle (ASA ≅ ASA) Method**
  Two triangles are congruent when two angles and the included side of one triangle are congruent to the corresponding parts of the second triangle.

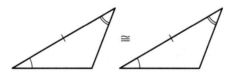

- **Angle-Angle-Side (AAS ≅ AAS) Method**
  Two triangles are congruent when two angles and the *not* included side of one triangle are congruent to the corresponding parts of the second triangle.

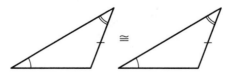

- **Hypotenuse-Leg (HL ≅ HL) Method**
  Two *right* triangles are congruent when the hypotenuse and a leg of one right triangle are congruent to the corresponding parts of the second right triangle.

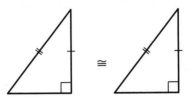

You may *not* conclude that two triangles are congruent when:

- Two sides and an angle that is *not* included of one triangle are congruent to the corresponding parts of the other triangle.

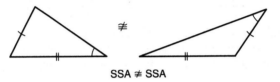

SSA ≇ SSA

- Three angles of one triangle are congruent to the corresponding parts of the other triangle.

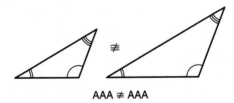

AAA ≇ AAA

Example

Using the parts marked in the accompanying figures as being congruent, determine in each case whether △I is congruent to △II. Give a reason for your answer.

(a)

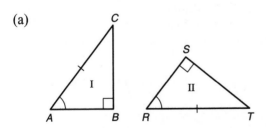

144

(b)

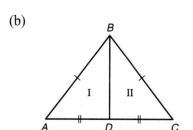

(c)

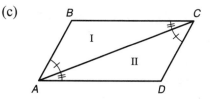

(d)

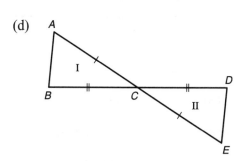

(e)

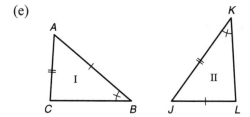

Solution: (a) $\triangle ABC \cong \triangle RST$

Apply the AAS method. Note that $\angle B$ and $\angle S$ are congruent since all right angles are congruent.

(b) $\triangle ADB \cong \triangle CDB$

Since $\overline{BD}$ is congruent to itself, the SSS method may be used.

(c) $\triangle ABC \cong \triangle CDA$

Since $\overline{AC}$ is congruent to itself, the ASA method may be used.

(d) $\triangle ABC \cong \triangle EDC$

Since the vertical angles are congruent, $\angle ACB$ and $\angle ECD$ are congruent. These angles are included by congruent pairs of corresponding sides, so the SAS method may be used.

(e) $\triangle ABC$ is *not* **congruent to** $\triangle JKL$

The pairs of angles that are congruent are *not* included by the congruent pairs of corresponding sides.

Using Corresponding Parts of Congruent Triangles

If you know that two triangles are congruent, you may conclude that *any* pair of corresponding angles are congruent and *any* pair of corresponding sides are congruent.

Check Your Understanding of Section 5.1

Multiple Choice

1. If $\triangle ABC \cong \triangle RST$, $AC = 3(x - 7)$ and $RT = 2(x - 6)$, what is the length of $\overline{AC}$?
   (1) 3   (2) 6   (3) 9   (4) 12

2. If $\triangle JKL \cong \triangle TOP$, m$\angle J = 101$, m$\angle K = 35$, and m$\angle P = 6x + 5$, what is the value of x?
   (1) 5   (2) 6.5   (3) 9.5   (4) 16

3. In the accompanying diagram of quadrilateral $ABCD$, diagonal $\overline{AC}$ bisects $\angle BAD$ and $\angle BCD$. Which statement can be used to prove $\triangle ABC \cong \triangle ADC$?
   (1) HL $\cong$ HL   (3) ASA $\cong$ ASA
   (2) SSS $\cong$ SSS   (4) SAS $\cong$ SAS

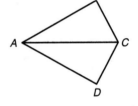

4. In the accompanying diagram of quadrilateral $QRST$, $\overline{RS} \cong \overline{ST}$, $\overline{SR} \perp \overline{QR}$, and $\overline{ST} \perp \overline{QT}$. Which method can be used to prove $\triangle QRS \cong \triangle QTS$?
   (1) HL $\cong$ HL   (3) ASA $\cong$ ASA
   (2) SAS $\cong$ SAS   (4) AAS $\cong$ AAS

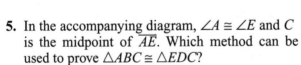

5. In the accompanying diagram, $\angle A \cong \angle E$ and C is the midpoint of $\overline{AE}$. Which method can be used to prove $\triangle ABC \cong \triangle EDC$?
   (1) SAS $\cong$ SAS   (3) SSS $\cong$ SSS
   (2) ASA $\cong$ ASA   (4) AAS $\cong$ AAS

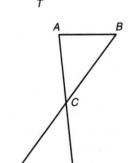

6. In the accompanying diagram, $\overline{RL} \perp \overline{LP}$, $\overline{LR}$ $\perp \overline{RT}$, and M is the midpoint of $\overline{TP}$. Which method can be used to prove $\triangle TMR \cong \triangle PML$?

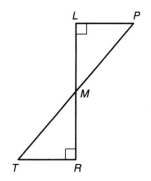

(1) SAS $\cong$ SAS   (3) SSS $\cong$ SSS
(2) HL $\cong$ HL   (4) AAS $\cong$ AAS

7. If the parts marked in each of the accompanying diagrams are congruent, in which diagram is $\triangle$ I not necessarily congruent to $\triangle$ II?

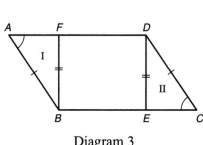

Diagram 1

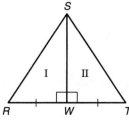

Diagram 2

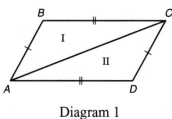

Diagram 3

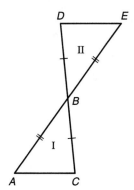

Diagram 4

(1) 1   (2) 2   (3) 3   (4) 4

147

5.2 ISOSCELES TRIANGLES

KEY IDEAS

In an isosceles triangle (Figure 5.2) the base angles have the same degree measure and are, therefore, congruent. Conversely, if two angles of a triangle have the same measure, then the sides opposite the angles have the same length.

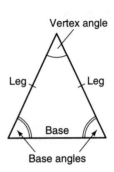

Figure 5.2 Isosceles Triangle

Congruent Sides Are Opposite Congruent Angles

An isosceles triangle has two sides, called the *legs*, with the same length. The remaining side, as shown in Figure 5.2, is called the *base*, and the angle opposite the base is the *vertex angle*. When a triangle in which two sides or two angles are congruent is folded along the bisector of the vertex angle, the two parts of the triangle will coincide. Therefore:

- If two sides of a triangle are equal in length, as in Figure 5.3, the angles opposite these sides are equal in measure.

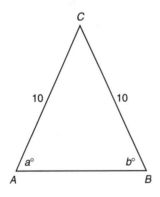

Figure 5.3 If $AC = BC$, $a = b$

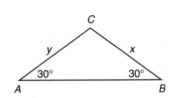

Figure 5.4 If $m\angle A = m\angle B$, $x = y$

- If two angles of a triangle are equal in measure, as in Figure 5.4, the sides opposite these angles are equal in length.

Example 1

If the degree measure of the vertex angle of an isosceles triangle is 80, what is the degree measure of each base angle?

Solution: **50**

Let x = measure of each base angle.

$$x + x + 80 = 180$$
$$2x = 180 - 80$$
$$x = \frac{100}{2}$$
$$= 50$$

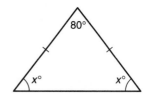

Example 2

If the degree measure of an exterior angle formed by extending the base of an isosceles triangle is 112, what is the degree measure of the vertex angle?

Solution: **44**

$m\angle A = 180 - 112 = 68$
Therefore, $m\angle B$ must also equal 68.

$$68 + 68 + m\angle C = 180$$
$$130 + m\angle C = 180$$
$$m\angle C = 180 - 136$$
$$= 44$$

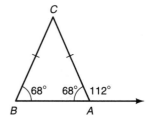

Example 3

The degree measure of a base angle of an isosceles triangle exceeds twice the degree measure of the vertex angle by 15. Find the degree measure of the vertex angle.

Solution: **30**

Let x = degree measure of the vertex angle.

Then $2x + 15$ = degree measure of a base angle.

$$(2x + 15) + (2x + 15) + x = 180$$
$$5x + 30 = 180$$
$$5x = 180 - 30$$
$$x = \frac{150}{5}$$
$$= 30$$

Equilateral Triangles

Since an equilateral triangle (Figure 5.5) has three equal sides, it is also an isosceles triangle, in which any pair of angles may be considered the base angles. Each angle of an equilateral triangle must, therefore, have the same measure.

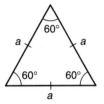

Figure 5.5 Equilateral Triangle

Congruent Angles Mean Congruent Opposite Sides

Although the converse of a true statement is *not necessarily* true, it is a fact that, if two angles of a triangle are congruent, the sides opposite these angles are also congruent.

Check Your Understanding of Section 5.2

A. Multiple Choice

1. If the degree measure of the vertex angle of an isosceles triangle is 38, the degree measure of an exterior angle at a vertex of a base angle is
   (1) 72   (4) 96   (3) 109   (4) 142

2. If the degree measure of an exterior angle at a vertex of a base angle in an isosceles triangle is 110, the degree measure of the vertex angle is
   (1) 70   (2) 35   (3) 40   (4) 20

3. If two angles of a triangle measure 56° and 68°, the triangle is
   (1) scalene   (2) isosceles   (3) obtuse   (4) right

4. The measure of a base angle of an isosceles triangle is four times the measure of the vertex angle. The degree measure of the vertex angle is
   (1) 20   (2) 30   (3) 36   (4) 135

5. If two isosceles triangles have congruent vertex angles, the two triangles must be
   (1) congruent   (2) right   (3) equiangular (4) equilateral

6. Which statements describe properties of an isoscles triangle?
  I. The median drawn to the base bisects the vertex angle.
  II. The altitude drawn to the base bisects the base.
  III. The vertex angle can never be obtuse.
  (1) I and III, only   (3) I and II, only
  (2) II and III, only   (4) I, II, and III

7. In the accompanying diagram, $\triangle ABC$ and $\triangle ACD$ are isosceles triangles with $AB = BC$ and $AD = CD$. If m$\angle ABC = 42$ and m$\angle ADC = 68$, what is m$\angle BAD$?
  (1) 115   (3) 125
  (2) 120   (4) 130

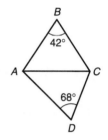

8. In the accompanying diagram, $\triangle ABC$ and $\triangle ACD$ are isosceles triangles with $AB = AC$ and $AD = AC$. If m$\angle ABC = 40$ and m$\angle ADC = 65$, what is m$\angle BAD$?

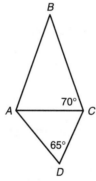

  (1) 80   (2) 90   (3) 120   (4) 135

9. In the accompanying diagram, $\triangle ABC$ and $\triangle DBE$ are isosceles triangles with $AB = BC$ and $DB = BE$. If m$\angle DBE = 26$ and m$\angle BCA = 65$, what is m$\angle ABD$?

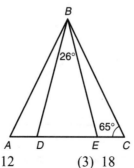

  (1) 9   (2) 12   (3) 18   (4) 24

B. *In each case, show how you arrived at your answer by clearly indicating all of the necessary steps, formula substitutions, diagrams, graphs, charts, etc.*

10. In isosceles triangle ABC the ratio of the measure of vertex angle A to the measure of $\angle B$ is 2 to 5. Find m$\angle C$.

11. Vertex angle A of isosceles triangle ABC measures 20° more than three times m$\angle B$. Find m$\angle C$.

5.3 INEQUALITIES IN A TRIANGLE

KEY IDEAS

In any triangle:

- the length of each side is less than the **sum** of the lengths of the other two sides and greater than the **difference** between these lengths.
- Unequal sides of a triangle lie opposite unequal angles so that the longest side lies opposite the angle with the largest measure and the shortest side lies opposite the angle with the smallest measure.

Testing Numbers as Possible Triangle Side Lengths

Suppose points A, B, and C are the three vertices of a triangle, as shown in Figure 5.6. Since the shortest distance between two points is a straight line, AC must be less than $AB + BC$.

To determine whether a set of any three positive numbers can represent the lengths of the sides of a triangle, check that *each* of the three numbers is less than the sum of the other two numbers.

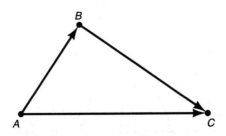

Figure 5.6 Testing Whether Numbers Are Possible Side Lengths

Example 1

Multiple Choice:
Which of the following sets of numbers can*not* represent the lengths of the sides of a triangle?

(1) 9, 40, 41   (2) 7, 7, 3   (3) 4, 5, 1   (4) 1.6, 1.4, 2.9

    Solution: **(3)**

For each choice, check that each of the three given numbers is less than the sum of the other two numbers. In choice (3), although $4 < 5 + 1$ and $1 < 4 + 5$, 5 is *not* less than $1 + 4$. Hence, the numbers 4, 5, and 1 cannot represent the lengths of the sides of a triangle.

Side-Length Limits

In a triangle, the range of possible lengths of a side depends on the lengths of the other two sides. If the lengths of two sides of a triangle are 4 and 9, you can conclude that the length, x, of the third side is greater than the difference, $9 - 4 = 5$, between the lengths of the other sides *and* less than their sum, $9 + 4 = 13$.

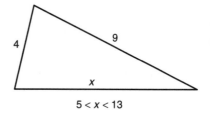

$5 < x < 13$

Example 2

If 3, 8, and x represent the lengths of the sides of a triangle, how many integer values for x are possible? Explain your reasoning.

    Solution: **5**

- The value of x must be greater than $8 - 3 = 5$ *and* less than $8 + 3 = 11$.
- Since $x > 5$ and $x < 11$, x must lie between 5 and 11.
- Because it is given that x is an integer, x can be equal to 6, 7, 8, 9, or 10.

Example 3

If the lengths of two sides of an isosceles triangle are 3 and 7, what is the length of the third side? Explain your reasoning.

    Solution: **7**

In an isosceles triangle two sides have the same length, so the length of the third side of the triangle is either 3 or 7. The third side must be greater than $7 - 3$ or 4. Hence, the length of the third side must be 7.

Example 4

Multiple Choice:
The shortest distance between city *A* and city *B* is 150 miles. The shortest distance between city *B* and city *C* is 350 miles. Which could be the shortest distance, in miles, between city *A* and city *C*?

(1) 175   (2) 200   (3) 250   (4) 300

Solution: **(3)**

The shortest distances between the three cities can be represented by the lengths of the sides of a triangle whose vertices are the three cities.

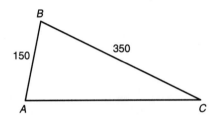

- The length of each side of a triangle is greater than the difference between the lengths of the other two sides.
  Hence, $AC > 350 - 150$, so $AC > 200$.
- Since the length of each side of a triangle is also less than the sum of the lengths of the other two sides, $AC < 150 + 350$, so $AC < 500$.
- Since AC must be between 200 and 500, the smallest distance in this range among the answer choices is 250.

Unequal Sides Opposite Unequal Angles

In a triangle, angles equal in measure are opposite sides with equal lengths. Similarly, angles that are unequal in measure are opposite sides that do not have the same length. Furthermore, the angle with the largest measure is opposite the longest side. These relationships are illustrated in Figure 5.7.

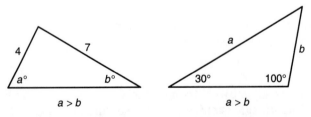

Figure 5.7 Unequal Side and Angle Relationships

154

Example 5

The degree measure of base angle *R* of isosceles triangle *RST* is 50. What is the longest side of the triangle?

Solution: $\overline{RT}$

Find the degree measures of the other two angles of the triangle. Since m∠*R* = 50, then m∠*T* = 50 and m∠*S* = 180 − 100 = 80. $\overline{RT}$ is the longest side since it lies opposite the largest angle (∠*S*).

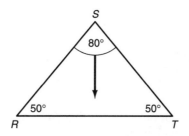

Check Your Understanding of Section 5.3

A. *Multiple Choice*

1. If the lengths of two sides of a triangular plot of land are 700 feet and 1200 feet, the number of feet in the length of the third side could be
   (1) 1600   (2) 1900   (3) 300   (4) 500

2. If the lengths of two sides of a triangle are 4 and 8, the length of the third side can*not* be
   (1) 5   (2) 6   (3) 7   (4) 4

3. Which set of numbers could represent the lengths of the sides of a triangle?
   (1) {3, 6, 3}   (2) {3, 9, 14}   (3) {3, 5, 7}   (4) {1, 2, 3}

4. If side $\overline{AB}$ of △*ABC* is extended from point *B* to point *D*, then m∠*DBC* is always
   (1) smaller than (m∠*A* + m∠*BCA*)
   (2) greater than (m∠*A* + m∠*BCA*)
   (3) equal to 2(m∠*A*)
   (4) greater than m∠*BCA*

5. In a triangle in which the lengths of two sides are 5 and 9, the length of the third side is an integer represented by x. Which statement is always true?
   (1) $5 \leq x \leq 13$ (3) $4 < x \leq 14$
   (2) $4 \leq x \leq 13$ (4) $4 \leq x < 14$

6. In $\triangle ABC$, m$\angle A = 55$ and m$\angle B = 60$. Which statement about $\triangle ABC$ is true?
   (1) All the sides have different lengths, and $\overline{AC}$ is the longest side.
   (2) All the sides have different lengths, and $\overline{AB}$ is the longest side.
   (3) Sides $\overline{AB}$ and $\overline{AC}$ have the same length and are longer than $\overline{BC}$.
   (4) Sides $\overline{AB}$ and $\overline{BC}$ have the same length and are longer than side $\overline{AC}$.

7. If the integer lengths of the three sides of a triangle are 4, x, and 9, what is the smallest possible perimeter of the triangle?
   (1) 18   (2) 19   (3) 20   (4) 21

8. In $\triangle ABC$, D is a point on $\overline{AC}$ such that $\overline{BD}$ bisects $\angle ABC$. If m$\angle ABC = 60$ and m$\angle C = 70$, which inequality is true?
   (1) $AD > AB$ (3) $AD > BD$
   (2) $BD > BC$ (4) $AB > AD$

9. A race course is being designed in the shape of a triangle in which one side is fixed at 3 kilometers and another side is fixed at 7 kilometers. If the remaining side of the course can be any integer distance, how many different courses can be designed?
   (1) 5   (2) 6   (3) 9   (4) 4

10. In $\triangle ABC$, m$\angle A = 64$ and m$\angle B = 58$. Which statement about $\triangle ABC$ is true?
    (1) All the sides have different lengths, and $\overline{AC}$ is the longest side.
    (2) All the sides have different lengths, and $\overline{BC}$ is the longest side.
    (3) Sides $\overline{AB}$ and $\overline{AC}$ have the same length and are longer than $\overline{BC}$.
    (4) Sides $\overline{AB}$ and $\overline{AC}$ have the same length and are shorter than side $\overline{BC}$.

11. In the accompanying diagram of isosceles triangle ABC, $\overline{AC} \cong \overline{BC}$, D is a point lying between A and B on base $\overline{AB}$, and $\overline{CD}$ is drawn. Which of the following inequality statements is true?
    (1) $AC > CD$
    (2) $CD > AC$
    (3) m$\angle A > m\angle ADC$
    (4) m$\angle B > m\angle BDC$

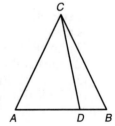

12. A box contains one 2-inch rod, one 3-inch rod, one 4-inch rod, and one 5-inch rod. What is the maximum number of different triangles that can be made using these rods as sides?

    (1) 1   (2) 2   (3) 3   (4) 4

13. In $\triangle ABC$, m$\angle B = 120$, m$\angle A = 55$, and D is a point on $\overline{AC}$ such that $\overline{BD}$ bisects $\angle ABC$. Which is the longest side of $\triangle ABD$?

    (1) $\overline{AB}$ (2) $\overline{AD}$ (3) $\overline{BD}$ (4) $\overline{DC}$

B. In each case, show how you arrived at your answer by clearly indicating all of the necessary steps, formula substitutions, diagrams, graphs, charts, etc.

14. In $\triangle ABC$, m$\angle A = 30$, and the degree measure of an exterior angle at B is 120. Which is the shortest side of the triangle?

15. In $\triangle RST$, $ST > RT$ and $RT > RS$.
    (a) If one of the angles of $\triangle RST$ is obtuse, which angle must it be? Give a reason for your answer.
    (b) If the degree measure of one of the angles of $\triangle RST$ is 60, which angle must it be? Give a reason for your answer.

16. If the lengths of the sides of a triangle are integers and their product is 105, what is a possible perimeter of the triangle? Explain how you arrived at your answer.

5.4 PARALLELOGRAMS AND TRAPEZOIDS

KEY IDEAS

A **parallelogram** is a quadrilateral in which both pairs of opposite sides are parallel. As a result, special relationships exist between the measures of the consecutive angles, opposite angles, and opposite sides of a parallelogram. A quadrilateral that has only one pair of parallel sides is called a **trapezoid**.

Angles of a Parallelogram

The notation $\square ABCD$ is read as "parallelogram $ABCD$. In $\square ABCD$, shown in Figure 5.8:

- The degree measures of the four angles add up to 360:

$$a + b + c + d = 360.$$

- The degree measures of any two consecutive angles add up to 180:

$$a+b=180 \quad \text{and} \quad b+c=180,$$
$$c+d=180 \quad \text{and} \quad a+d=180.$$

- Opposite angles have equal degree measures:

$$a=c \quad \text{and} \quad b=d$$

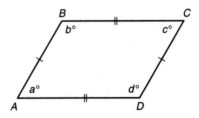

Figure 5.8   Parallelogram *ABCD*

Sides of a Parallelogram

In □*ABCD*, shown in Figure 5.9:
- Opposite sides are parallel:

$$\overline{AD} \parallel \overline{BC} \quad \text{and} \quad \overline{AB} \parallel \overline{CD}.$$

- Opposite sides have the same lengths:

$$AD = BC \quad \text{and} \quad AB = CD.$$

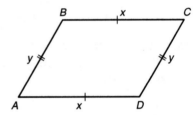

Figure 5.9   Side Relationships in a Parallelogram

Diagonals of a Parallelogram

The diagonals of a parallelogram separate the parallelogram into two congruent triangles. As shown in Figure 5.10, diagonal $\overline{AC}$ forms congruent alternate interior angles with each pair of parallel sides.

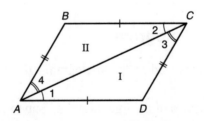

Figure 5.10   △I ≅ △II by ASA ≅ ASA

The diagonals of a parallelogram bisect each other, as shown in Figure 5.11.

$$AE = EC \quad \text{and} \quad DE = EB.$$

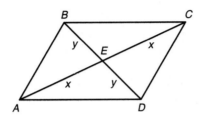

Figure 5.11   Point *E* Is the Midpoint of Both Diagonals

Example 1

In $\square ABCD$, $m\angle B = 5x - 43$ and $m\angle D = 3x - 7$. What is $m\angle A$?

Solution: **133**

- First find the value of x. Since angles B and D are opposite angles of a parallelogram, they have equal measures.

$$m\angle B = m\angle D$$
$$5x - 43 = 3x - 7$$
$$5x = 3x - 7 + 43$$
$$5x = 3x + 36$$
$$5x - 3x = 36$$
$$2x = 36$$
$$x = \frac{36}{2} = 18$$

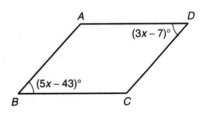

- Since $x = 18$:

$$m\angle B = 5x - 43$$
$$= 5(18) - 43$$
$$= 90 - 43$$
$$= 47$$

- Consecutive angles of a parallelogram are supplementary. Thus:

$$m\angle A = 180 - m\angle B$$
$$= 180 - 47$$
$$= 133$$

Example 2

In the accompanying figure, *ABCD* is a parallelogram. If *EB* = *AB* and m∠*CBE* = 57, what is the value of *x*?

Solution: **66**

- Opposite sides of a parallelogram are parallel, so $\overline{BC} \parallel \overline{AE}$. Since alternate interior angles formed by parallel lines have equal measures, m∠*CBE* = m∠*AEB* = 57.
- Because *EB* = *AB*, m∠*A* = m∠*AEB* = 57.
- Since the measures of the angles of a triangle add up to 180:

$$x + 57 + 57 = 180$$
$$x + 114 = 180$$
$$x = 180 - 114$$
$$= 66$$

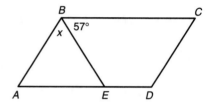

Trapezoids

Unlike a parallelogram, a **trapezoid** is a quadrilateral with only one pair of parallel sides. As shown in Figure 5.12, the parallel sides are called **bases** and the nonparallel sides are **legs**. In any trapezoid:

- The degree measures of the four angles add up to 360:

    $$a + b + c + d = 360.$$

- The lower and upper base angles are supplementary:

    $$a + b = 180 \quad \text{and} \quad c + d = 180.$$

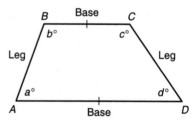

Figure 5.12 Trapezoid *ABCD*

An **isosceles trapezoid** (Figure 5.13) is a trapezoid in which the legs are congruent. In an isosceles trapezoid:

- The lower base angles have equal degree measures and the upper base angles have equal degree measures:

    $$a = d \quad \text{and} \quad b = c$$

- The diagonals have equal lengths:

    $$AC = BD$$

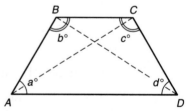

Figure 5.13 Isosceles Trapezoid *ABCD*

Check Your Understanding of Section 5.4

A. *Multiple Choice*

1. Which statement about a parallelogram is *not always* true?
   (1) Diagonals are perpendicular.
   (2) Opposite sides are congruent.
   (3) Opposite angles are congruent.
   (4) Consecutive angles are supplementary.

2. If quadrilateral *ABCD* is a parallelogram, which statement *must* be true?
   (1) $\overline{AC} \perp \overline{BD}$
   (2) $\overline{AC} \cong \overline{BD}$
   (3) $\overline{AC}$ bisects $\angle DAB$ and $\angle BCD$.
   (4) $\overline{AC}$ and $\overline{BD}$ bisect each other.

   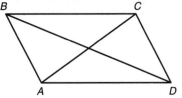

3. In the accompanying diagram of parallelogram *ABCD*, $\overline{EC} \perp \overline{DC}$, $\angle B \cong \angle E$, and m$\angle A = 100$. What is m$\angle CDE$?
   (1) 10   (3) 30
   (2) 20   (4) 80

   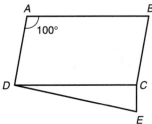

4. In the accompanying diagram of parallelogram *MATH*, m$\angle T = 100$ and $\overline{SH}$ bisects $\angle MHT$. What is m$\angle HSA$?
   (1) 80   (3) 120
   (2) 100   (4) 140

   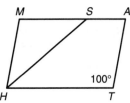

5. In isosceles trapezoid *ABCD*, $\overline{AB} \parallel \overline{CD}$, and $AD = BC$. What is m$\angle A +$ m$\angle C$?
   (1) 45   (2) 90   (3) 180   (4) 360

6. In the accompanying diagram of parallelogram *ABCD*, *EB* = *CD*, and m∠*CBE* = 57. What is m∠*ABE*?

   (1) 57   (2) 66   (3) 114   (4) 123

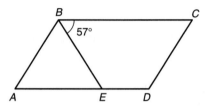

7. In the accompanying diagram of parallelogram *ABCD*, m∠*ABC* = 125, $\overline{AD}$ is extended to *E*, and *AB* = *CE*. What s m∠*DCE*?

   (1) 55   (2) 60   (3) 70   (4) 75

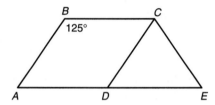

8. In the accompanying diagram of parallelogram *ABCD*, $\overline{DE} \perp \overline{AC}$, m∠*DCA* = 40, and m∠*ADE* = 70. What is m∠*ABC*?

   (1) 100   (3) 120
   (2) 110   (4) 140

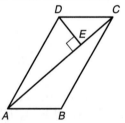

9. In the accompanying diagram of isosceles trapezoid *ABCD*, $\overline{DEC} \parallel \overline{AB}$, $\overline{AD} \cong \overline{DE}$, and m∠*ADC* = 110. What is m∠*ABC*?

   (1) 35   (2) 40   (3) 45   (4) 80

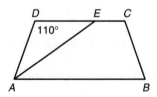

10. In the accompanying diagram of isosceles trapezoid $ABCD$, $\overline{AB} \parallel \overline{DC}$ and diagonals $\overline{DB}$ and $\overline{AC}$ intersect at E. Which statement is *not* true?

(1) $\overline{AC} \cong \overline{BD}$

(2) $\angle CDB \cong \angle DBA$

(3) $\triangle ADC \cong \triangle ABC$

(4) $\triangle CBA \cong \triangle DAB$

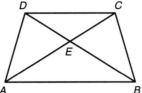

11. In the accompanying diagram of parallelogram $ABCD$, $\text{m}\angle A = x$, $\text{m}\angle DEF = y$, and $\text{m}\angle DFE = z$. Which statement is correct?

(1) $x < y + z$ (3) $x > y + z$

(2) $x = y + z$ (4) $360 - x = y + z$

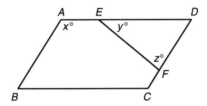

12. In parallelogram $MATH$, the degree measure of $\angle T$ exceeds two times the degree measure of $\angle H$ by 30. What is the degree measure of the largest angle of the parallelogram?

(1) 50   (2) 60   (3) 120   (4) 130

13. In parallelogram $TRIG$, $\text{m}\angle R = 2x + 19$ and $\text{m}\angle G = 4x - 17$. What is $\text{m}\angle T$?

(1) 48   (2) 55   (3) 125   (4) 132

14. In parallelogram $RSTW$, diagonals $\overline{RT}$ and $\overline{SW}$ intersect at point A. If $SA = x - 3$ and $AW = 2x - 37$, what is the length of $\overline{SW}$?

(1) 31   (2) 34   (3) 68   (4) 62

5.5 SPECIAL PARALLELOGRAMS

KEY IDEAS

A parallelogram may be equiangular (rectangle), equilateral (rhombus), or both equiangular and equilateral (square).

Rectangle

A **rectangle** is a parallelogram with four right angles. A rectangle has these properties:

- All of the properties of a parallelogram.
- Four right angles. In Figure 5.14, angles 1, 2, 3, and 4 are right angles.
- Congruent diagonals. In Figure 5.14, $\overline{AC} \cong \overline{DB}$.

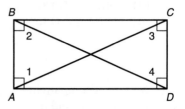

Example 1

Figure 5.14 Rectangle

In rectangle $ABCD$, diagonals $\overline{AC}$ and $\overline{BD}$ intersect at point E. If $AE = 2x - 9$ and $CE = x + 7$, find BD.

Solution: **46**

Since the diagonals of a rectangle bisect each other:

$$AE = CE$$
$$2x - 9 = x + 7$$
$$2x = x + 16$$
$$2x - x = 16$$
$$x = 16$$

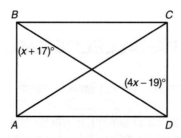

Hence $CE = x + 7 = 16 + 7 = 23$. Also, $AE = 23$. Therefore, $AC = 23 + 23 = 46$. Since the diagonals of a rectangle are congruent,

$$BD = AC = 46$$

Example 2

In the accompanying diagram of rectangle $ABCD$, m$\angle ABD = x + 17$ and m$\angle BDC = 4x - 19$. What is m$\angle ADB$?

Solution: **61**

- The opposite sides of a rectangle are parallel. Since $\overline{AB} \parallel \overline{CD}$, the measures of alternate interior angles formed by transversal $\overline{BD}$ are equal. Thus:

$$m\angle BDC = m\angle ABD$$
$$4x - 19 = x + 17$$
$$4x - x = 19 + 17$$
$$3x = 36$$
$$x = \frac{36}{3} = 12$$

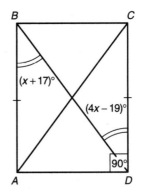

- Because $x = 12$, $m\angle BDC = 4x - 19 = 4(12) - 19 = 48 - 19 = 29$.
- Since each of the four angles of a rectangle is a right angle, $m\angle ADC = 90$. Hence:

$$m\angle ADB + m\angle BDC = 90$$
$$m\angle ADB + 29 = 90$$
$$m\angle ADB = 90 - 29 = 61$$

Rhombus

A **rhombus** is a parallelogram with four congruent sides. A rhombus has these properties:

- All of the properties of a parallelogram.
- Four sides that have the same length. In Figure 5.15,

$$AB = BC = CD = AD.$$

- Diagonals that intersect at right angles.
  In Figure 5.15, since $\overline{AC} \perp \overline{BD}$ at E, there are four right angles at E.
- Diagonals that bisect opposite pairs of angles.
  In Figure 5.15:

$$\angle 1 \cong \angle 2 \quad \text{and} \quad \angle 3 \cong \angle 4,$$
$$\angle 5 \cong \angle 6 \quad \text{and} \quad \angle 7 \cong \angle 8.$$

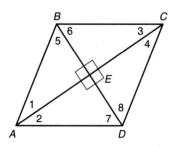

Figure 5.15 Rhombus

Example 3

Given that *ABCD* is a rhombus and m∠1 = 40. Find the degree measure of each of the following angles:
(a) m∠2
(b) m∠3
(c) m∠*ADC*

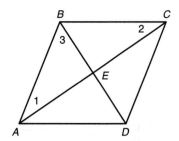

Solution: (a) **40**

Triangle *ABC* is isosceles since $\overline{AB} \cong \overline{BD}$. Hence, the base angles of the triangle must be congruent: m∠1 = m∠2 = 40.

(b) **50**

In △*AEB*, ∠*AEB* is a right angle since the diagonals of a rhombus are perpendicular to each other. Since the sum of the degree measures of the angles of a triangle is 180, m∠3 must be 50.

(c) **100**

Since the diagonals of a rhombus bisect the angles of the rhombus, if m∠3 = 50, then m∠*ABC* = 100. Since opposite angles of a rhombus are equal in measure, m∠*ADC* must also equal 100.

Example 4

The perimeter of rhombus *ABCD* is 20, and m∠*A* = 60. What is the length of the shorter diagonal of the rhombus?

Solution: **5**

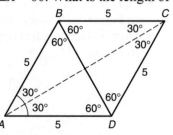

- Each side of a rhombus has the same length. Therefore, if the perimeter of the rhombus is 20, the length of each side is $\dfrac{20}{4} = 5$.

166

- Since consecutive angles of a rhombus are supplementary, m∠B = m∠D = 120.
- The shorter diagonal is $\overline{BD}$ since it lies opposite a 60° angle, while $\overline{AC}$ is the longer diagonal since it lies opposite a 120° angle.
- Diagonal $\overline{BD}$ bisects ∠B and ∠D, so △ABD is equiangular and, as a result, is also equilateral. Hence, $BD = AB = AD = 5$.

Square

A **square** is a parallelogram with four right angles and four sides that have the same length. A square has all of the special properties of a rectangle and a rhombus, as shown in Figure 5.16.

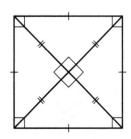

Figure 5.16 Square

Check Your Understanding of Section 5.5

A. Multiple Choice

1. A parallelogram must be a rectangle if its diagonals
   (1) bisect each other
   (2) bisect the angles to which they are drawn
   (3) are perpendicular to each other
   (4) are congruent

2. The diagonals of a rhombus do *not*
   (1) bisect each other
   (2) bisect the angles to which they are drawn
   (3) intersect at right angles
   (4) have the same length

3. In the accompanying diagram, *ABCD* is a rectangle, *E* is a point on $\overline{CD}$, m∠DAE = 30, and m∠CBE = 20. What is the value of *x*?
   (1) 40 (2) 50
   (3) 60 (4) 70

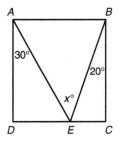

4. If the diagonals of a parallelogram are perpendicular but not congruent, then the parallelogram is
   (1) a rectangle
   (2) a rhombus
   (3) a square
   (4) an isosceles trapezoid

5. In rhombus *PQRS*, diagonals $\overline{PR}$ and $\overline{QS}$ intersect at *T*. Which statement is *always* true?
   (1) Quadrilateral *PQRS* is a square.
   (2) Triangle *RTQ* is a right triangle.
   (3) Triangle *PQS* is equilateral.
   (4) Diagonals $\overline{PR}$ and $\overline{QS}$ are congruent.

6. Which statements describe the properties of the diagonals of a rectangle?
   I. The diagonals are congruent.
   II. The diagonals are perpendicular.
   III. The diagonals bisect each other.
   (1) II and III, only
   (2) I and II, only
   (3) I and III, only
   (4) I, II, and III

7. An example of a quadrilateral whose diagonals are congruent but do not bisect each other is
   (1) a square
   (2) an isosceles trapezoid
   (3) a rhombus
   (4) a rectangle

8. If the lengths of two consecutive sides of a rhombus are represented by $3x - 6$ and $x + 14$, the perimeter of the rhombus is
   (1) 10   (2) 24   (3) 72   (4) 96

B. *In each case, show how you arrived at your answer by clearly indicating all of the necessary steps, formula substitutions, diagrams, graphs, charts, etc.*

9. In rectangle *ABCD*, if $AC = 3x - 5$ and $BD = x + 7$, what is the value of *x*?

10. In rhombus *STAR*, $ST = 4y - 9$ and $TA = 2y + 5$. What is the perimeter of the rhombus?

11. In the accompanying diagram of rectangle *ABCD*, diagonals $\overline{AC}$ and $\overline{BD}$ intersect at *E*.
(a) Is $\triangle BAD \cong \triangle CDA$? Give a reason for your answer.
(b) If m$\angle CAD = 3x - 22$ and m$\angle BDA = x + 14$, what is m$\angle AED$?

12. In the accompanying diagram, if m$\angle BCA = 2x - 9$ and m$\angle DAC = x + 13$, what is m$\angle ACD$?

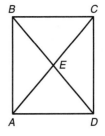

Exercises 11 and 12

Unit Three ALGEBRAIC METHODS

CHAPTER 6

POLYNOMIAL ARITHMETIC AND FACTORING

6.1 CLASSIFYING AND COMBINING POLYNOMIALS

=== **KEY IDEAS** ===

Algebraic terms such as $7x$, $-3xy$, and $\frac{1}{2}ab^2$ are called *monomials*. A **monomial** is a single number, a variable, or the *product* of one or more numbers and variables with positive exponents. The prefix *poly* means "many." A **polynomial** is an expression that contains one or more monomials: an example is $3x^2 - 5x + 1$. Since polynomials represent real numbers, arithmetic operations can be performed with polynomials using the properties of real numbers.

Like Monomials

The monomials $3xy^2$ and $4xy^2$ are **like monomials** since they differ only in their numerical coefficients. Like monomials can be combined by adding or subtracting their numerical coefficients, as in

$$3xy^2 + 4xy^2 = (3+4)xy^2 = 7xy^2$$
$$2ac - 3ac = (2-3)ac = -ac$$

Polynomials in One Variable

A polynomial in one variable is in **standard form** when its monomial terms are arranged so that the exponents decrease from left to right. The polynomial $2x - 5x^2 + 3$ is in standard form when its terms are rearranged and are written as $-5x^2 + 2x + 3$. The **degree** of a polynomial in one variable is the largest exponent of the variable. Thus, $-5x^2 + 2x + 3$ is a second-degree polynomial since the largest exponent is 2. The polynomial $y^3 + 3y^2 - 4y + 9$ is a third-degree polynomial.

Naming Polynomials

Sometimes it is convenient to refer to a polynomial by the number of terms it contains. The names of polynomials with one, two, and three terms are shown in the accompanying table.

Example	Number of Terms	Name of Polynomial
$4x^2$	One	Monomial
$7x - 3y$	Two	Binomial
$5n^2 - n - 2$	Three	Trinomial

Adding Polynomials

To add two polynomials such as $2x^2 - 5x - 1$ and $x^2 + 3x - 7$:

- Write one polynomial underneath the other polynomial so that like terms are aligned in the same column:

$$2x^2 - 5x - 1$$
$$x^2 + 3x - 7$$

- Combine like terms in each column:

$$\overline{3x^2 - 2x - 8}$$

Example 1

The lengths of the sides of a triangle are represented by $3x + 5y$, $2x - 9y$, and $5x + y$.
    What is the perimeter of the triangle in terms of x and y?

   Solution: **$10x - 3y$**

The perimeter of a triangle is the sum of the lengths of its three sides. Thus:

$$3x + 5y$$
$$2x - 9y$$
$$5x + \ y$$
$$\text{Perimeter} = \overline{10x - 3y}$$

Subtracting Polynomials

To subtract a polynomial from another polynomial, add the *opposite* of each term of the polynomial that is being subtracted.

Example 2

Express as a binomial: $x - (3x - 4)$.

   Solution: **$-2x + 4$**

Remove the parentheses in $x - (3x - 4)$ by rewriting each term inside the parentheses with its opposite sign. Then combine like terms.

$$x - (3x - 4) = x - 3x + 4$$
$$= -2x + 4$$

Example 3

Subtract $3a - 2b - 9c$ from $5a + 7b - 4c$.

Solution: **$2a + 9b + 5c$**

Method 1: Do horizontally.

$$(5a + 7b - 4c) - (3a - 2b - 9c)$$
$$= (5a + 7b - 4c) + (-3a + 2b + 9c)$$
$$= (5a - 3a) + (7b + 2b) + (-4c + 9c)$$
$$= \quad 2a \quad + \quad 9b \quad + \quad 5c$$

Method 2: Do vertically. Write the polynomial being subtracted on the line underneath the other polynomial, aligning like terms in the same column. Then change to an addition example by replacing the sign of each term in the polynomial on the second line with its opposite sign, and combine like terms.

$$
\begin{array}{cc}
\text{Original Subtraction} & \text{Equivalent Addition} \\
\text{example} & \text{example} \\
5a + 7b - 4c & 5a + 7b - 4c \\
-\underline{3a - 2b - 9c} & +\underline{-3a + 2b + 9c} \\
& 2a + 9b + 5c
\end{array}
$$

$\rightarrow$

Check Your Understanding of Section 6.1

A. Multiple Choice

1. What is the sum of $4a^2 - 7a - 5$ and $-6a^2 - 2a + 7$?
   (1) $-2a^2 - 9a + 2$ (3) $-10a^2 + 5a + 12$
   (2) $2a^2 - 5a + 2$ (4) $2a^4 - 9a^2 + 2$

2. If $2y^2 - 7y + 6$ is subtracted from $3y^2 - 2y + 5$, the result is
   (1) $5y^2 - 9y + 11$ (3) $y^2 + 5y - 1$
   (2) $-y^2 - 5y + 1$ (4) $y^2 - 9y + 11$

3. When $a^2 + a - 3$ is subtracted from $3a^2 - 5$, the result is
   (1) $2a^2 - a - 2$ (3) $2a^2 - a + 2$
   (2) $-2a^2 + a + 2$ (4) $4a^2 + a - 8$

B. *In each case, show how you arrived at your answer by clearly indicating all of the necessary steps, formula substitutions, diagrams, graphs, charts, etc.*

4–7. Combine like terms and express in simplest form.

4. $(5y - 8) + (-2 + 3y)$ **6.** $(-x^3 + 7x^2 - 9) + (3x^3 + x^2 - 6x)$

5. $(4n^2 - 11) - (7n^2 - 6)$ **7.** $(2x - 5y + 4z) - (-3x + 2y - 3z)$

8. What is the difference when $5x^2 - 2x + 3$ is subtracted from $3x^2 - 4x - 3$?

9. From the sum of $(2x^3 + 6x^2 - 3)$ and $(x^3 - 9x + 7)$, subtract $-x^3 + 5x - 2$.

10. What is the additive inverse of $4x^3 - 5x^2 + 13$?

11. The value of $ax^2 + bx$ is 5 when $x = 1$, and is equal to 0 when $x = -1$. What are the values of a and b? Explain how you arrived at your answer.

6.2 MULTIPLYING AND DIVIDING POLYNOMIALS

⌃ KEY IDEAS

When multiplying or dividing monomials, keep in mind that:

- powers of the same base are *multiplied* by *adding* their exponents. For example:

$$x^5 \cdot x^3 = x^{5+3} = x^8$$

- powers of the same base are *divided* by *subtracting* their exponents. For example:

$$\frac{x^5}{x^3} = x^{5-3} = x^2 \quad \text{and} \quad \frac{x}{x^3} = \frac{x^1}{x^3} = x^{1-3} = x^{-2} = \frac{1}{x^2}$$

Multiplying Monomials

To multiply two monomials, group like factors together. Then multiply the like factors. For example:

$$\left(6x^3y\right)\left(\frac{1}{3}xy^2\right) = \left(6 \cdot \frac{1}{3}\right)\left(x^3 \cdot x\right)\left(y \cdot y^2\right) = 2x^4y^3$$

Dividing Monomials

To divide a monomial by another monomial, group the quotients of like factors together. Then divide the like factors. For example,

$$\frac{-12a^6b}{3a^4b^2} = \left(\frac{-12}{3}\right)\left(\frac{a^6}{a^4}\right)\left(\frac{b}{b^2}\right) = -4a^2b^{-1}$$

Change any negative exponent to a positive exponent by inverting the base:

$$= 4a^2\frac{1}{b^1} \quad \text{or} \quad \frac{4a^2}{b}$$

Multiplying a Polynomial by a Monomial

To multiply a polynomial by a monomial, multiply each term of the polynomial by the monomial. For example:

$$3a^2\left(a^2 - 4a + 5\right) = 3a^2\left(a^2\right) + 3a^2(-4a) + 3a^2(5)$$
$$= \quad 3a^4 \quad - \quad 12a^3 \quad + \quad 15a^2$$

Dividing a Polynomial by a Monomial

To divide a polynomial by a monomial, divide each term of the polynomial by the monomial. For example:

$$\frac{72x^3 - 32x^2 + 8x}{8x} = \frac{72x^3}{8x} + \frac{-32x^2}{8x} + \frac{8x}{8x}$$
$$= 9x^{3-1} - 4x^{2-1} + 1$$
$$= \quad 9x^2 \quad - \quad 4x \quad + 1$$

Example 1

Gary notices that, when he adds $2 + 4 + 6$, the sum is 12, which is divisible by 3. Gary tells Tom that, when *any* three consecutive even integers are added together, the sum is *always* divisible by 3. State whether Gary is correct or incorrect and explain your answer.

Solution: **Correct**

Let x, $x + 2$, and $x + 4$ represent a set of *any* three consecutive even integers. Then add these three numbers, and divide the sum by 3 to see whether there is a remainder:

$$\frac{x+(x+2)+(x+4)}{3} = \frac{3x+6}{3}$$

$$= \frac{3x}{3} + \frac{6}{3}$$

$$= x+2$$

Since there is no remainder, the sum of *any* three consecutive even integers is always divisible by 3.

Multiplying a Polynomial by a Polynomial

To multiply a polynomial by another polynomial, write the second polynomial underneath the first polynomial. Then multiply each term of the second polynomial by the polynomial above it in much the same way as two multidigit numbers are multiplied. For example:

$$
\begin{array}{rll}
& 2x & + 7 \\
& x & + 3 \\
\hline
3(2x+7) = & 6x & +21 \\
x(2x+7) = 2x^2 & + 7x & \\
\hline
\end{array}
$$

Add like terms
in each column: $2x^2 + 13x + 21 \leftarrow$ Final product

Multiplying Binomials Using "FOIL"

You can multiply two binomials such as $(2x + 7)$ and $(x + 3)$ together *horizontally* by forming the sum of these four products:

Product of **F**irst terms: $(\underline{2x} + 7)(\underline{x} + 3) = (2x)(x) +$
Product of **O**uter terms: $(\underline{2x} + 7)(x + \underline{3}) = \quad 2x^2 \quad + (2x)(3) +$
Product of **I**nner terms: $(2 + \underline{7})(\underline{x} + 3) = \quad 2x^2 \quad + \quad 6x \quad + (7)(x) +$
Product of **L**ast terms: $(2x + \underline{7})(x + \underline{3}) = \quad 2x^2 \quad + \quad 6x \quad + \quad 7x \quad + (7)(3)$
$\qquad\qquad\qquad\qquad\qquad = \quad 2x^2 \quad + \quad 13x \quad + \quad 21$

You can remember the four products that are needed by remembering the word "FOIL."

Modeling Algebraic Processes

The process of multiplying two binomials such as $(2x + 7)$ by $(x + 3)$ can be modeled geometrically as shown in the accompanying diagram:

$$2x + 7$$

	F $2x^2$	I $7x$	
x $+$ 3			x $+$ 3
	O $6x$	L 21	

Since the area of the big rectangle must be equal to the sum of the areas of the four smaller rectangles:

$$\overbrace{(2x+7)(x+3)}^{\text{Area of big rectangle}} = \overbrace{2x^2+6x+7x+21}^{\text{Sum of areas of four small rectangles}}$$

Example 2

Multiply $(x - 6)$ and $(x + 2)$.

Solution: $x^2 - 4x - 12$

$$(x-6)(x+2) = \overbrace{x \cdot x}^{F} + \overbrace{2 \cdot x}^{O} + \overbrace{(-6)(x)}^{I} + \overbrace{(-6)(+2)}^{L}$$
$$= x^2 \quad +[2x - 6x] \quad -12$$
$$= x^2 \quad -4x \quad -12$$

Multiplying a Special Pair of Binomials

The product of two binomials is not always a trinomial. For example:

$$(2y+5)(2y-5) = \overbrace{2y \cdot 2y}^{F} + \overbrace{2y \cdot (-5)}^{O} + \overbrace{(5)(2y)}^{I} + \overbrace{(5)(-5)}^{L}$$
$$= 4y^2 \quad +[-10y+10y] \quad -25$$
$$= 4y^2 \quad -25$$

Notice that, when the sum and difference of the same two terms are multiplied together, as in $(2y + 5)(2y - 5)$, the sum of the products of the outer and inner terms is 0. The result is a *binomial* that can be formed by simply taking the difference of the *squares* of the two terms that are being added and subtracted:

$$(2y+5)(2y-5) = (2y)^2 - (5)^2 = 4y^2 - 25$$

======== **MATH FACTS** ========

The binomials in a pair formed by taking the sum and difference of the same two terms, such as $(A + B)$ and $(A - B)$, are called **conjugate binomials**. The product of conjugate binomials is always another binomial that is the difference between the squares of the first and last terms of the conjugate binomials:

$$(A + B)(A - B) = A^2 - B^2$$

For example:

$$(3x - 4)(3x + 4) = (3x)^2 - (4)^2 = 9x^2 - 16$$

Check Your Understanding of Section 6.2

A. *Multiple Choice*

1. The product $\left(\dfrac{2a^3}{5b}\right)\left(\dfrac{3a^2}{7b}\right)$ is

   (1) $\dfrac{5a^5}{12b^2}$ (2) $\dfrac{5a^6}{12b}$ (3) $\dfrac{6a^5}{35b^2}$ (4) $\dfrac{6a^6}{35b}$

2. What is the quotient of $\dfrac{26x^4y^2}{13xy}$?

   (1) $2x^4y^2$ (2) $13x^5y^3$ (3) $2x^3y$ (4) $13x^3y$

3. The product of $-3xy^2$ and $5x^2y^3$ is
   (1) $-8x^3y^5$ (2) $-15x^3y^5$ (3) $-15x^2y^5$ (4) $-15x^3y^6$

4. What is the product of $1.45(xy^2)^3$ and $2.6xy^3$?
   (1) $3.77x^4y^9$ (2) $4.05x^4y^9$ (3) $3.77x^3y^{18}$ (4) $4.05x^3y^9$

5. If $14x^3 - 35x^2 + 7x$ is divided by $7x$ $(x \neq 0)$, the quotient is
   (1) $2x^2 - 5x$ (3) $2x^3 - 5x^2 + x$
   (2) $2x^2 - 5x + 1$ (4) $2x^2 - 5x + x$

6. Chad had a garden that was in the shape of a rectangle. Its length was twice its width. He decided to make a new garden that was 2 feet longer and 2 feet wider than the first garden. If x represents the original width of the garden, which expression represents the difference between the area of the new garden and the area of the original garden?
   (1) $6x + 4$ (2) $2x^2$ (3) $x^2 + 3x + 2$ (4) 8

B. *In each case, show how you arrived at your answer by clearly indicating all of the necessary steps, formula substitutions, diagrams, graphs, charts, etc.*

7–21. Find each product or quotient.

7. $\left(-\dfrac{1}{2}a^2b\right)(-8ab^3)$ **15.** $5y(y^3 - 8y - 4)$

8. $(-2x)(-3x^2)(-4x^3)$ **16.** $(3x + 7)(2x - 9)$

9. $(0.4y^3)(-0.15xy^2)$ **17.** $(5w - 8)(5w + 8)$

10. $\dfrac{xy^2}{x^3y}$ **18.** $(4b - 3)(b + 2)$

11. $\dfrac{3x^5}{-27x^4}$ **19.** $(0.3y^2 + 1)(0.3y^2 - 1)$

12. $\dfrac{8a^2b^3}{12a^2b^5}$ **20.** $(1 - 3x)(2 + x)$

13. $1.05x^5y^3 \div 0.35x^2y^4$ **21.** $(2x - 3)^2$

14. $\left(\dfrac{1}{2xy}\right) \div \left(\dfrac{1}{6x^3y^2}\right)$

22–24. Find each quotient.

22. $\dfrac{18r^4 - 27r^3s^2}{9rs}$ **23.** $\dfrac{21c^3 - 12c^2 + 3c}{-3c}$ **24.** $\dfrac{0.14a^3 - 1.05a^2b}{0.7a}$

25. How does the square of a number, x, compare to the product obtained by multiplying together the two numbers obtained by increasing x by 6 and decreasing x by 6?

26. Roberto claims that the sum of any five consecutive integers is always evenly divisible by 5. Explain why you agree or disagree with Roberto.

27. In the accompanying diagram, the width of the inner rectangle is represented by x and the length by $2x - 1$. The width of the outer rectangle is represented by $x + 3$ and the length by $x + 5$.
(a) Express the area of the shaded region as a trinomial in terms of x.
(b) If the perimeter of the outer rectangle is 24, what is the value of x?
(c) What is the area, in square units, of the shaded region?
(d) The area of the inner rectangle is what percent, correct to the *nearest tenth* of a percent, of the area of the outer rectangle?

179

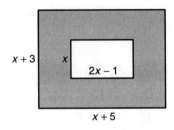

6.3 FACTORING POLYNOMIALS

KEY IDEAS

Sometimes it is useful to know what quantities were multiplied together to obtain a certain product. The process of reversing multiplication is called **factoring**.

Operation	Example
Multiplication	$3x(x + 2) = 3x^2 + 6x$
Factoring	$3x^2 + 6x = 3x(x + 2)$

Finding the Greatest Common Factor (GCF)

The **greatest common factor** of a polynomial is the greatest monomial that divides evenly into *each* term of the polynomial. For example, the GCF of $21a^5 + 14a^3$ is $7a^3$ since 7 is the largest integer that divides evenly into 21 *and* 14, and a^3 is the greatest power of a that is contained in a^5 and a^3.

Factoring a Polynomial by Removing the GCF

Factoring a polynomial means writing the polynomial as the product of two or more lower degree polynomials each of which is called a **factor** of the original polynomial. If you know the GCF of a polynomial, you can find the other factor by dividing the polynomial by the GCF. For example, if you know that 5 is a factor of 30, you divide 30 by 5 to obtain the corresponding factor, 6: $30 = 5 \times 6$.

To factor a polynomial such as $21a^5 + 14a^3$, you need to know the GCF and then use it to find the other factor.

- Since $7a^3$ is the GCF of $21a^5 + 14a^3$, find the corresponding factor by dividing the original polynomial by $7a^3$:

$$\frac{21a^5}{7a^3} + \frac{14a^3}{7a^3} = 3a^2 + 2$$

- Write $21a^5 + 14a^3$ in factored form:

$$21a^5 + 14a^3 = 7a^3(3a^2 + 2)$$

- Check by multiplying the two factors together to make sure the product is the original polynomial:

$$7a^3(3a^2 + 2) = 7a^3 \cdot 3a^2 + 7a^3 \cdot 2 = 21a^5 + 14a^3$$

Not all polynomials can be factored. A polynomial such as $3x^2 + 5$ that cannot be factored except by writing it as the product of itself and 1 (or as the product of its opposite and -1) is called a **prime polynomial**.

Example 1

Factor $6a^2b - 21ab$.

Solution: **$3ab(2a - 7)$**

Since the GCF of $6a^2b - 21ab$ is $3ab$, the corresponding factor is

$$\frac{6a^2b}{3ab} - \frac{21ab}{3ab} = 2a - 7$$

Hence, $6a^2b - 21ab = 3ab(2a - 7)$.

Example 2

If $xz = y - x$, what is x in terms of y and z?

Solution: $x = \dfrac{y}{z+1}$

- Write the given equation: $xz = y - x$
- Add x to each side: $xz + x = y$
- Factor out x: $x(z + 1) = y$

- Divide by $z + 1$: $x = \dfrac{y}{z+1}$

Check Your Understanding of Section 6.3

A. *Multiple Choice*

1. If $A = p + prt$, then $p =$

   (1) $\dfrac{A}{rt}$
   (2) $\dfrac{A}{1+rt}$
   (3) $A - rt$
   (4) $\dfrac{A}{1-rt}$

2. If $ay - c = d + by$, then $y =$

   (1) $\dfrac{c+d}{a+b}$
   (2) $\dfrac{a-b}{c+d}$
   (3) $\dfrac{c+d}{a-b}$
   (4) $\dfrac{a+c}{d-b}$

B. *In each case, show how you arrived at your answer by clearly indicating all of the necessary steps, formula substitutions, diagrams, graphs, charts, etc.*

3–17. Factor.

3. $15x^2 - 6x$

4. $7p^2 + 7q^2$

5. $x^3 + x^2 - x$

6. $3y^7 - 6y^5 + 12y^3$

7. $-4a - 4b$

8. $8u^5w^2 - 40u^2w^5$

9. $y^2 - 144$

10. $0.24x^2 - 0.36xy$

11. $\dfrac{1}{4}a^2b - \dfrac{3}{4}ab^3$

12. $81 - x^2$

13. $p^2 - \dfrac{1}{9}$

14. $b^2 - 0.36$

15. $\dfrac{4}{9}c^2 - 1$

16. $100a^2 - 49b^2$

17. $4x^2(x-1) - 7(x-1)$

6.4 SPECIAL FACTORING TECHNIQUES

∧ KEY IDEAS ∠ ⟍

Some special binomials and trinomials can be factored as the product of two binomials.

Factoring the Difference Between Two Squares

When two binomials have the form $A + B$ and $A - B$, they can be quickly multiplied together by writing $A^2 - B^2$. For example:

$$(x+5)(x-5) = x^2 - 5^2 = x^2 - 25$$

Therefore, whenever you encounter a binomial that has the form $A^2 - B^2$, you can reverse the multiplication by factoring the binomial as $(A + B)(A - B)$. For instance:

$$x^2 - 25 = (x)^2 - (5)^2$$
$$= (x+5)(x-5)$$

==================== **MATH FACTS** ====================

A binomial that is the *difference* between two squares ($A^2 - B^2$) can be factored as the product of the sum ($A + B$) and the difference ($A - B$) of the terms that are being squared:

$$A^2 - B^2 = (A + B)(A - B)$$

_____

Example 1

Factor $4a^2 - 25b^2$.

    Solution: **(2a + 5b)(2a – 5b)**

$$4a^2 - 25b^2 = (2a)^2 - (5b)^2 = (2a+5b)(2a-5b)$$

Example 2

Factor $0.16y^4 - 0.09$.

    Solution: **(0.4y² + 0.3)(0.4y² – 0.3)**

$$0.16y^4 - 0.09 = (0.4y^2)^2 - (0.3)^2 = (0.4y^2 + 0.3)(0.4y^2 - 0.3)$$

If you are not sure what decimal number has as its square 0.16 (or 0.09), use your calculator to find the square root of the given number.

Factoring $x^2 + bx + c$

Since $(x + 2)(x + 5) = x^2 + 7x + 10$, you know that $x^2 + 7x + 10 = (x + 2)(x + 5)$. Notice that the binomial factors contain 2 and 5 since these are the only two integers that when multiplied together give 10, the last term in

$x^2 + 7x + 10$, *and* when added together equal 7, the coefficient of the x-term in $x^2 + 7x + 10$.

$$x^2 + \underbrace{7x}_{\text{sum of 2 and 5}} + \overbrace{10}^{\text{product of 2 and 5}} = (x+2)(x+5)$$

MATH FACTS

To factor a quadratic trinomial that has the form $x^2 + bx + c$:

- Think of two numbers that when multiplied together give c (the number term) *and* when added together give b (the coefficient of the x-term).
- Place these two numbers, call them p and q, inside the parentheses of the binomial factors:

$$x^2 + bx + c = (x + p)(x + q)$$

- Check that the factoring is correct by multiplying the two binomial factors together and comparing the product to the original quadratic expression.

Example 3

Factor $y^2 - 7y + 12$ as the product of two binomials.

   Solution: **$(y - 3)(y - 4)$**

- Think: "What two numbers when multiplied together give $+12$, and when added together give -7?" Because the product of the two numbers you are seeking is positive $(+12)$ and their sum is negative (-7), both numbers must be negative. Since $(-3) \times (-4) = +12$ and $(-3) + (-4) = -7$, the correct factors of $+12$ are -3 and -4.
- Do the factoring: $y^2 - 7y + 12 = (y - 3)(y - 4)$
- Check the factoring:

$$(y - 3)(y - 4) = y \cdot y - 3y - 4y + (-3)(-4)$$
$$= y^2 - 7y + 12$$

Example 4

Factor $n^2 - 5n - 14$ as the product of two binomials.

   Solution: **$(n + 2)(n - 7)$**

- Think: "What two numbers when multiplied together give -14, and when added together give -5?" The two factors of -14 must have opposite signs. Since $(+2) \times (-7) = -14$ and $(+2) + (-7) = -5$, the correct factors of -14 are $+2$ and -7.
- Do the factoring: $n^2 - 5n - 14 = (n + 2)(n - 7)$
- Check the factoring:

$$(n+2)(n-7) = n \cdot n + 2n - 7n + (+2)(-7)$$
$$= n^2 - 5n - 14$$

Factoring $ax^2 + bx + c(a > 1)$

Factoring a quadratic trinomial becomes more difficult when the numerical coefficient of the x^2-term is different from 1.

Example 5

Factor $3x^2 + 10x + 8$.

Solution: **$(3x + 4)(x + 2)$**

Use the reverse of FOIL.

- First factor the x^2-term:

$$3x^2 + 10x + 8 = (3x + ?)(x + ?)$$

- Identify possibilities for the missing numerical terms in the binomial factors. The missing terms are the two integers whose product is $+8$, the last term of $3x^2 + 10x + 8$, and that make the sum of the outer and inner products of the terms of the binomial factors equal to $+10x$. The two factors of $+8$ must have the same sign. Because the coefficient of the x-term in $3x^2 + 10x + 8$ is positive, the factors of 8 are positive. Thus, the possible integer factors of $+8$ are limited to 1 and 8; 2 and 4.
- Use trial and elimination to find the factors of $+8$ that work:

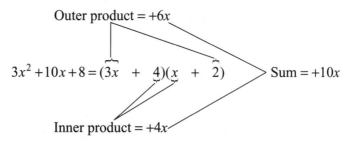

Keep in mind that the placement of the factors of 8 matters. Although $(3x + 2)(x + 4)$ contains the correct factors of 8, they are not placed correctly since the sum of the outer and inner products is $12x + 2x = 14x$ rather than $10x$.

Factoring Completely

A polynomial is **factored completely** when each of its factors cannot be factored further. Sometimes it is necessary to use more than one factoring method in order to factor a polynomial completely.

Example 6

Factor $2x^3 - 50x$ completely.

> *Solution:* $\mathbf{2x(x + 5)(x - 5)}$
> - Factor out the GCF, $2x$: $2x^3 - 50x = 2x(x^2 - 25)$
> - Factor the binomial: $= 2x(x + 5)(x - 5)$

Example 7

Factor $3t^3 + 18t^2 - 48t$ completely.

> *Solution:* $\mathbf{3t(t + 8)(t - 2)}$
> - Factor out the GCF, $3t$: $3t^3 + 18t^2 - 48t = 3t(t^2 + 6t - 16)$
> - Factor the trinomial: $= 3t(t + 8)(t - 2)$

Check Your Understanding of Section 6.4

A. Multiple Choice

1. Which is a factor of $y^2 + y - 30$?
   (1) $(y - 6)$ (2) $(y + 6)$ (3) $(y - 3)$ (4) $(y + 3)$

2. If $ax^2 + bx + c = (2x - 3)(x + 5)$, what is the value of b?
   (1) -15 (2) 2   (3) 7   (4) 4

3. Which expression is a factored form of $3x^2 + 16x - 12$?
   (1) $(3x + 2)(x - 6)$ (3) $(3x - 2)(x + 6)$
   (2) $(3x + 6)(x - 2)$ (4) $(3x - 1)(x + 12)$

4. Which expression is a factored form of $0.04y^2 - 9$?
   (1) $(0.02y + 3)(0.02y - 3)$ (3) $(0.2y + 3)(0.2y - 3)$
   (2) $(0.2y + 9)(0.2y - 1)$ (4) $(4y + 3)(0.01y - 3)$

5. If the area of a rectangle is represented by $3x^2 - 7x - 20$ and its width is represented by $x - 4$, which expression represents the length of this rectangle?
   (1) $3x + 5$ (2) $3x - 5$ (3) $x + 15$ (4) $x - 15$

6. If $x^2 + 2x + k = (x + 5)(x + p)$, then
   (1) $p = 3$ and $k = -5$ (3) $p = -5$ and $k = -3$
   (2) $p = -3$ and $k = 15$ (4) $p = -3$ and $k = -15$

7. Which product is a factored form of $y^3 - 4y$?
   (1) $y(y - 1)(y - 2)$ (3) $y(y - 4)(y + 4)$
   (2) $(y^2 + 1)(y - 4)$ (4) $y(y + 2)(y - 2)$

8. Which product is a factored form of $2x^2 - 10x - 12$?
   (1) $2(x + 2)(x - 2)$ (3) $2(x + 6)(x - 1)$
   (2) $2(x + 3)(x - 2)$ (4) $2(x + 1)(x - 6)$

9. Expressed in factored form, the binomial $2x^2y - 4xy^3$ is equivalent to
   (1) $2xy(x - 2y)$ (3) $2xy(xy - 4y)$
   (2) $2xy(x - 2y^2)$ (4) $2x^2y^3(y - 2)$

B. *In each case, show how you arrived at your answer by clearly indicating all of the necessary steps, formula substitutions, diagrams, graphs, charts, etc.*

10–18. Factor each trinomial as the product of two binomials.

10. $x^2 + 8x + 15$ 13. $a^2 - 4a - 45$ 16. $3x^2 + 2x - 21$

11. $x^2 - 10x + 21$ 14. $b^2 + 3b - 40$ 17. $4n^2 + 11n - 3$

12. $y^2 + 6y + 9$ 15. $w^2 - 13w + 42$ 18. $5s^2 - 14s - 3$

19–27. In each case, factor completely.

19. $2y^3 - 50y$ 24. $2x^3 + 2x^2 - 112x$

20. $-5t^2 + 5$ 25. $10y^4 + 50y^3 - 500y^2$

21. $4m^2 - 4n^2$ 26. $18x^3 - 50xy^2$

22. $8xy^3 - 72xy$ 27. $\dfrac{1}{2}x^3 - 18x$

23. $-2y^2 - 14x - 20$

187

6.5 SOLVING QUADRATIC EQUATIONS BY FACTORING

$$\wedge$$
KEY IDEAS
$$\diagup\quad\diagdown$$

Equations such as $x^2 = 16$, $x^2 - 5x = 0$, and $x^2 + 4x = 5$ are called **quadratic equations** because in each equation the greatest exponent of any variable is 2. A quadratic equation that has the **standard form**

$$ax^2 + bx + c = 0 \quad (a \neq 0)$$

can be solved by breaking it down into two linear equations. To do this, factor the quadratic expression and then set each factor equal to 0.

Zero Product Rule

According to the **zero product rule**, if the product of two real numbers is 0, then at least one of these two numbers must be equal to 0. This rule provides a method for solving a quadratic equation in which one side is 0 and the other side contains a factorable quadratic expression. For example, if $x^2 - 5x = 0$, then $x(x - 5) = 0$. According to the zero product rule, either $x = 0$ or $x - 5 = 0$. Hence, the two roots of $x^2 - 5x = 0$ are $x = 0$ and $x = 5$.

A Strategy for Solving Factorable Quadratic Equations

Before solving a quadratic equation, make sure all of the nonzero terms are on the same side of the equation and 0 is isolated on the other side. For example, to solve $x^2 + 4x = 5$, you must first rearrange the terms by subtracting 5 from each side of the equation so that 0 remains alone on the right side. Thus, to solve the equation $x^2 + 4x = 5$:

- Put the quadratic equation into the standard form $ax^2 + bx + c = 0$: $\qquad x^2 + 4x - 5 = 0$
- Factor the quadratic expression: $\qquad (x + 5)(x - 1) = 0$
- Set each factor equal to 0: $\qquad x + 5 = 0 \quad$ or $\quad x - 1 = 0$
- Solve each first-degree equation: $\qquad x = -5 \quad$ or $\qquad x = 1$
- Check each root in the original equation:

$$\text{Let } x = -5. \qquad\qquad \text{Let } x = 1.$$
$$x^2 + 4x = 5 \qquad\qquad x^2 + 4x = 5$$
$$(-5)^2 + 4(-5) \overset{?}{=} 5 \qquad\qquad (1)^2 + 4(1) \overset{?}{=} 5$$
$$25 - 20 \overset{✔}{=} 5 \qquad\qquad 1 + 4 \overset{✔}{=} 5$$

Example 1

Solve for y and check: $6y^2 + 18y + 12 = 0$.

Solution: $y = -2$ or $y = -1$

Since 6 is a common factor of each term, simplify the equation before factoring the quadratic trinomial by dividing each term of the equation by 6:

$$\frac{6y^2}{6} + \frac{18y}{6} + \frac{12}{6} = 0$$
$$y^2 + 3y + 2 = 0$$
$$(y+2)(y+1) = 0$$
$$y + 2 = 0 \quad \text{or} \quad y + 1 = 0$$
$$y = -2 \qquad\qquad y = -1$$

The check is left for you.

Example 2

Raymond said that $x = -2$ and $x = 1$ are the roots of the quadratic equation $x^2 - 3x = -2$ because:

$$x^2 - 3x = -2$$
$$x(x - 3) = -2$$
$$x = -2 \quad \text{or} \quad x - 3 = -2$$
$$x = -2 + 3$$
$$= 1$$

State whether Raymond's answer is correct or incorrect, and explain your answer.

Solution: **Incorrect**

When $x(x - 3) = -2$, it is *not* correct to set each factor equal to -2 because, if the product of two numbers is -2, each number may be equal to a number different from -2. For example, one number could be -0.001 and the other number could be 2000 since $-0.001 \times 2000 = -2$. Only when the product of two numbers is 0 can you assume that one or both of these numbers are equal to 0. You can also tell that -2 is not a root by checking -2 in the original equation:

$$(-2)^2 - 3(-2) \overset{?}{=} -2$$
$$4 + 6 \neq -2$$

Example 3

Find the solution set and check: $28 - x^2 = 3x$.

Solution: **{−7, 4}**

- Collect all of the nonzero terms on the left side of the equation by subtracting $3x$ from each side: $\qquad 28 - x^2 - 3x = 0$
- Write the quadratic trinomial in standard form: $\qquad -x^2 - 3x + 28 = 0$
- Make the coefficient of the x^2-term positive by multiplying each member of the equation by -1. This is equivalent to changing the sign of each term to its opposite: $\qquad x^2 + 3x - 28 = 0$
- Factor the quadratic trinomial: $\qquad (x + 7)(x - 4) = 0$
- Set each binomial factor equal to 0: $\qquad x + 7 = 0 \quad$ or $\quad x - 4 = 0$
- Solve each first-degree equation: $\qquad x = -7 \quad$ or $\qquad x = 4$
- Wrtie the solution set: $\qquad \{-7, 4\}$

The check is left for you.

TIP

You can solve $28 - x^2 = 3x$ with a graphing calculator by letting $Y_1 = 28 - x^2$ and $Y_2 = 3x$, and then graphing Y_1 and Y_2 on the same set of axes in an appropriate viewing window. The x-coordinates of the points of intersection of the two graphs represent the solutions to the quadratic equation. Pages 316, 353–354, 384–385, and 394–395 discuss the use of the graphing calculator.

Example 4

Find the positive root of $\dfrac{x+5}{2x} = \dfrac{x-2}{3}$.

Solution: **5**

- Eliminate the fractions by making the cross-products equal:
$$2x(x - 2) = 3(x + 5)$$
$$2x^2 - 4x = 3x + 15$$
- Put the quadratic equation into standard form by subtracting $3x$ and 15 from each side:
$$2x^2 - 4x - 3x - 15 = 0$$
$$2x^2 - 7x - 15 = 0$$
- Write the binomial factors:
$$(2x + ?)(x + ?) = 0$$
- Find the two missing numbers of the binomial factors by finding the two integers whose product is -15 *and* that make the sum of the outer

190

and inner products equal to $-7x$. The possible pairs of factors of -15 are 3 and -5; -3 and 5; 1 and -15; and -1 and 15. Use trial and elimination to find that the correct factors of -15 are 3 and -5:

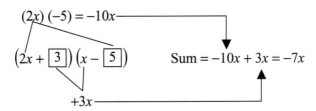

$(2x)\,(-5) = -10x$

$\left(2x + \boxed{3}\right)\left(x - \boxed{5}\right)$ Sum $= -10x + 3x = -7x$

$+3x$

- With the quadratic expression is factored form, apply the zero product rule:

$$(2x+3)(x-5) = 0$$
$$2x+3=0 \quad \text{or} \quad x-5=0$$
$$x = -\frac{3}{2} \quad \text{or} \quad x = 5$$

Since the problem asks for the positive root, the answer is $x = 5$.

Check Your Understanding of Section 6.5

A. *Multiple Choice*

1. If $(x - 3)(x + 2) = 0$, which is the larger of the two roots?
   (1) -2 (2) 2   (3) 3   (4) -3

2. If $x^2 + 11x + 30 = 0$, which is the smaller of the two roots?
   (1) -6 (2) -5 (3) 5   (4) 6

3. If one of the roots of the equation $x^2 - x + q = 0$ is 3, what is the other root?
   (1) -2 (2) 2   (3) -1 (4) -4

4. For what value(s) of x is the fraction $\dfrac{x+4}{x^2 - 2x - 3}$ *not* defined?
   (1) $-1, 3$ (2) $1, -3$ (3) $-3, -1$ (4) -4

B. *In each case, show how you arrived at your answer by clearly indicating all of the necessary steps, formula substitutions, diagrams, graphs, charts, etc.*

5–16. Find the solution set, and check.

5. $x(2x - 1) = 0$

6. $y^2 + 3y + 2 = 0$

7. $x^2 + 4 = 5x$

8. $x^2 = x + 12$

9. $13n - n^2 = 0$

10. $5r + 3 = 2r^2$

11. $5b - 2b^2 + 18 = 0$

12. $9x^2 - 12x + 4 = 0$

13. $3x = \dfrac{x^2}{2}$

14. $6t^2 = 7t + 3$

15. $3p^2 + 14p = 5$

16. $\dfrac{x^2}{2} + \dfrac{23x}{10} - 1 = 0$

17–19. Solve each equation and check.

17. $\dfrac{x+5}{x+1} = \dfrac{x-1}{4}$

18. $\dfrac{x-3}{x-2} = \dfrac{x+3}{2x}$

19. $\dfrac{x-2}{x} = \dfrac{x+4}{3x}$

20. The perimeter of a certain rectangle is 24 inches. When the length of the rectangle is doubled and the width is tripled, the area of the rectangle is increased by 160 square inches. Find the dimensions of the original rectangle.

6.6 SOLVING WORD PROBLEMS

KEY IDEAS

The relationships between quantities described in some word problems may lead to quadratic equations. The solutions of these quadratic equations should be checked to make sure they fit the conditions of the word problem.

Area- and Volume-Related Problems

When solving area- and volume-related word problems, carefully label diagrams with the key items of information that you are given. Reject any negative solutions that represent dimensions of a figure.

Example 1

A rectangular photograph is 7 inches long and 3 inches wide. The photograph is enlarged by increasing its length and width by the same amount. If the area of the enlarged photograph is 96 square inches, what are the new dimensions of the photograph?

Solution: **Length = 12 in, width = 8 in**

Let x = number of inches by which the length and width are increased.

Method 1: Use an algebraic model.

$$\overbrace{(7+x)}^{\text{New length}} \times \overbrace{(3+x)}^{\text{New width}} = \overbrace{96}^{\text{New area}}$$

Thus, $x^2 + 10x + 21 = 96$ so $x^2 + 10x - 75 = 0$, which makes $(x + 15)(x - 5) = 0$. Since the only positive root of this equation is $x = 5$, the length and width are each increased by 5 in. Therefore, the new length is $7 + 5 = 12$ in, and the new width is $3 + 5 = 8$ in.

Method 2: Guess, check, and revise.

	Guess		
x	New Length	New Width	New Area
1	$7 + 1 = 8$	$3 + 1 = 4$	$8 \times 4 = 32 \leftarrow$ Much too low
4	$7 + 4 = 11$	$3 + 4 = 7$	$11 \times 7 = 77 \leftarrow$ Still too low
5	$7 + 5 = 12$	$3 + 5 = 8$	$12 \times 8 = 96 \leftarrow$ This is the answer!

Method 3: Work backwards.

- The original photograph is 7 in long and 3 in wide. Since the length and width are increased by the same amount, the dimensions of the enlarged photograph must also differ by 4 in.
- Work backwards from the new area of 96 in² by thinking of two positive integers whose product is 96 and whose difference is 4.
- Test different pairs of factors of 96 until you find that $12 \times 8 = 96$ and $12 - 8 = 4$. Hence, the enlarged photograph is 12 in long and 8 in wide.

Example 2

Four equal squares are cut off at each corner of a 15-inch by 12-inch rectangular piece of cardboard, as shown in the diagram. The corners are then turned straight up to form a rectangular box that is open at the top.

(a) What size square must be cut off at each corner so that the area of the base of the box will be 60% of the area of the original rectangle?

(b) What is the volume of the box?

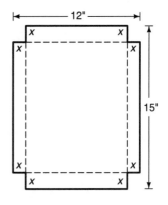

Solution: (a) $\frac{3}{2}$ **in**

Lex x represent the length of the side of each of the squares. After the four equal squares are cut off and the corners turned up, the dimensions of the rectangular base are $15 - 2x$ by $12 - 2x$.

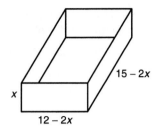

Hence:

$$\overbrace{(15-2x)(12-2x)}^{\text{Area of base of box}} = \overbrace{0.60(15\times 12)}^{\substack{\text{60\% of area of} \\ \text{original rectangle}}}$$

$$180 - 30x - 24x + 4x^2 = 0.60(180)$$

$$4x^2 - 54x + 180 = 108$$

$$4x^2 - 54x + 72 = 0$$

$$\frac{4x^2}{2} - \frac{54x}{2} + \frac{72}{2} = \frac{0}{2}$$

$$2x^2 - 27x + 36 = 0$$

$$(2x-3)(x-12) = 0$$

$$2x - 3 = 0 \quad \text{or} \quad x - 12 = 0$$

$$x = \frac{3}{2} \quad \text{or} \qquad x = 12 \leftarrow \text{Reject!}$$

The width of the base of the box, $12 - 2x$, must be greater than 0. This means that x must be less than 6 and greater than 0. You must, therefore, reject the solution $x = 12$ because it doesn't fit the conditions of the problem.

    (b) **162 in³**

The volume of a rectangular box is equal to the area of its base times its height. For this box:

$$\text{Height} = x = \frac{3}{2} \text{ in}$$

$$\text{Length} = 15 - 12x = 15 - 2\left(\frac{3}{2}\right) = 15 - 3 = 12 \text{ in}$$

$$\text{Width} = 12 - 2x = 12 - 2\left(\frac{3}{2}\right) = 12 - 3 = 9 \text{ in}$$

Hence:

$$\text{Volume} = 12 \text{ in} \times 9 \text{ in} \times \frac{3}{2} \text{ in} = 108 \times 1.5 = 162 \text{ in}^3$$

Example 3

The length of a rectangular garden is twice its width. The garden is surrounded by a rectangular concrete walk having a uniform width of 4 feet. If the area of the garden and the walk is 330 square feet, what are the dimensions of the garden?

   Solution: **Width = 7 ft, length = 14 ft**

Let x = width of the garden.
   Then $2x$ = length of the garden. Draw a diagram.

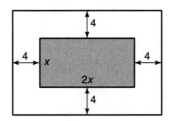

In the accompanying diagram the innermost rectangle represents the garden. Since the walk has a uniform width, the width of the larger (outer) rectangle is $4 + x + 4 = x + 8$. The length of the larger rectangle is $4 + 2x + 4 = 2x + 8$. The area of the larger rectangle is given as 330 ft^2. Hence:

$$\text{Length} \times \text{Width} = \text{Area}$$
$$(2x+8)(x+8) = 330$$
$$2x^2 + 24x + 64 = 330$$
$$2x^2 + 24x + 64 - 330 = 0$$
$$2x^2 + 24x - 266 = 0$$
$$\frac{2x^2}{2} + \frac{24x}{2} - \frac{266}{2} = \frac{0}{2}$$
$$x^2 + 12x - 133 = 0$$
$$(x-7)(x+19) = 0$$
$$x - 7 = 0 \quad \text{or} \quad x + 19 = 0$$
$$x = 7 \quad \text{or} \qquad x = -19$$

Reject -19 since the width must be a positive number. The width, x, of the garden is 7 ft, and the length, $2x$, is 14 ft.

Number-Related Problems

Some word problems involve the relationships between two or more numbers that are represented in terms of the same variable.

195

Example 4

Find the largest of three consecutive odd integers if the product of the first and the third integers is 6 more than three times the second integer. [*Only an algebraic solution will be accepted.*]

Solution: **7**

If the three consecutive odd integers are represented by x, $x + 2$, and $x + 4$, then

$$\underbrace{x(x+4)}_{\substack{\text{Product of the first} \\ \text{and third integers}}} = \underbrace{3(x+2)}_{\substack{\text{3 times the} \\ \text{second integer}}} \underbrace{+6}_{\text{by 6}}$$

$$x^2 + 4x = 3x + 6 + 6$$

$$x^2 + 4x = 3x + 12$$

$$x^2 + 4x - 3x - 12 = 0$$

$$x^2 + x - 12 = 0$$

$$(x - 3)(x + 4) = 0$$

$$x - 3 = 0 \quad \text{or} \quad x + 4 = 0$$

$$x = 3 \qquad\qquad x = -4 \leftarrow \text{Reject since } x \text{ must be odd.}$$

Then $x + 2 = 5$

and $x + 4 = 7$

Hence, 7 is the largest of the three consecutive odd integers.

Check Your Understanding of Section 6.6

In each case, show how you arrived at your answer by clearly indicating all of the necessary steps, formula substitutions, diagrams, graphs, charts, etc. In each case, only an algebraic solution will be accepted.

1. A rectangular flower garden has an area of 180 square feet. If the width of the garden is 3 feet less than the length, what is the minimum number of feet of fencing needed to completely enclose the garden?

2. The height of a rectangular box is 2 inches, and the length of the box exceeds its width by 5 inches. If the volume of the box is 208 cubic inches, find the number of inches in the length and the width of the box.

3. A builder needs to extend a rectangular floor measuring 6 feet by 8 feet so that the number of square feet in the area of the floor increases by 72

square feet. If each of the original dimensions is increased by the same number of feet, what are the dimensions of the new floor?

4. A rectangular plot of land is subdivided into a square flower garden, a rectangular vegetable garden, and a rectangular patio, as shown in the accompanying diagram.

Flower garden	Vegetable garden	Patio

The length of the longer side of the vegetable garden exceeds the length of the flower garden by 4 feet, and the length of the longer side of the patio is twice the length of the flower garden. If the total area of the original plot of land is 360 square feet, find the numbers of square feet in the areas of each garden and the patio.

5. Find two consecutive positive integers such that the square of the first decreased by 25 equals three times the second.

6. If the second of three positive consecutive integers is added to the product of the first and the third, the result is 71. Find the three integers.

7. A positive number is 1 more than twice another number. If the difference of their squares is 40, find the larger of the two numbers.

8. The side of a certain square is 3 feet longer than the side of another square. If the sum of the areas of the squares is 117 square feet, find the length of a side of the smaller square.

9. A rectangular piece of cardboard is twice as long as it is wide. From each of its four corners a square piece 3 inches on a side is cut out. The flaps at each corner are then turned up to form an open box. If the volume of the box is 168 cubic inches, what were the original dimensions of the piece of cardboard?

10. If the length of one side of a square garden in increased by 3 feet, and the length of an adjacent side is increased by 2 feet, the are of the garden increases to 72 square feet. What is the length of a side of the original garden?

11. A rectangular picture 30 centimeters wide and 50 centimeters long is surrounded by a frame having a uniform width. If the combined area of the picture and the frame is 2016 square centimeters, what is the width of the frame?

12. The art staff at Central High School is determining the dimensions of paper to be used in the senior yearbook. The area of each sheet is to be 432 square centimeters. The staff has agreed on margins of 3 centimeters on each side and 4 centimeters on top and bottom. If the printed matter is to occupy 192 centimeters on each page, what must be the overall length and width of the paper?

13. A rectangular picture 24 inches by 32 inches is surrounded by a border of uniform width. If the area of the border is 528 square inches less than the area of the picture, find the width of the border.

14. The sum, S, of a list of n consecutive positive integers beginning with 1 can be determined by evaluating the formula $S = \dfrac{n}{2}(n+1)$. How many consecutive positive integers beginning with 1 must be added together so that the sum is 351? [*Only an algebraic solution will be accepted.*]

15. Alexi throws a pebble into an unused well. The distance, d, the pebble falls after t seconds can be described by the equation $d = 16t^2 + 56t$ where d is measured in feet. If the water in the well is 240 feet below ground level, how many seconds will it take for the pebble to hit the water?

CHAPTER 7

OPERATIONS WITH ALGEBRAIC FRACTIONS

7.1 SIMPLIFYING ALGEBRAIC FRACTIONS

KEY IDEAS

A fraction is in **lowest terms** when its numerator and denominator have no factor in common other than 1 or –1. Thus, the fraction $\frac{12}{15}$ is *not* in lowest terms since 3 is a factor of both 12 and 15. To write $\frac{12}{15}$ in lowest terms, factor the numerator and factor the denominator. Then cancel any factor that appears in both the numerator and the denominator since the quotient of these identical factors is 1:

$$\frac{12}{15} = \frac{4 \times \overset{1}{\cancel{3}}}{5 \times \cancel{3}} = \frac{4}{5} \times 1 = \frac{4}{5}$$

Fractions that contain variables are reduced to lowest terms in much the same way.

Algebraic Fractions

Algebraic fractions are fractions such as $\frac{2x}{3}$, $\frac{y}{4}$, and $\frac{x+y}{x^2-y^2}$ that include variables. Since division by 0 is not defined, always assume that a variable in the denominator of a fraction can never have a value that will make the denominator of the fraction evaluate to 0. For example, the denominator of the fraction $\frac{3}{x+1}$ has a value of 0 when $x = -1$. Therefore, when working with this fraction, assume that x cannot equal -1.

Writing Algebraic Fractions in Lowest Terms

To write an algebraic fraction in lowest terms:

- Factor the numerator and factor the denominator, where possible.
- Divide out (cancel) any factor that is found in both the numerator *and* the denominator.

199

- Multiply the remaining factors in the numerator, and multiply the remaining factors in the denominator.

Example 1

Write $\dfrac{4x+12}{x^2+3x}$ in lowest terms.

Solution: $\dfrac{4}{x}$

- Factor the numerator and the denominator: $\dfrac{4x+12}{x^2+3x} = \dfrac{4(x+3)}{x(x+3)}$

- Cancel common factors: $= \dfrac{\overset{1}{4\cancel{(x+3)}}}{x\cancel{(x+3)}}$

- Multiply the remaining factors: $= \dfrac{4}{x}$

Example 2

When working on her homework, Carla wrote

$$\dfrac{\overset{1}{\cancel{3}a}+b^2}{\cancel{3}a} = 1+b^2$$

Did Carla simplify the fraction correctly? Explain your answer.

Solution: **No**

Carla made a mistake, just as it would be a mistake to simplify $\dfrac{3+4}{3}$ by writing

$$\dfrac{\overset{1}{\cancel{3}}+4}{\cancel{3}} = 1+4 = 5 \leftarrow \text{Wrong!}$$

Only factors of a *product* common to both the numerator and the denominator can be canceled. Since, in $\dfrac{3a+b^2}{3a}$, $3a$ is being *added* to b^2, it is not a factor of the numerator, and so cannot be divided out.

Example 3

Write $\dfrac{3-3y^2}{y^2+4y-5}$ in lowest terms.

   Solution: $\dfrac{-3(1+y)}{y+5}$

- Factor the numerator and factor the denominator completely:

$$\frac{3-3y^2}{y^2+4y-5}=\frac{3\left(1-y^2\right)}{(y-1)(y+5)}$$

$$=\frac{3(1-y)(1+y)}{(y-1)(y+5)}$$

- Rewrite $3\,(1-y)$ as $-3\,(y-1)$

$$=\frac{-3(y-1)(1+y)}{(y-1)(y+5)}$$

- Cancel common factors:

$$=\frac{\overset{1}{-3\cancel{(y-1)}}(1+y)}{\cancel{(y-1)}(y+5)}$$

- Multiply the remaining factors:

$$=\frac{-3(1+y)}{y+5}$$

Example 4

Simplify: $\dfrac{10a^2-15ab}{4a^2-9b^2}$.

   Solution: $\dfrac{5a}{2a+3b}$

- Factor the numerator and factor the denominator completely:

$$\frac{10a^2-15ab}{4a^2-9b^2}=\frac{5a(2a-3b)}{(2a-3b)(2a+3b)}$$

- Cancel common factors:

$$=\frac{\overset{1}{5a\cancel{(2a-3b)}}}{\cancel{(2a-3b)}(2a+3b)}$$

- Multiply the remaining factors:

$$=\frac{5a}{2a+3b}$$

Check Your Understanding of Section 7.1

A. *Multiple Choice*

1. What is the value of $\dfrac{(x+1)^2}{x^2-1}$ when $x = 1.002$?

   (1) 4004   (2) 1001   (3) 200.2   (4) 4008.004

2. Expressed in simplest form, $\dfrac{x^2-x-6}{x^2-9}$ $(x \neq \pm 3)$ is equivalent to

   (1) $\dfrac{x+2}{x-3}$ (2) $\dfrac{x+2}{x+3}$ (3) $\dfrac{x-2}{x-3}$ (4) $\dfrac{x-2}{x+3}$

3. Which fraction is expressed in simplest form?

   (1) $\dfrac{x-1}{x^2-1}$ (3) $\dfrac{x+1}{x^2-1}$

   (2) $\dfrac{x-1}{x^2-2x+1}$ (4) $\dfrac{x+1}{x^2+1}$

B. *In each case, show how you arrived at your answer by clearly indicating all of the necessary steps, formula substitutions, diagrams, graphs, charts, etc.*

4–15. Simplify by writing each fraction in lowest terms.

4. $\dfrac{2x-16}{x^2-64}$

5. $\dfrac{a^2-16}{3a+12}$

6. $\dfrac{2ab^2-2a^2b}{4ab}$

7. $\dfrac{0.48xy-0.16y}{0.8y}$

8. $\dfrac{21r^2s-7r^3s}{14rs}$

9. $\dfrac{10xy+30x^2}{xy^2-9x^3}$

10. $\dfrac{10-5x}{x^2-x-2}$

11. $\dfrac{2x^2-50}{2x^2+14x+20}$

12. $\dfrac{x^2-y^2}{(x-y)^2}$

13. $\dfrac{x^2-x-42}{x^2+7x+6}$

14. $\dfrac{4x^2-9y^2}{10x^2y+15xy^2}$

15. $\dfrac{24x^2y-6y}{4x^2+14x+6}$

7.2 MULTIPLYING AND DIVIDING ALGEBRAIC FRACTIONS

==================== **KEY IDEAS** ====================

To multiply two fractions, first divide out any common factors in the numerators and the denominators of the fractions. Then multiply together the remaining factors in the numerators, and multiply together the remaining factors in the denominators. For example:

$$\frac{4}{9} \times \frac{3}{10} = \frac{\overset{2}{\cancel{4}}}{\underset{3}{\cancel{9}}} \times \frac{\overset{1}{\cancel{3}}}{\underset{5}{\cancel{10}}} \Rightarrow \overbrace{\frac{4 \div 2 = 2}{9 \div 3 = 3}}^{\textit{Think:}} \times \frac{3 \div 3 = 1}{10 \div 2 = 5} = \frac{2 \times 1}{3 \times 5} = \frac{2}{15}$$

To divide one fraction by a second fraction, change to a multiplication example by multiplying the first fraction by the reciprocal of the second fraction.

Multiplying Algebraic Fractions

Before multiplying fractions that contain polynomials, factor and divide out any matching pairs of factors in the numerators and the denominators.

Example 1

Write the product $\dfrac{12y^2}{x^2+7x} \cdot \dfrac{x^2-49}{2y^5}$ in lowest terms.

Solution: $\dfrac{6(x-7)}{xy^3}$

- Factor where possible:

$$\frac{12y^2}{x^2+7x} \cdot \frac{x^2-49}{2y^5} = \frac{12y^2}{x(x+7)} \cdot \frac{(x+7)(x-7)}{2y^5}$$

- Cancel any factor that appears in both a numerator and a denominator:

$$= \frac{\overset{6}{\cancel{12}y^2}}{x\cancel{(x+7)}} \cdot \frac{\overset{1}{\cancel{(x+7)}}(x-7)}{\underset{y^3}{\cancel{2y^5}}}$$

- Multiply together the remaining factors in the numerators, and multiply together the remaining factors in the denominator:

$$= \frac{6(x-7)}{xy^3}$$

203

Dividing Algebraic Fractions

Algebraic fractions, like fractions in arithmetic, are divided by inverting the second fraction and then multiplying.

Example 2

Write the quotient $\dfrac{8m^2}{3} \div \dfrac{6m^3}{3m-12}$ in lowest terms.

Solution: $\dfrac{4(m-4)}{3m}$

- Change to a multiplication example:

$$\frac{8m^2}{3} \div \frac{6m^3}{3m-12} = \frac{8m^2}{3} \cdot \frac{3m-12}{6m^3}$$

- Factor:

$$= \frac{8m^2}{3} \cdot \frac{3(m-4)}{6m^3}$$

- Cancel common factors:

$$= \frac{\overset{4}{\cancel{8m^2}}}{\cancel{3}} \cdot \frac{\overset{1}{\cancel{3}}(m-4)}{\underset{3m}{\cancel{6m^3}}}$$

- Multiply the remaining factors:

$$= \frac{4(m-4)}{3m}$$

Example 3

Write the quotient $\dfrac{x^2-2x-8}{x^2-25} \div \dfrac{x^2-4}{2x+10}$ in lowest terms.

Solution: $\dfrac{2(x-4)}{(x-5)(x-2)}$

- Change to a multiplication example:

$$\frac{x^2 - 2x - 8}{x^2 - 25} \div \frac{x^2 - 4}{2x + 10} = \frac{x^2 - 2x - 8}{x^2 - 25} \cdot \frac{2x + 10}{x^2 - 4}$$

- Factor:

$$= \frac{(x-4)(x+2)}{(x-5)(x+5)} \cdot \frac{2(x+5)}{(x+2)(x-2)}$$

- Cancel common factors:

$$= \frac{(x-4)\overset{1}{\cancel{(x+2)}}}{(x-5)\cancel{(x+5)}} \cdot \frac{2\overset{1}{\cancel{(x+5)}}}{\cancel{(x+2)}(x-2)}$$

- Simplify:

$$= \frac{2(x-4)}{(x-5)(x-2)}$$

Check Your Understanding of Section 7.2

In each case, show how you arrived at your answer by clearly indicating all of the necessary steps, formula substitutions, diagrams, graphs, charts, etc.

1–10. For all values of the variables for which each expression is defined, perform the indicated operation and express the result in simplest form.

1. $\dfrac{3b}{4a} \cdot \dfrac{8a^2 - 4a}{9b^2}$

2. $\dfrac{2x+6}{8xy} \div \dfrac{x+3}{2y^2}$

3. $\dfrac{3x^2 y}{x^2 - y^2} \cdot \dfrac{2x+2y}{6xy^2}$

4. $\dfrac{x^2 - x - 6}{3x - 9} \cdot \dfrac{2}{x+2}$

5. $\dfrac{x^2 - 9}{x^2 + 6x + 9} \cdot \dfrac{x^2 + 3x}{x^2}$

6. $\dfrac{x^2 + 3x - 4}{5x - 5} \cdot \dfrac{10x^2 - 40x}{x^2 - 16}$

7. $\dfrac{y^2 - 7y + 10}{25y - y^2} \div \dfrac{y^2 - 4}{25y^3}$

8. $\dfrac{y^2 - 49}{y^2 - 3y - 28} \cdot \dfrac{3y + 12}{y^2 + 5y - 14}$

9. $\dfrac{x^2 - 3x}{x^2 + 2x} \div \dfrac{x^2 - 5x + 6}{x^2 - 4}$

10. $\dfrac{x^2 + 4x + 4}{2x - 3} \div \dfrac{x^2 - 4}{2x^2 - 7x + 6}$

7.3 CONVERTING UNITS OF MEASUREMENT

Sometimes it is necessary to convert from one unit of measurement to another. This conversion can be accomplished by multiplying the original unit by a fractional conversion factor or by writing a proportion.

Multiplication Property of Conversion Factors

Since 60 minutes = 1 hour, 3 hours can be changed into an equivalent number of minutes by multiplying 3 hours by the conversion factor $\frac{60 \text{ minutes}}{1 \text{ hour}}$. For example:

$$3 \text{ hr} \times \frac{60 \text{ min}}{1 \text{ hr}} = 3 \times 60 \text{ min} = 180 \text{ min}$$

Because the conversion factor $\frac{60 \text{ minutes}}{1 \text{ hour}}$ has the same amount of time in both its numerator and its denominator, it is numerically equal to 1. A conversion factor is always formed so that its numerical value is 1. This guarantees that multiplying by an appropriate conversion factor changes the *unit* of measurement of a quantity, but not its amount.

Finding a Conversion Factor

A conversion factor between two related units of measurement is obtained by writing an equation that defines their numerical relationship. For instance, 5280 feet = 1 mile. Hence:

- Dividing both sides of the equation 5280 feet = 1 mile by 1 mile gives the conversion factor $\frac{5280 \text{ feet}}{1 \text{ mile}}$ for changing from miles into an equivalent number of feet. For example, to change 1.5 miles into an equivalent number of feet, write:

$$1.5 \text{ mi} \times \frac{5280 \text{ ft}}{1 \text{ mi}} = 1.5 \times 5280 \text{ ft} = 7920 \text{ ft}$$

- Dividing both sides of the equation 5280 feet = 1 mile by 5280 feet gives the conversion factor $\frac{1 \text{ mile}}{5280 \text{ feet}}$ for changing from feet into an equivalent number of miles. For example, to change 3960 feet into an equivalent number of miles, write:

$$3960 \cancel{ft} \times \frac{1 \text{ mi}}{5280 \cancel{ft}} = \frac{3960}{5280} \text{ mi} = 0.75 \text{ mi}$$

Notice that, when converting from one unit to another, like units that appear in a numerator and in a denominator are canceled so that the final answer is in the correct unit of measurement. If the answer is *not* in the expected unit of measurement, you may have used the wrong conversion factor.

Example 1

If an object is moving at an average rate of 18 kilometers per minute, what is its average rate of speed in meters per second?

Solution: $\mathbf{300 \dfrac{m}{s}}$

Since 1 km = 1000 m, to convert from kilometers to meters use the conversion factor $\frac{1000 \text{ m}}{1 \text{ km}}$. Thus:

$$18 \frac{\text{km}}{\text{min}} = 18 \frac{\text{km}}{\text{min}} \times \frac{1000 \text{ m}}{1 \text{ km}} \times \frac{?}{?}$$

The numerator for the conversion factor for time must contain minutes so that it cancels the minutes in the denominator of $18 \frac{\text{km}}{\text{min}}$.

Hence, use the conversion factor $\frac{1 \text{ min}}{60 \text{ s}}$:

$$18 \frac{\text{km}}{\text{min}} = 18 \frac{\cancel{\text{km}}}{\cancel{\text{min}}} \times \frac{1000 \text{ m}}{1 \cancel{\text{km}}} \times \frac{1 \cancel{\text{min}}}{60 \text{ s}}$$

$$= \frac{18 \times 1000 \text{ m}}{60 \text{ s}}$$

$$= 300 \frac{\text{m}}{\text{s}}$$

Using Proportions to Convert Units of Measurement

Since the conversion ratio of one unit to another is fixed, you can use a proportion to convert between units of measurement. For example, to convert 13,200 feet into an equivalent number of miles, write the extended proportion

$$\frac{\text{mi}}{\text{ft}} = \frac{x}{13,200 \text{ ft}} = \frac{1 \text{ mi}}{5280 \text{ ft}}$$

where one of the ratios is the conversion factor, and x represents the number of miles equivalent to 13,200 feet. Hence:

$$x = \frac{1\,\text{mi}}{5280\,\text{ft}} \times 13,200\,\text{ft}$$

$$= \frac{13,200}{5280}\,\text{mi}$$

$$= 2.5\,\text{mi}$$

Check Your Understanding of Section 7.3

A. Multiple Choice

1. A car is traveling at an average rate of 60 miles per hour. How many miles per minute is the car traveling?

   (1) 1   (2) $\dfrac{1}{60}$ (3) $\dfrac{1}{360}$ (4) 3600

2. A car is traveling at an average rate of 45 miles per hour. How many feet per second is the car traveling?

   (1) 48   (2) 60   (3) 66   (4) 88

3. In which number is 72 meters per hour expressed as kilometers per hour?

   (1) 7.2×10^{-2} (2) 7.2×10^{2} (3) 7.2×10^{-3} (4) 7.2×10^{3}

4. Which expression could be used to change 8 kilometers per hour to meters per minute?

   (1) $\dfrac{8\,\text{km}}{\text{h}} \times \dfrac{1\,\text{km}}{1000\,\text{m}} \times \dfrac{1\,\text{h}}{60\,\text{min}}$ (3) $\dfrac{8\,\text{km}}{\text{h}} \times \dfrac{1000\,\text{m}}{1\,\text{km}} \times \dfrac{1\,\text{h}}{60\,\text{min}}$

   (2) $\dfrac{8\,\text{km}}{\text{h}} \times \dfrac{1000\,\text{m}}{1\,\text{m}} \times \dfrac{60\,\text{min}}{1\,\text{h}}$ (4) $\dfrac{8\,\text{km}}{\text{h}} \times \dfrac{1\,\text{km}}{1000\,\text{m}} \times \dfrac{60\,\text{min}}{1\,\text{h}}$

5. There are 12 players on a basketball team. Before a game, both ankles of each player are taped. Each roll of tape will tape three ankles. Which product can be used to determine the number of rolls of tape needed to tape all the players' ankles?

   (1) $12\,\text{players} \cdot \dfrac{1\,\text{player}}{2\,\text{ankles}} \cdot \dfrac{3\,\text{ankles}}{1\,\text{roll}}$

   (2) $12\,\text{players} \cdot 2\,\text{ankles} \cdot \dfrac{3\,\text{rolls}}{1\,\text{ankle}}$

(3) $12 \text{ players} \cdot \dfrac{2 \text{ ankles}}{1 \text{ player}} \cdot \dfrac{1 \text{ roll}}{3 \text{ ankles}}$

(4) $12 \text{ players} \cdot \dfrac{1 \text{ roll}}{3 \text{ players}} \cdot \dfrac{3 \text{ ankles}}{1 \text{ roll}}$

B. *In each case, show how you arrived at your answer by clearly indicating all of the necessary steps, formula substitutions, diagrams, graphs, charts, etc.*

6. At a party, six 1-liter bottles of soda are completely emptied into 8-ounce cups. What is the least number of cups that are needed? [Assume 1 L is approximately equal to 1.1 qt.]

7. A rectangular fish tank is 75 centimeters in length, 45 centimeters in width, and 48 centimeters in height. What is the greatest number of quarts of water, to the *nearest tenth*, that the fish tank can hold? [1 L = $100 \text{ cc} \approx 1.1 \text{ qt}$]

8. A carpet costs $27 per square yard. What is the cost, in dollars, of a roll of carpeting that is 12 feet in width and 25 feet in length?

9. One knot is 1 nautical mile per hour, and 1 nautical mile is 6080 feet. If a cruiser ship has an average speed of 3.5 knots, what is the average speed, in miles per hour, of the ship?

10. Roger bought a generator that will run for 1.5 hours on 1 liter of gas. If the gas tank has the shape of a rectangular box that is 25 centimeters by 20 centimeters by 16 centimeters, how long will the generator run on a full tank of gas? [1 L = 1000 cc]

7.4 ADDING AND SUBTRACTING ALGEBRAIC FRACTIONS

KEY IDEAS

To add (or subtract) fractions that have the same denominators, write the sum (or difference) of the numerators over the common denominator. For example:

$$\frac{2}{7} + \frac{3}{7} = \frac{2+3}{7} = \frac{5}{7}$$

If the fractions have different denominators, first change each fraction to an equivalent fraction having the lowest common denominator (LCD) as its denominator.

Combining Algebraic Fractions Having the Same Denominator

Algebraic fractions are added and subtracted in much the same way as are fractions in arithmetic.

Example 1

Write the difference $\dfrac{5a+b}{10ab} - \dfrac{3a-b}{10ab}$ in simplest form.

Solution: $\dfrac{a+b}{5ab}$

Write the difference of the numerators over the common denominator. Then combine like terms and simplify:

$$\frac{5a+b}{10ab} - \frac{3a-b}{10ab} = \frac{5a+b-(3a-b)}{10ab}$$
$$= \frac{5a+b-3a+b}{10ab}$$
$$= \frac{2a+2b}{10ab}$$
$$= \frac{2(a+b)}{10ab}$$
$$= \frac{a+b}{5ab}$$

Combining Algebraic Fractions Having Different Denominators

When adding and subtracting algebraic fractions with unlike denominators, first determine the LCD of the fractions. Then use the multiplication property of 1 to change each fraction into an equivalent fraction with the LCD as its denominator.

Example 2

Express the sum of $\dfrac{3}{10x}$ and $\dfrac{4}{15x}$ as a single fraction in simplest form.

Solution: $\dfrac{17}{30x}$

The LCD of $10x$ and $15x$ is $30x$, which is the smallest term into which both $10x$ and $15x$ divide evenly. Multiply the first fraction by 1 in the form of $\frac{3}{3}$ in

210

order to change it into an equivalent fraction that has $30x$ as its denominator. Change the second fraction into an equivalent fraction that has $30x$ as its denominator by multiplying it by 1 in the form of $\frac{2}{2}$.

Thus:

$$\frac{3}{10x}+\frac{4}{15x}=\frac{3}{3}\cdot\left(\frac{3}{10x}\right)+\frac{2}{2}\cdot\left(\frac{4}{15x}\right)$$

$$=\frac{9}{30x}\quad+\frac{8}{30x}$$

$$=\frac{9+8}{30x}$$

$$=\frac{17}{30x}$$

Example 3

Express $\dfrac{x}{2}-\dfrac{5x}{12}+\dfrac{x}{4}$ as a single fraction in simplest form.

Solution: $\dfrac{x}{3}$

The LCD of 2, 12, and 4 is 12, which is the smallest positive integer into which 2, 12, and 4 divide evenly. Change the first fraction into an equivalent fraction that has 12 as its denominator by multiplying it by 1 in the form of $\frac{6}{6}$. The second fraction already has the LCD as its denominator. Multiply the last fraction by 1 in the form of $\frac{3}{3}$ so that it becomes an equivalent fraction with 12 as its denominator. Thus:

$$\frac{x}{2}-\frac{5x}{12}+\frac{x}{4}=\frac{6}{6}\cdot\left(\frac{x}{2}\right)-\frac{5x}{12}+\frac{3}{3}\cdot\left(\frac{x}{4}\right)$$

$$=\frac{6x}{12}\quad-\frac{5x}{12}+\frac{3x}{12}$$

$$=\frac{6x-5x+3x}{12}$$

$$=\frac{4x}{12}$$

$$=\frac{x}{3}$$

Example 4

Write the sum $\dfrac{2x+1}{6y} + \dfrac{3x-5}{9y}$ in simplest form.

Solution: $\dfrac{12x-7}{18y}$

The LCD of $6y$ and $9y$ is $18y$, which is the smallest expression into which $6y$ and $9y$ divide evenly. Change the fractions into equivalent fractions that have $18y$ as their denominator by multiplying the first fraction by 1 in the form of $\frac{3}{3}$, and by multiplying the second fraction by 1 in the form of $\frac{2}{2}$. Thus:

$$\frac{2x+1}{6y} + \frac{3x-5}{9y} = \frac{3}{3} \cdot \left(\frac{2x+1}{6y}\right) + \frac{2}{2} \cdot \left(\frac{3x-5}{9y}\right)$$

$$= \frac{3(2x+1) \quad +2(3x-5)}{18y}$$

$$= \frac{(6x+3)+(6x-10)}{18y}$$

$$= \frac{12x-7}{18y}$$

Example 5

Express the difference $\dfrac{7}{2x} - \dfrac{3}{x+2}$ as a single fraction.

Solution: $\dfrac{x+14}{2x^2+4x}$

The LCD of $2x$ and $x + 2$ is $2x(x + 2)$. Change the fractions into equivalent fractions that have $2x(x + 2)$ as their common denominator by multiplying the first fraction by 1 in the form of $\frac{x+2}{x+2}$, and by multiplying the second fraction by 1 in the form of $\frac{2x}{2x}$:

$$\frac{7}{2x} - \frac{3}{x+2} = \frac{x+2}{x+2} \cdot \left(\frac{7}{2x}\right) - \frac{2x}{2x} \cdot \left(\frac{3}{x+2}\right)$$

$$= \frac{7(x+2) - 2x(3)}{2x(x+2)}$$

$$= \frac{7x+14-6x}{2x(x+2)}$$

$$= \frac{x+14}{2x^2+4x}$$

Example 6

Multiple Choice:

Expressed as a single fraction, what is $\dfrac{1}{x}+\dfrac{1}{x+1}$ $(x \neq 0, -1)$?

(1) $\dfrac{2}{2x+1}$ (2) $\dfrac{2}{x^2+x}$ (3) $\dfrac{2x+1}{x^2+x}$ (4) $\dfrac{3x}{x+2}$

Solution: **(3)**

Method 1: Use algebraic reasoning.

$$\frac{1}{x}+\frac{1}{x+1} = \frac{1}{x}\left(\frac{x+1}{x+1}\right)+\left(\frac{1}{x+1}\right)\cdot\frac{x}{x}$$

$$= \frac{x+1}{x^2-x} \quad + \quad \frac{x}{x^2+x}$$

$$= \frac{2x+1}{x^2+x}$$

Method 2: Look for a pattern using particular numbers.

Substitute an easy number for x and perform the arithmetic. Keep substituting easy numbers until you find a pattern.

- Suppose $x = 1$. Then

$$\frac{1}{1}+\frac{1}{2} = \frac{2}{2}+\frac{1}{2} = \frac{2+1}{2}.$$

- Suppose $x = 2$. Then

$$\frac{1}{2}+\frac{1}{3} = \frac{3}{6}+\frac{2}{6} = \frac{3+2}{6}.$$

- Suppose $x = 3$. Then

$$\frac{1}{3}+\frac{1}{4} = \frac{4}{12}+\frac{3}{12} = \frac{4+3}{12} = \frac{\text{sum of original denominators}}{\text{product of original denominators}}.$$

Hence:

$$\frac{1}{x}+\frac{1}{x+1} = \frac{\text{sum of denominators}}{\text{product of denominators}}$$

$$= \frac{x+(x+1)}{x(x+1)}$$

$$= \frac{2x+1}{x^2+x}$$

Check Your Understanding of Section 7.4

A. *Multiple Choice*

1. Expressed as a single fraction, what is $\dfrac{x-7}{2}+\dfrac{x+2}{6}$?

   (1) $\dfrac{2x-5}{8}$ (2) $\dfrac{4x-19}{6}$ (3) $\dfrac{8x-5}{12}$ (4) $\dfrac{x^2-14}{12}$

2. Which expression is equivalent to $\dfrac{a}{x}+\dfrac{3b}{2x}$?

   (1) $\dfrac{2a+3b}{3x}$ (2) $\dfrac{2a+6b}{3x}$ (3) $\dfrac{2a+3b}{2x}$ (4) $\dfrac{5ab}{2x}$

3. What is the sum of $\dfrac{x-2}{3}$ and $\dfrac{x-3}{2}$?

   (1) $\dfrac{2x-5}{5}$ (2) $\dfrac{5x-5}{6}$ (3) $\dfrac{2x-5}{6}$ (4) $\dfrac{5x-13}{6}$

4. Expressed as a single fraction, what is $\dfrac{2}{x}-\dfrac{1}{x+1}(x\neq 0,-1)$?

   (1) $\dfrac{3x-1}{x^2+x}$ (2) $\dfrac{x+2}{x^2+x}$ (3) $\dfrac{1}{2x+1}$ (4) $\dfrac{1}{x}$

5. Expressed as a single fraction, what is $\dfrac{1}{x}+\dfrac{1}{1-x}(x\neq 0,-1)$?

   (1) $\dfrac{1}{x(1-x)}$ (2) $\dfrac{-1}{x(x+1)}$ (3) $-\dfrac{2}{x}$ (4) $\dfrac{1}{x(x-1)}$

6. The sum of $\dfrac{y-4}{2y}$ and $\dfrac{3y-5}{5y}$ is

   (1) $\dfrac{11y-30}{10y}$ (2) $\dfrac{4y-9}{10y}$ (3) $11y-30$ (4) $\dfrac{4y-9}{7y}$

7. Expressed as a single fraction, what is $\dfrac{x^2}{(x+1)^2}-\dfrac{x-1}{x+1}(x\neq-1)$?

   (1) $\dfrac{2x^2+1}{(x+1)^2}$ (2) $\dfrac{1}{(x+1)^2}$ (3) $\dfrac{1-2x^2}{(x+1)^2}$ (4) $\dfrac{x-1}{x+1}$

8. Expressed as a single fraction, what is $\dfrac{2y}{y^2-9}+\dfrac{6}{9-y^2}(y\neq 3,-3)$?

(1) $\dfrac{-2}{y-3}$

(3) $\dfrac{-4y}{(y+3)(y-3)}$

(2) $\dfrac{y+3}{y-3}$

(4) $\dfrac{2}{y+3}$

B. *In each case, show how you arrived at your answer by clearly indicating all of the necessary steps, formula substitutions, diagrams, graphs, charts, etc.*

9–20. Express each sum or difference as a single fraction in simplest form.

9. $\dfrac{4b}{5x}-\dfrac{3b}{10x}$

15. $\dfrac{a^2+1}{a^2-1}-\dfrac{a}{a+1}$

10. $\dfrac{3y-5}{5xy}-\dfrac{1}{10x}$

16. $\dfrac{3x-9}{x^2-9}-\dfrac{1}{x+3}$

11. $\dfrac{b-5}{10b}+\dfrac{b+10}{15b}$

17. $\dfrac{7x-2}{4}-x$

12. $\dfrac{3x-1}{7x}+\dfrac{x+9}{14x}$

18. $\dfrac{3}{10xy}-\dfrac{10x-y}{5xy^2}$

13. $\dfrac{3y-5}{5y}-\dfrac{y-4}{2y}$

19. $\dfrac{5y+7}{y^2-16}-\dfrac{3}{4y+16}$

14. $\dfrac{10x+1}{4x}-\dfrac{x+5}{6x}$

20. $\dfrac{y^2}{x^2-4y^2}+\dfrac{y}{2x+4y}$

7.5 SOLVING EQUATIONS WITH FRACTIONS

 KEY IDEAS

To solve an equation that contains fractions, change the equation into an equivalent equation that does not have any fractions. To eliminate the fractions, multiply each side of the equation by the lowest common denominator of all of its fractional terms.

Example 1

Solve for y: $\dfrac{3y}{4}+\dfrac{7}{12}=\dfrac{y}{6}$.

Solution: **−1**

The LCD of the denominators 4, 12, and 6 is 12, which is the smallest positive integer into which 4, 12, and 6 divide evenly. Eliminate the fractions by multiplying each term on both sides of the equation by 12:

$$\overset{3}{\cancel{12}}\left(\dfrac{3y}{\cancel{4}}\right)+\overset{1}{\cancel{12}}\left(\dfrac{7}{\cancel{12}}\right)=\overset{2}{\cancel{12}}\left(\dfrac{y}{\cancel{6}}\right)$$

$$9y+7=2y$$
$$7=2y-9y$$
$$7=-7y$$
$$\dfrac{7}{-7}=\dfrac{-7y}{-7}$$
$$-1=y \text{ or } y=-1$$

Example 2

Solve for x: $\dfrac{x+1}{2}=\dfrac{5x-3}{3}+5$.

Solution: **−3**

- Write the equation:
- Clear the equation of fractions by multiplying each term on both sides by 6, the LCD of 2 and 3:
  Simplify:
- Remove the parentheses:

$$\dfrac{x+1}{2}=\dfrac{5x-3}{3}+5$$

$$6\left(\dfrac{x+1}{2}\right)=6\left(\dfrac{5x-3}{3}\right)+6(5)$$

$$3(x+1)=2(5x-3)+30$$

$$3\cdot x+3\cdot 1=2\cdot 5x-2\cdot 3+30$$

$$3x+3=10x-6+30$$

$$=10x+24$$

- Collect like x-terms on the left side by subtracting $10x$ from each side:

$$3x-(10x)+3=10x-(10x)+24$$
$$-7x+3=24$$

- Subtract 3 from each side:

$$-7x+3-(3)=24-(3)$$
$$-7x=21$$
$$\dfrac{-7x}{-7}=\dfrac{21}{-7}$$
$$x=-3$$

Example 3

Solve for the positive value of n: $\dfrac{3}{5} + \dfrac{n-2}{3} = \dfrac{14}{5n}$.

Solution: **3**

Since $15n$ is the smallest expression into which 5, 3, and $5n$ divide evenly, clear the equation of its fractions by multiplying each term on both sides by $15n$:

$$15n\left(\frac{3}{5}\right) + 15n\left(\frac{n-2}{3}\right) = 15n\left(\frac{14}{5n}\right)$$

$$9n + 5n(n-2) = 42$$

$$9n + 5n^2 - 10n = 42$$

$$5n^2 - n - 42 = 0$$

$$(5n+14)(n-3) = 0$$

$$5n + 14 = 0 \qquad \text{or } n - 3 = 0$$

$$n = -\frac{14}{5} \quad \text{or} \quad n = 3$$

Check Your Understanding of Section 7.5

Show how you arrived at your answer by clearly indicating all of the necessary steps, formula substitutions, diagrams, graphs, charts, etc.

1–9. In each case, solve for the variable and check.

1. $\dfrac{n}{2} - 3 = \dfrac{n}{5}$

2. $\dfrac{x+1}{4} - \dfrac{2}{3} = \dfrac{1}{12}$

3. $\dfrac{3b}{4} = \dfrac{2b}{5} + \dfrac{21}{20}$

4. $\dfrac{y-2}{2} + \dfrac{2y-1}{20} = \dfrac{y}{4}$

5. $\dfrac{8}{3x} - \dfrac{x-1}{12} = \dfrac{1}{6x}$

6. $\dfrac{4}{x} - 1 = \dfrac{x-4}{2}$

7. $\dfrac{3x-1}{6x} + \dfrac{x+1}{4} = \dfrac{1}{3}$

8. $\dfrac{3}{x} + \dfrac{6}{x+3} = 2$

9. $\dfrac{1}{y} - \dfrac{y+1}{8} = \dfrac{y-1}{4y}$

10. If $\frac{1}{2}$ is added to the reciprocal of a number, the result is 1 less than twice the reciprocal of the original number. Find the number.

11. If the reciprocal of a number is multiplied by 3, the result exceeds the reciprocal of the original number by $\frac{1}{3}$. Find the number.

12. If the reciprocal of a number is multiplied by 1 less than the original number, the result exceeds $\frac{1}{2}$ the reciprocal of the original number by $\frac{5}{8}$. Find the number.

13. If $x = \dfrac{a}{a^2 - b^2} + \dfrac{b}{a^2 - b^2}$, solve for a in terms of b and x.

14. On Tuesday, CityEx delivered 7 less than two times the number of packages it had delivered on Monday. The number of packages delivered on Wednesday was equal to 10 more than the average number of packages delivered on Monday and Tuesday. If the total number of packages that was delivered over the 3 days was 661, how many packages were delivered on Monday?

15. A student got 75% of the questions on a test correct. If he answered 11 of the first 13 questions correctly, and two-thirds of the remaining questions correctly, how many questions were on the test?

CHAPTER 8

RADICALS AND RIGHT TRIANGLES

8.1 SQUARE ROOTS AND REAL NUMBERS

KEY IDEAS

To find the length of a side of a square with an area of 36, we need to answer this question: "The product of which two identical positive numbers is 36?" The answer is the *square root* of 36, denoted as $\sqrt{36}$, which is 6 since $6 \times 6 = 36$.

Definition of Square Root

A **square root** of a nonnegative number N is one of two equal numbers whose product is N. Every positive number has two square roots. The square roots of 4 are $+2$ and -2 since $(+2)(+2) = 4$, and also $(-2)(-2) = 4$. The **radical sign**, $\sqrt{}$, is used to indicate the *principal square root* of the number written underneath it, called the **radicand**.

The **principal square root** of a positive number is its positive square root. Thus, $\sqrt{4} = +2$, where 4 is the *radicand* and $+2$ is the *principal square root* of 4. When we need to refer to the *negative* square root of 4, we write $-\sqrt{4}$. Thus, $-\sqrt{4}$ indicates -2.

Rational Numbers and Perfect Squares

A number that can be written as a fraction with an integer numerator and a nonzero integer denominator is called a **rational number**. For example, 9 is a rational number since $9 = \dfrac{9}{1}$. The square root of a number may or may not be rational. The square roots of numbers such as 1, 4, 9, and $\frac{16}{25}$ are rational. Any number whose square root is rational is called a **perfect square**. For instance, the smallest perfect square integer greater than 100 is 121 since $\sqrt{121} = 11$.

Irrational Numbers

When you use a calculator to evaluate the square roots of numbers such as 2, 3, and 79, you will find that the answers are decimal numbers that have more decimal digits than can fit in any calculator display. The numbers $\sqrt{2}$, $\sqrt{3}$, and $\sqrt{79}$ are **irrational numbers** since they are nonrepeating decimal numbers that do not end. For example, $\sqrt{2} = 1.4142136 \ldots$.

The Set of Real Numbers and Its Subsets

The **set of real numbers** consists of the sets of *rational* numbers and *irrational* numbers.

- The **set of rational numbers** includes these types of numbers:
  1. Integers since these numbers can be written in fractional form with a denominator of 1. For example, 3, 0, and −2 are rational numbers since $3 = \dfrac{3}{1}$, $0 = \dfrac{0}{1}$, and $-2 = \dfrac{-2}{1}$.
  2. Decimal numbers that end, such as $1.25 \left(= \dfrac{5}{4} \right)$.
  3. Nonending decimal numbers in which one or more nonzero digits repeat endlessly, as in $0.3333 \ldots \left(= \dfrac{1}{3} \right)$ and $0.636363 \ldots \left(= \dfrac{7}{11} \right)$.

     Sometimes the repeating part of a nonending decimal is indicated by using a short-hand notation in which a bar is placed over the digit or set of digits that repeat, as in $0.3333 \ldots = 0.\overline{3}$ and $0.636363 \ldots = 0.\overline{63}$.
- The **set of irrational numbers** includes numbers such as

$$\pi \, (= 3.14159\ldots) \quad \text{and} \quad \sqrt{3} \, (= 1.73205\ldots)$$

that cannot be expressed as the quotient of two integers. Thus, an irrational number can only be *approximated* by a decimal number that ends or by a fraction with an integer numerator and a nonzero integer denominator.

Example 1

Multiple Choice:
Which is an irrational number?

(1) $\sqrt{49}$ (2) $\sqrt{8}$ (3) $\sqrt{\dfrac{9}{25}}$ (4) $0.171717\ldots$

Solution: **(2)**

- Choice (1): $\sqrt{49}$ is rational since $\sqrt{49} = 7$.
- Choice (2): Since 8 is not a perfect square, $\sqrt{8}$ is irrational. $\Big\}$ ← This is the correct answer.

220

- Choice (3): $\sqrt{\dfrac{9}{25}}$ is rational since $\sqrt{\dfrac{9}{25}} = \dfrac{3}{5}$.
- Choice (4): $0.171717\ldots$ is rational since the digits 17 repeat endlessly.

Example 2

Express the repeating decimal $0.171717\ldots$ as the quotient of two integers.

Solution: $\dfrac{17}{99}$

Subtract $0.171717\ldots$ from a multiple of itself in order to eliminate the repeating decimal part of the number. Subtracting $0.171717\ldots$ from 100 times that same number, N, will make the decimal parts of the two numbers exactly the same, so their difference will be 0:

$$100N = 17.1717\ldots$$
$$-N = 0.1717\ldots$$
$$99N = 17.0000\ldots$$
$$N = \dfrac{17}{99}$$

Simplifying Radicals

To simplify a radical, remove any perfect square factors from the radicand so that the resulting radicand does not contain any perfect square factors greater than 1. For example, to write $\sqrt{80}$ in simplest form:

- Factor the radicand, if possible, so that one of its factors is the *largest* perfect square factor of 80:
- Write the radical over each factor:
- Evaluate the square root of the perfect square:

$\sqrt{80} = \sqrt{16 \cdot 5}$
$= \sqrt{16} \cdot \sqrt{5}$
$= 4\sqrt{5}$

Example 3

Simplify: $\sqrt{28a^4b^2}$.

Solution: $2\sqrt{7}a^2b$

To simplify $\sqrt{28a^4b^2}$, distribute the radical over each factor of $28a^4b^2$:

$$\sqrt{28a^4b^2} = \sqrt{28} \cdot \sqrt{a^4} \cdot \sqrt{b^2}$$
$$= \sqrt{4} \cdot \sqrt{7} \cdot a^2 \cdot b$$
$$= 2\sqrt{7}a^2b$$

Solving Quadratic Equations with No Middle Term

To solve a quadratic equation such as $2x^2 - 50 = 0$, in which the x-term is missing, solve for the x^2-term. Then take the square root of both sides of the equation. Thus:

- $2x^2 - 50 = 0$ becomes $2x^2 = 50$, so $x^2 = \dfrac{50}{2} = 25$.
- Since the solution set of $x^2 = 25$ includes the two square roots of 25, then $x = +\sqrt{25} = +5$ or $x = -\sqrt{25} = -5$. Sometimes the notation $x = \pm\sqrt{25}$ is used to abbreviate the statement $x = +\sqrt{25}$ or $x = -\sqrt{25}$.

Check Your Understanding of Section 8.1

A. Multiple Choice

1. Which is an irrational number?

   (1) $\sqrt{\dfrac{1}{81}}$

   (2) $-\sqrt{\dfrac{9}{25}}$

   (3) $0.101001000100001\ldots$

   (4) $0.454545\ldots$

2. Which is an irrational number?

   (1) $\sqrt{121}$

   (2) $\sqrt{\dfrac{25}{9}}$

   (3) $\sqrt{40}$

   (4) $0.373737\ldots$

3. If $\sqrt{7-2x}$ represents a rational number, x cannot be equal to

   (1) 1   (2) $\dfrac{3}{2}$ (3) 3   (4) -1

4. If $4x^2 - 48 = x^2$, then $x =$

   (1) -4 (2) -2 (3) ± 4 (4) $+4, -2$

5. If $\dfrac{1}{2}\sqrt{96} = k\sqrt{6}$, then k is equal to

   (1) 8   (2) 2   (3) 3   (4) 4

6. What is the value of $\sqrt{\dfrac{x+4}{2}}$ when $x = \dfrac{1}{2}$?

(1) $\dfrac{3}{2}$ (2) 2   (3) $\dfrac{3}{4}$ (4) $\dfrac{5}{2}$

7. If $n^2 < n < \sqrt{n}$, then n could be

(1) 1   (2) 0   (3) $\dfrac{1}{9}$ (4) 9

8. If $\dfrac{1}{\sqrt{k}} < \sqrt{k} < k$, then k could be

(1) $\dfrac{1}{4}$ (2) 1   (3) 0   (4) 4

9. If $0 < t < 1$, which expression is *not* true?

(1) $t < \dfrac{1}{t}$ (2) $t^2 < \sqrt{t}$ (3) $t^2 > t$ (4) $\sqrt{t} > t$

B. *In each case, show how you arrived at your answer by clearly indicating all of the necessary steps, formula substitutions, diagrams, graphs, charts, etc.*

10–17. Write each expression in simplest form.

10. $\sqrt{0.04} + \sqrt{0.36}$ **14.** $\sqrt{\dfrac{9}{25}}$ **16.** $\dfrac{\sqrt{600}}{5}$

11. $\sqrt{48}$

12. $-\sqrt{63}$ **15.** $\dfrac{1}{2}\sqrt{192}$ **17.** $3 - 2\sqrt{4}$

13. $2\sqrt{108}$

18–20. Express each repeating decimal number as the ratio of two integers.

18. $0.4141\overline{41}$ **19.** $0.3\overline{8}$ **20.** $1.72\overline{72}$

21–26. Simplify.

21. $\sqrt{32r^2s^6}$ **23.** $a\sqrt{75a^4}$ **25.** $\sqrt{\dfrac{40a^7b^2}{5a^3b^2}}$

22. $2\sqrt{50x^2y^4}$ **24.** $\sqrt{12s^4t}$ **26.** $\sqrt{\dfrac{54x^5y^2}{3x^3y^6}}$

8.2 ARITHMETIC OPERATIONS WITH RADICALS

=== KEY IDEAS ===

Since radicals are real numbers, we can add, subtract, multiply, and divide them using the properties of real numbers. Square root radicals such as $7\sqrt{3}$ and $5\sqrt{3}$ are called **like radicals** since they have the same radicand. Like radicals can be added or subtracted in the same way that like algebraic terms are combined. For example:

$$7\sqrt{3} + 5\sqrt{3} = (7+5)\sqrt{3} = 12\sqrt{3}$$

Multiplying Radicals

To multiply radicals, use the rule

$$a\sqrt{x} \cdot b\sqrt{y} = ab\sqrt{xy}$$

provided that x and y are nonnegative numbers. For example:

$$3\sqrt{7} \cdot 2\sqrt{5} = (3 \times 2)\sqrt{7 \times 5} = 6\sqrt{35}$$

Dividing Radicals

To divide radicals, use the rule

$$\frac{a\sqrt{x}}{b\sqrt{y}} = \frac{a}{b}\sqrt{\frac{x}{y}}$$

provided that x and y are nonnegative numbers, and the denominator of each fraction is not 0. For example:

$$\frac{20\sqrt{6}}{4\sqrt{2}} = \frac{20}{4}\sqrt{\frac{6}{2}} = 5\sqrt{3}$$

Squaring Radicals

Squaring a radical eliminates the radical. For example:

$$\left(\sqrt{9}\right)^2 = \sqrt{9} \cdot \sqrt{9} = \sqrt{81} = 9$$

Thus, $\left(\sqrt{9}\right)^2 = 9$. In general, the product of the square roots of the *same* nonnegative number is equal to that number (radicand).

═══════════ **MATH FACTS** ═══════════

If N stands for any nonnegative number, then:
$$\sqrt{N} \times \sqrt{N} = N$$

Example 1

Write $\left(2\sqrt{7}\right)^2$ in simplest form.

Solution: **28**

$$\begin{aligned}
\left(2\sqrt{7}\right)^2 &= \left(2\sqrt{7}\right) \times \left(2\sqrt{7}\right) \\
&= (2 \times 2) \times \left(\sqrt{7} \times \sqrt{7}\right) \\
&= 4 \quad\times\quad 7 \\
&= 28
\end{aligned}$$

Rationalizing Denominators

If a radical appears in the denominator of a fraction, the denominator can be "rationalized" by changing the fraction into an equivalent fraction that does not have a radical in its denominator. To rationalize the denominator of $\frac{15}{\sqrt{3}}$, multiply the fraction by 1 in the form of $\frac{\sqrt{3}}{\sqrt{3}}$:

$$\begin{aligned}
\frac{15}{\sqrt{3}} &= \frac{15}{\sqrt{3}} \cdot \left(\frac{\sqrt{3}}{\sqrt{3}}\right) \\
&= \frac{15\sqrt{3}}{\sqrt{3} \cdot \sqrt{3}} \\
&= \frac{\overset{5}{\cancel{15}}\sqrt{3}}{\underset{1}{\cancel{3}}} \\
&= 5\sqrt{3}
\end{aligned}$$

Adding and Subtracting Radicals

A radical may need to be simplified before it can be combined with another radical. For example, to add unlike radicals $5\sqrt{3}$ and $\sqrt{48}$, first rewrite $\sqrt{48}$ as follows:

$$\begin{aligned}
5\sqrt{3} + \sqrt{48} &= 5\sqrt{3} + \overset{\sqrt{48}}{\overbrace{\sqrt{16} \cdot \sqrt{3}}} \\
&= 5\sqrt{3} + 4\sqrt{3} \\
&= 9\sqrt{3}
\end{aligned}$$

It is not always possible to combine radicals. For example, the sum $\sqrt{2} + \sqrt{5}$ cannot be simplified further.

Example 2

If $3\sqrt{20} - \sqrt{5} + \dfrac{1}{2}\sqrt{80} = x\sqrt{5}$, what is the value of x?

  Solution: **7**

Change the first and third radicals into equivalent radicals having 5 as their radicands. Then combine the three radicals.

$$3\sqrt{20} - \sqrt{5} + \frac{1}{2}\sqrt{80} = 3\sqrt{4} \cdot \sqrt{5} - \sqrt{5} + \frac{1}{2}\sqrt{16} \cdot \sqrt{5}$$

$$= (3 \cdot 2)\sqrt{5} - \sqrt{5} + \left(\frac{1}{2} \cdot 4\right)\sqrt{5}$$

$$= 6\sqrt{5} - \sqrt{5} \quad + \quad 2\sqrt{5}$$

$$= 5\sqrt{5} \quad\quad + \quad 2\sqrt{5}$$

$$= 7\sqrt{5}$$

Since $3\sqrt{20} - \sqrt{5} + \dfrac{1}{2}\sqrt{80} = x\sqrt{5} = 7\sqrt{5}$, $x = 7$.

Check Your Understanding of Section 8.2

A. *Multiple Choice*

1. The expression $2\sqrt{2} + \sqrt{50}$ is equivalent to
   (1) $2\sqrt{52}$ (2) $3\sqrt{52}$ (3) $7\sqrt{2}$ (4) $27\sqrt{2}$

2. The expression $5\sqrt{2} - \sqrt{18}$ is equivalent to
   (1) $2\sqrt{2}$ (2) $-2\sqrt{2}$ (3) $8\sqrt{2}$ (4) $-8\sqrt{2}$

3. The expression $4\sqrt{12} + 2\sqrt{27}$ is equivalent to
   (1) $6\sqrt{39}$ (2) $34\sqrt{3}$ (3) $14\sqrt{3}$ (4) $14\sqrt{6}$

4. If the length of a rectangle is $5\sqrt{2}$ and the width is $2\sqrt{3}$, what is the area of the rectangle?
   (1) $10\sqrt{6}$ (2) $7\sqrt{6}$ (3) $7\sqrt{5}$ (4) $10\sqrt{5}$

5. If the length of a rectangle is $\sqrt{27}$ and the width is $\sqrt{12}$, what is the perimeter in simplest radical form?
   (1) $5\sqrt{3}$ (2) $10\sqrt{3}$ (3) $5\sqrt{6}$ (4) $26\sqrt{3}$

6. If the dimensions of a rectangular box are $\sqrt{2}$, $\sqrt{3}$, and $4\sqrt{6}$, what is the volume of the box in simplest form?
   (1) 24   (2) $4\sqrt{11}$ (3) $4\sqrt{30}$ (4) $8\sqrt{3}$

B. In each case, show how you arrived at your answer by clearly indicating all of the necessary steps, formula substitutions, diagrams, graphs, charts, etc.

7–18. In each case, perform the indicated operation, and write the answer in simplest radical form.

7. $\sqrt{11} - 8\sqrt{11}$

13. $\left(5\sqrt{14}\right) \cdot \left(3\sqrt{10}\right)$

8. $2\sqrt{8} + 8\sqrt{2}$

14. $\dfrac{12\sqrt{54}}{4\sqrt{3}}$

9. $\dfrac{1}{2}\sqrt{20} + \dfrac{1}{3}\sqrt{45}$

15. $\sqrt{75} + 2\sqrt{12} - \sqrt{48}$

10. $\left(3\sqrt{5}\right)^2$

16. $\dfrac{\sqrt{50} - \sqrt{8}}{\sqrt{2}}$

11. $\sqrt{18} + \left(\sqrt{2}\right)^3$

17. $\sqrt{2}\left(3\sqrt{2} - \sqrt{8}\right)$

12. $\dfrac{3\sqrt{252}}{12\sqrt{7}}$

18. $\dfrac{\sqrt{27} + 4\sqrt{3}}{\sqrt{12}}$

19. If $\sqrt{18} - \sqrt{200} + \sqrt{72} = x\sqrt{2}$, what is the value of x?

20. If a and b are positive numbers, explain how the value of $\left(\sqrt{a} + \sqrt{b}\right)^2$ compares with the value of $a + b$.

8.3   THE PYTHAGOREAN THEOREM

========================= **KEY IDEAS** =========================

In a *right triangle*, the longest side faces the right angle and is called the **hypotenuse**. The other two sides are called **legs**. The Pythagorean theorem relates the *squares* of the lengths of the three sides of a right triangle:

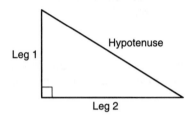

$$(\text{Leg 1})^2 + (\text{Leg 2})^2 = (\text{Hypotenuse})^2$$

Side Relationships in a Right Triangle

The Pythagorean relationship can be used to find the length of *any* side of a right triangle when the lengths of the other two sides are known. If in right triangle *ACB*, shown in the accompanying diagram, leg $a = 2$ and hypotenuse $c = \sqrt{13}$, then:

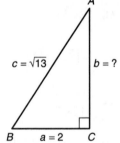

$$a^2 + b^2 = c^2$$
$$2^2 + b^2 = \left(\sqrt{13}\right)^2$$
$$4 + b^2 = 13$$
$$b^2 = 13 - 4$$
$$= 9$$

Because $b^2 = 9$ and b must be positive, leg $b = +\sqrt{9} = 3$.

Example 1

If the length of a diagonal of a square is 10, what is the length of a side of the square?

Solution: **$5\sqrt{2}$**

- Draw square $ABCD$ and diagonal $\overline{AC}$, as shown in the accompanying diagram.

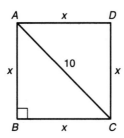

- Since a square has four sides of equal length, let x represent the length of each side.
- In right triangle ABC, use the Pythagorean relationship:

$$x^2 + x^2 = 10^2$$
$$2x^2 = 100$$
$$x^2 = \frac{100}{2} = 50$$
$$x = \sqrt{50} = \sqrt{25} \cdot \sqrt{2} = 5\sqrt{2}$$

Example 2

The cross section of an attic is in the shape of an isosceles trapezoid, as shown in the accompanying figure. The height, BE, of the attic is 10 feet, AD = 31 feet, and BC = 19 feet. What is the length of $\overline{AB}$ to the *nearest foot*?

Solution: **11.7**

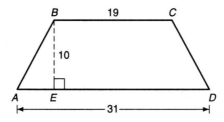

- Drop altitude CF from C to $\overline{AD}$, thus forming rectangle $EBCF$ as shown in the accompanying diagram.

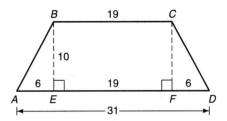

- In rectangle $EBCF$, $EF = BC = 19$. Thus, $AE + DF = 31 - 19 = 12$.
- Since right triangle $AEB \cong$ right triangle DFC,

$$AE = DF = \frac{1}{2} \times 12 = 6.$$

- In right triangle AEB use the Pythagorean relationship:

$$
\begin{aligned}
(AB)^2 &= (AE)^2 + (BE)^2 \\
&= (6)^2 \quad + (10)^2 \\
&= 36 \quad + 100 \\
&= 136 \\
AB &= \sqrt{136} \approx 11.7 \text{ ft}
\end{aligned}
$$

Pythagorean Triples

A **Pythagorean triple** is a set of three positive integers, a, b, and c, that are related so that $a^2 + b^2 = c^2$.

- The numbers 3, 4, and 5 form a Pythagorean triple since

$$\underbrace{3^2 + 4^2}_{9+16} = \underbrace{5^2}_{25}$$

- The numbers 5, 12, and 13 form a Pythagorean triple since

$$\underbrace{5^2 + 12^2}_{25+144} = \underbrace{13^2}_{169}$$

- The numbers 8, 15, and 17 form a Pythagorean triple since

$$\underbrace{8^2 + 15^2}_{64+225} = \underbrace{17^2}_{289}$$

Any whole-number multiple of a Pythagorean triple is also a Pythagorean triple. For example, multiplying each member of the set {3, 4, 5} by 2 produces another Pythagorean triple set, 6, 8, 10.

Examples 3, 4, and 5 make use of Pythagorean triples.

Example 3

A ladder 13 feet in length rests against a vertical building, as shown in the accompanying diagram. The foot of the ladder is 5 feet from the building. How far up the building does the ladder reach?

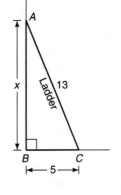

Solution: **12 ft**

- The ladder forms the hypotenuse of right triangle *ABC*. Notice that the lengths of the sides of the right triangle form a 5-12-13 Pythagorean triple. Therefore, the ladder reaches 12 ft up the building.
- If you did not recognize that the lengths form a Pythagorean triple, you could have solved this problem by letting $x = AB$ and proceeding as follows:

$$x^2 + 5^2 = 13^2$$
$$x^2 + 25 = 169$$
$$x^2 = 169 - 25$$
$$x = \sqrt{144} = 12\,\text{ft}$$

Example 4

The perimeter of an isosceles triangle is 50, and the length of the base is 16. What is the length of the altitude drawn to the base?

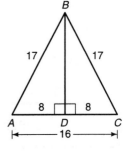

Solution: **15**

- Draw isosceles triangle *ABC* with altitude $\overline{BD}$, as shown in the accompanying diagram.
- The perimeter of $\triangle ABC$ is 50 and $AC = 16$, so $AB + BC = 50 - 16 = 34$. Thus,

$$AB = BC = \frac{1}{2} \times 34 = 17.$$

- The altitude drawn to the base of an isosceles triangle bisects the base. Since $AD = 8$, the lengths of the sides of right triangle *ADB* form a $8 - 15 - 17$ Pythagorean triple in which $BD = 15$.

Example 5

If the lengths of the diagonals of a rhombus are 18 and 24, what is the length of a side of the rhombus?

Solution: **15**

- Draw rhombus $ABCD$. Diagonals $\overline{AC}$ and $\overline{BD}$ bisect each other and intersect at right angles, as shown in the accompanying diagram.
- The lengths of the legs of right triangle AED are 9 and 12. Since $9 = 3 \times 3$ and $12 = 3 \times 4$, the length of the hypotenuse, AD, is $3 \times 5 = 15$. In other words, the lengths of the sides of right triangle AED form a 9-12-15 Pythagorean triple in which $AD = 15$.

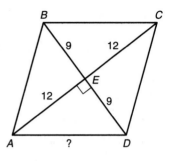

Converse of the Pythagorean Relationship

Is $\triangle ABC$ a right triangle if $AC = 5$, $BC = \sqrt{11}$, and $AB = 6$? A triangle is a right triangle if the square of the length of the longest side of the triangle is equal to the sum of the squares of the lengths of the other two sides. Since the longest side of $\triangle ABC$ is $\overline{AB}$:

$$(AB)^2 \overset{?}{=} (AC)^2 + (BC)^2$$

$$(6)^2 \overset{?}{=} (5)^2 + (\sqrt{11})^2$$

$$36 \overset{?}{=} 25 + 11$$

$$36 \overset{\checkmark}{=} 36$$

Yes, $\triangle ABC$ is a right triangle.

Check Your Understanding of Section 8.3

A. Multiple Choice

1. If the lengths of the legs of a right triangle are 4 and 7, what is the length of the hypotenuse?
   (1) $\sqrt{3}$ (2) $\sqrt{11}$ (3) $\sqrt{33}$ (4) $\sqrt{65}$

2. The length of the hypotenuse of a right triangle is 20 centimeters, and the length of one leg is 12 centimeters. The number of centimeters in the length of the other leg is
   (1) 8   (2) 16   (3) 32   (4) $\sqrt{544}$

3. What is the length of a side of a square whose diagonal measures $4\sqrt{2}$?
   (1) $8\sqrt{2}$ (2) 8   (3) $16\sqrt{2}$ (4) 4

4. Which of the following sets of numbers can *not* represent the lengths of the sides of a right triangle?
   (1) 7, 24, 25   (2) 2, 6, $3\sqrt{5}$ (3) 9, 40, 41   (4) 2, $\sqrt{5}$, 3

5. What is the area of a rectangle whose diagonal measures 10 and whose width measures 6?
   (1) 48   (2) 60   (3) 80   (4) 120

6. If the lengths of the diagonals of a rhombus are 6 and 8, the perimeter of the rhombus is
   (1) 14   (2) 20   (3) 28   (4) 40

7. In rectangle *ABCD*, *AD* = 10 and diagonal *AC* = 26. What is the perimeter of rectangle *ABCD*?
   (1) 34   (2) 68   (3) 240   (4) 260

8. At 9:00 A.M. a car starts at point *A* and travels north for 1 hour at an average rate of 60 miles per hour. Without stopping, the car then travels east for 2 hours at an average rate of 45 miles per hour. At noon, what is the best approximation of the distance, in miles, of the car from point *A*?
   (1) 100   (2) 105   (3) 108   (4) 115

B. In each case, show how you arrived at your answer by clearly indicating all of the necessary steps, formula substitutions, diagrams, graphs, charts, etc.

9. In the accompanying diagram of isosceles trapezoid *ABCD*, $\overline{DE} \perp \overline{AB}$, $\overline{CF} \perp \overline{AB}$, *AB* = 18, *CD* = 10, and *AD* = *BC* = 5. What is the perimeter of quadrilateral *EDCF*?

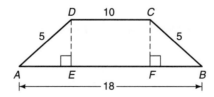

10. In an isosceles trapezoid, the length of each leg is 19 feet, and the lengths of the bases are 11 feet and 25 feet. What is the number of feet, correct to the nearest *tenth*, in the height of the trapezoid?

11. What is the length, to the *nearest tenth of an inch*, of the longest board that can fit through a rectangular doorway 6 feet and 8 inches in height and 3 feet in width?

12. What is the length of a diagonal of a square whose perimeter is $8\sqrt{5}$?

13. A ladder 25 feet in length rests against a vertical building. If the foot of the ladder is 7 feet from the building, how far up the building does the ladder reach?

14. The diagonal of a rectangle exceeds the length by 1. The width of the rectangle is 1 less than one-third of the length. Find the length of the diagonal.

15. Linda and Sue are raising money for a charity by participating in a walkathon. Their sponsors have pledged $5 for each mile each girl walks. Sue begins at point A and walks 5 miles north, then 6 miles east, and finally 3 miles north again to point B. If Linda walks directly in a straight line from A to B, how many more dollars does Sue raise than Linda?

16. The perimeter of a right triangle is 180. If the length of the hypotenuse is 82, find the length of the shorter leg of the triangle.

17. In isosceles triangle ABC, $\overline{AB} \cong \overline{BC}$, and $\overline{BD} \perp \overline{AC}$. If $BD = x$, $AB = 2x - 1$, and $AC = 2x + 2$, find the length of $\overline{BD}$.

18. What is the number of inches in the longest board with negligible width that can fit inside a rectangular box 6 inches wide, 8 inches long, and 2 feet high? Explain how you arrived at your answer.

8.4 SIDE RELATIONSHIPS IN SPECIAL RIGHT TRIANGLES

KEY IDEAS

Knowing the side relationships in special right triangles in which the two acute angles measure 45° and 45° or 30° and 60° can simplify your work.

Special Right Triangle Relationships

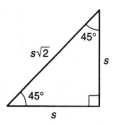
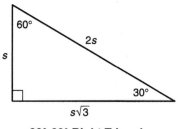

45°-45° (Isosceles) Right Triangle   **30°-60° Right Triangle**

The diagrams above summarize the relationships between the lengths of the sides in two special right triangles.

Example 1

In isosceles trapezoid $ABCD$, $\overline{BC} \parallel \overline{AD}$, and m$\angle A$ = m$\angle D$ = 45. The lengths of the bases are 5 inches and 19 inches. What is the exact number of inches in the length of each leg of the trapezoid?

   Solution: $7\sqrt{2}$

- Drop altitudes from B and C, as shown in the accompanying figure. Then find the length of $\overline{AB}$.

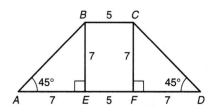

- Since $BEFC$ is a rectangle, $EF = BC = 5$. Therefore, $AE + DF = 19 - 5 = 14$.

- Right triangle $AEB \cong$ right triangle DFC. Hence, $AE = DF = \dfrac{1}{2} \times$ $14 = 7$.
- In right triangle AEB, $AB = BE \times \sqrt{2} = 7\sqrt{2}$.

Example 2

What is the length of the altitude drawn to a side of an equilateral triangle whose perimeter is 36?

Solution: **6√3**

If the perimeter of an equilateral triangle is 36, then the length of each side is $\dfrac{36}{3} = 12$. Draw equilateral triangle *ABC*, and drop altitude $\overline{BD}$ from *B* to $\overline{AC}$.

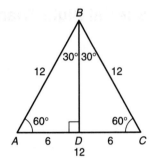

Method 1: Use 30°-60° right triangle relationships.

- Since an altitude drawn to a side of an equilateral triangle divides the triangle into two congruent triangles, *AD* = *CD* = 6 and m∠1 = m∠2 = 30.
- In 30°-60° right triangle *ADB*, $\overline{BD}$ is the leg opposite the 60° angle, so

$$BD = \frac{1}{2} \times AB \times \sqrt{3}$$
$$= \frac{1}{2} \times 12 \times \sqrt{3}$$
$$= 6\sqrt{3}$$

Method 2: Use the Pythagorean theorem.

$$(BD)^2 + (AD)^2 = (AB)^2$$
$$(BD)^2 + (6)^2 = (12)^2$$
$$(BD)^2 + 36 = 144$$
$$(BD)^2 = 108$$
$$BD = \sqrt{108} = \sqrt{36} \cdot \sqrt{3} = 6\sqrt{3}$$

Check Your Understanding of Section 8.4

A. Multiple Choice

1. What is the number of centimeters in the length of an altitude of an equilateral triangle whose side measures 4 centimeters?

   (1) $2\sqrt{3}$ (2) 2   (3) $4\sqrt{3}$ (4) 4

2. In right triangle *ABC*, $\angle C$ is the right angle and m$\angle B = 30$. What is the ratio of *AC* to *BC*?

(1) $2:\sqrt{3}$ (2) $1:2$ (3) $1:\sqrt{3}$ (4) $1:1$

3. What is the perimeter of an equilateral triangle if the length of an altitude is $5\sqrt{3}$?

(1) $15\sqrt{3}$ (2) 15 (3) 30 (4) $30\sqrt{3}$

4. The lengths of two adjacent sides of a parallelogram are 6 and 15. If the measure of an included angle is 60°, what is the length of the shorter diagonal of the parallelogram?

(1) $\sqrt{171}$ (2) $\sqrt{148}$ (3) $\sqrt{153}$ (4) $\sqrt{261}$

5. In the accompanying figure of right triangles *ABC* and *EDC*, $\overline{BCD}$, $\overline{AC} \perp \overline{EC}$, *BD* = 10, *CE* = 8, and m$\angle ACB = 30$. What is the length of $\overline{AB}$?

(1) $2\sqrt{3}$ (2) 3 (3) $3\sqrt{3}$ (4) 6

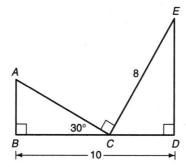

B. *In each case, show how you arrived at your answer by clearly indicating all of the necessary steps, formula substitutions, diagrams, graphs, charts, etc.*

6. In the accompanying diagram of $\triangle RTS$, m$\angle SRT = 60$, *RS* = 8, *RT* = 14, and $\overline{SH} \perp \overline{RT}$. What is the length of $\overline{ST}$ correct to the *nearest tenth*?

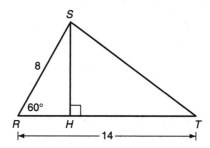

7 and 8. The lengths of the bases of an isosceles trapezoid are 9 and 25.

7. Find the perimeter if the measure of each of the lower base angles is 30°.

8. Find the perimeter if the measure of each of the lower base angles is 45°.

9. The length of each side of a rhombus is 10, and the measure of an angle is 60°. What is the length of:
   (a) the shorter diagonal?   (b) the longer diagonal?

10. A 12-foot ladder leans against the side of a building, making an angle of 30° with the ground. To increase the reach of the ladder against the building, the ladder is moved so that it makes an angle of 60° with the ground, as shown in the accompanying diagram. To the *nearest tenth of a foot*, how much further up the building does the ladder now reach?

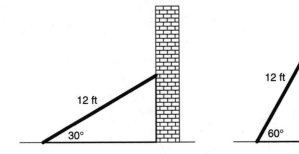

AREA, VOLUME, AND MEASUREMENT

CHAPTER 9

AREA, CIRCUMFERENCE, AND VOLUME

9.1 AREAS OF TRIANGLES AND SPECIAL QUADRILATERALS

KEY IDEAS

The **area** of a figure is the number of 1-by-1-unit square boxes that the figure can enclose. The areas of triangles and special quadrilaterals can be calculated by using the correct formulas.

Area of a Parallelogram = base × height

In a parallelogram, any side may be considered the base, b. The height, h, of the parallelogram is the altitude drawn to that base from the opposite side. A parallelogram has the same area as a rectangle with the same base and height, as shown in the accompanying figure.

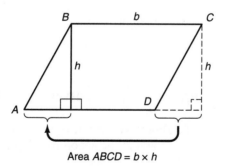

Area $ABCD = b \times h$

Example 1

The lengths of a pair of adjacent sides of parallelogram $ABCD$ are 6 and 10 centimeters. If the degree measure of their included angle is 30, find the area of the parallelogram.

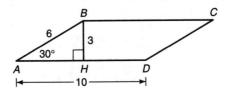

Solution: **30 cm²**

In parallelogram *ABCD*, altitude *BH* = 3 since the length of the leg opposite a 30° angle in a 30°-60° right triangle is one-half the length of the hypotenuse (side $\overline{AB}$).

$$\text{Area of parallelogram } ABCD = \text{base} \times \text{height}$$
$$= AD \times BH$$
$$= 10 \times 3 = 30 \text{ cm}^2$$

Example 2

Find the area of rectangle *ABCD*, whose diagonal is 10 and whose base is 8.

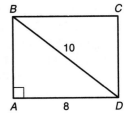

Solution: **48 square units**

The height (width) is 6 since $\triangle BAD$ is a 6-8-10 right triangle. Therefore:

$$\text{Area of rectangle} = \text{base} \times \text{height}$$
$$= 8 \times 6 = 48 \text{ square units}$$

Area of a Rhombus = $\dfrac{1}{2}$ × (diagonal₁ × diagonal₂)

Because the diagonals of a rhombus meet at right angles, a rhombus can be enclosed by a rectangle, as shown in the accompanying figure.

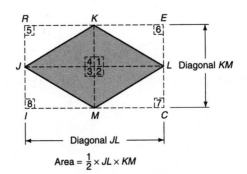

240

- Rectangle *RECT* contains 8 right triangles that have the same area.
- Since rhombus *JKLM* contains 4 of the 8 right triangles, the area of rhombus *JKLM* is $\frac{4}{8}$ or $\frac{1}{2}$ the area of rectangle *RECT*.
- The area of rectangle *RECT* is equal to $JL \times KM$. Hence, the area of rhombus *JKLM* is $\frac{1}{2} \times JL \times KM$ or, equivalently, one-half the product of the lengths of its two diagonals.

Area of a Square = $\frac{1}{2} \times$ (diagonal)²

Since a square, shown in the accompanying diagram, is a special rhombus in which the two diagonals have the same length, the area of a square can be found by squaring the length of either diagonal and then multiplying the result by $\frac{1}{2}$.

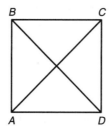

Area = $\frac{1}{2} \times (AC)^2 = \frac{1}{2} \times (BD)^2$

Example 3

The length of the longer diagonal of a rhombus is 24, and the length of a side is 13. Find the area of the rhombus.

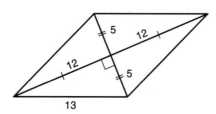

Solution: **120 square units**

As the accompanying diagram illustrates, the triangle that contains one-half of the length of the shorter diagonal is a 5-12-13 right triangle, so the length of the shorter diagonal is 5 + 5 or 10.

241

$$\text{Area of rhombus} = \frac{1}{2} \times \text{Diagonal}_1 \times \text{Diagonal}_2$$

$$= \frac{1}{2} \times 10 \qquad \times 24$$

$$= \frac{1}{2}(240) = 120 \text{ square units}$$

Area of a Triangle $= \dfrac{1}{2} \times$ base $\times$ height

Since the diagonal of a parallelogram divides the parallelogram into two congruent triangles, the area of each triangle is one-half of the area of the parallelogram, as shown in the accompanying figure.

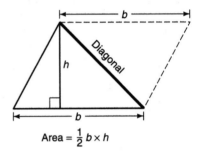

Area $= \frac{1}{2} b \times h$

Example 4

Find the area of each of the following triangles:

(a)

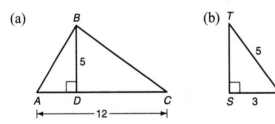

(b)

(c)

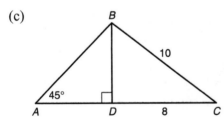

Solution: (a) **30 square units**

In $\triangle ABC$, base $= AC = 12$ and height $= BD = 5$. Hence:

$$\text{Area of } \triangle ABC = \frac{1}{2} \times AC \times BD$$

$$= \frac{1}{2} \times 12 \times 5$$

$$= 30 \text{ square units}$$

(b) **6 square units**

The lengths of the sides of right triangle *RST* form a 3-4-5 right triangle in which leg $ST = 4$.

$$\text{Area of right triangle } RST = \frac{1}{2} \times RS \times ST$$

$$= \frac{1}{2} \times 3 \times 4$$

$$= 6 \text{ square units}$$

(c) **42 square units**

- The lengths of the sides of right triangle *CDB* form a 6-8-10 right triangle in which leg $BD = 6$.
- In right triangle *ADB*, $m\angle A = 45$, so $AD = BD = 6$. Thus, $AC = AD + CD = 6 + 8 = 14$. Thus:

$$\text{Area } \triangle ABC = \frac{1}{2} \times AC \times BD$$

$$= \frac{1}{2} \times 14 \times 6$$

$$= 42 \text{ square units}$$

Example 5

In the accompanying figure, what is the area of $\triangle ABC$?

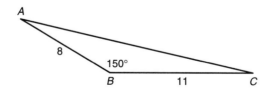

Solution: **22 square units**

In an obtuse triangle, the altitude drawn to either side of the obtuse angle will fall outside the triangle, as shown in the accompanying figure, in which base $\overline{BC}$ is extended to H.

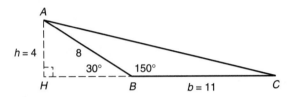

- In right triangle AHB,

$$\mathrm{m}\angle ABH = 180 - 150 = 30$$

- Since altitude $\overline{AH}$ is the leg opposite the 30° angle in a right triangle,

$$AH = \frac{1}{2}(AB) = \frac{1}{2}(8) = 4 \cdot$$

- Thus:

$$\text{Area of } \triangle ABC = \frac{1}{2} \times BC \times AH$$

$$= \frac{1}{2} \times 11 \times 4$$

$$= 22 \text{ square units}$$

Area of a Trapezoid = height × $\left(\dfrac{\textbf{sum of bases}}{\textbf{2}}\right)$

In a trapezoid, the bases are the parallel sides and the height is the common perpendicular segment to the bases. A diagonal divides a trapezoid into two triangles with same height, as shown in the accompanying figure. Hence, the area of trapezoid $TRAP$ is the sum of the areas of $\triangle TRP$ and $\triangle RPA$, which is $\frac{1}{2}b_1 h + \frac{1}{2}b_2 h$ or, after factoring, $h \times \left(\dfrac{b_1 + b_2}{2}\right)$.

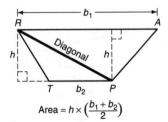

$$\text{Area} = h \times \left(\frac{b_1 + b_2}{2}\right)$$

Example 6

In the accompanying figure, ABCD is a trapezoid with $\overline{AB} \parallel \overline{CD}$, $\overline{BA} \perp \overline{AD}$, $\overline{CD} \perp \overline{AD}$, $AB = 10$, $BC = 17$, and $AD = 15$. Find the area of trapezoid ABCD.

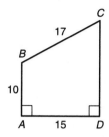

Solution: **210 square units**

First find the length of base $\overline{CD}$ by drawing altitude $\overline{BH}$ to $\overline{CD}$. Since parallel lines are everywhere equidistant, $BH = AD = 15$. The lengths of the sides of right triangle BHC form an 8-15-17 Pythagorean triple, where $CH = 8$. Thus, $CD = CH + HD = 8 + 10 = 18$.

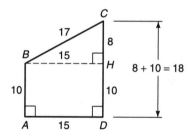

$$\text{Area of trapezoid } ABCD = BH \times \frac{AB + CD}{2}$$

$$= 15 \ \times \frac{10 + 18}{2}$$

$$= 15 \ \times 14$$

$$= 420 = 210 \text{ square units}$$

Example 7

What is the area of an isosceles trapezoid in which the bases measure 8 inches and 20 inches and each of the lower base angles measures 45°?

Solution: **84 in²**

- $HK = BC = 8$. Hence, $AH + DK = 20 - 8 = 12$.
- Because trapezoid $ABCD$ is isosceles, $AH = DK$, $AH = \dfrac{1}{2}(12) = 6$.
- Since $\overline{BH}$ is the leg opposite the 45° angle in a 45°-45° right triangle, $BH = AH = 6$. Thus:

$$\text{Area of isosceles trapezoid } ABCD = BH \times \frac{AD + BC}{2}$$

$$= 6 \quad \times \frac{20 + 8}{2}$$

$$= 6 \quad \times 14$$

$$= 84 \text{ in}^2$$

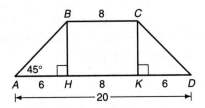

Check Your Understanding of Section 9.1

A. *Multiple Choice*

1. What is the number of square units in the area of a square whose diagonal is 8?
   (1) 16   (2) $16\sqrt{2}$ (3) 32   (4) $32\sqrt{2}$

2. What is the number of square units in the area of a parallelogram in which the lengths of two sides are 12 and 15 and the measure of their included angle is 60°?
   (1) $45\sqrt{2}$ (2) $60\sqrt{2}$ (3) 90   (4) $90\sqrt{3}$

3. An altitude is drawn from vertex B of rhombus $ABCD$, intersecting $\overline{AD}$ at point H. If $AB = 13$ and $HD = 8$, what is the area of rhombus $ABCD$?
   (1) 48   (2) 65   (3) 78   (4) 156

4. In the accompanying figure of quadrilateral $ABCD$, $\overline{DA} \perp \overline{AB}$, $\overline{DB} \perp \overline{BC}$, $AD = 4$, $AB = 3$, and $CD = 13$. What is the number of square units in the area of quadrilateral $ABCD$?

(1) 28   (2) 32   (3) 36   (4) 42

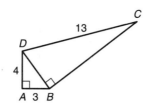

5. In the accompanying diagram of square $ABCD$, $AB = x - 1$ and $BC = 5x - 13$. What is the number of square units in the area of square $ABCD$?

(1) 1   (2) 9   (3) 16   (4) 4

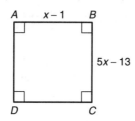

6. What is the area of rhombus $ABCD$ if $AB = 13$ inches and diagonal $AC = 10$ inches?

(1) 60 in^2 (2) 65 in^2 (3) 120 in^2 (4) 130 in^2

7. In the accompanying diagram of isosceles trapezoid $ABCD$, $BC = 7$ centimeters and $AD = 19$ centimeters. If the perimeter of $ABCD$ is 46 centimeters, what is the area?

(1) 525 cm^2 (3) 112 cm^2
(2) 104 cm^2 (4) 124 cm^2

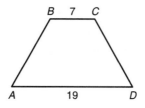

8. In the accompanying diagram of quadrilateral $ABCD$, $\overline{CD} \perp \overline{BC}$, $\overline{AB} \perp \overline{BC}$, m$\angle A = 45$, $BC = 6$, and $CD = 2$. What is the number of square units in the area of quadrilateral $ABCD$?

(1) 24   (2) 30   (3) 32   (4) 36

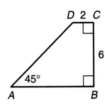

9. What is the area of a triangle in which the lengths of a pair of sides are 6 centimeters and 14 centimeters and the degree measure of their included angle is 135?

(1) 21 cm^2 (2) $21\sqrt{2}$ cm^2 (3) 42 cm^2 (4) $42\sqrt{2}$ cm^2

10. The perimeter of rhombus $ABCD$ is 68 inches and the length of one of its diagonals is 16 inches. What is the area of rhombus $ABCD$?

(1) 60 in^2 (2) 120 in^2 (3) 240 in^2 (4) 544 in^2

11. In isosceles triangle ABC, $\overline{AB} \cong \overline{BC}$, m$\angle ABC = 120$, and $AB = 20$. What is the number of square units in the area of $\triangle ABC$?

(1) 100   (2) $100\sqrt{3}$ (3) 200   (4) $200\sqrt{3}$

12. What is the area of an equilateral triangle in which one side measures 12 centimeters?

(1) $18\sqrt{3}$ cm^2 (2) $36\sqrt{3}$ cm^2 (3) 72 cm^2 (4) $72\sqrt{3}$ cm^2

13. What is the number of centimeters in the perimeter of a square whose area is 32 square centimeters?

(1) $16\sqrt{2}$ (2) $32\sqrt{2}$ (3) 24   (4) 64

14. In right triangle ABC, $\angle C$ is a right angle and $AC = 10$ inches. If the area of $\triangle ABC$ is 150 square inches, what is the number of inches in the length of $\overline{AB}$?

(1) 15   (2) $10\sqrt{10}$ (3) $10\sqrt{15}$ (4) $20\sqrt{2}$

15. In the accompanying diagram of rectangle $ABCD$, E is a point on $\overline{DC}$, $EC = 5$, $BE = 13$, and $AB = 20$. What is the number of square units in the area of trapezoid $ABED$?

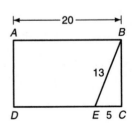

248

16. In the accompanying diagram, $\overline{AC}$ and $\overline{DE}$ are perpendicular to $\overline{BC}$, $AC = BC = 10$, and $DE = 8$. What is the area of trapezoid $ACED$?

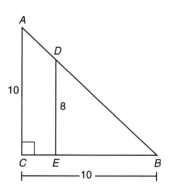

17. In the accompanying figure of right triangle ABC, $\angle C$ is a right angle, $AB = 25$, $DB = 17$, and $CD = 8$. What is the area of $\triangle ADB$?

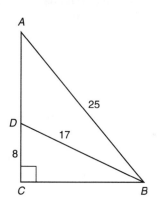

18. A corner is cut off a 9-inch by 9-inch square piece of paper. The cut is made x inches from a corner, as shown in the accompanying diagram. After the corner is removed, the area of the figure that remains is $\frac{7}{8}$ the area of the original square. What is the value of x?

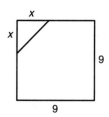

19. In the accompanying figure of trapezoid $ABCD$, $\overline{CD} \parallel \overline{AB}$. Triangles AOD, DOC, and COB are equilateral, $\overline{OE} \perp \overline{CD}$, and $OA = 6$ inches.

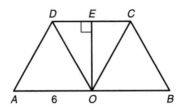

(a) Find the area of $\triangle DOC$.
(b) Find the area of trapezoid $ABCD$.
(v) Find the area of trapezoid $OECB$.

20. Mr. Carlin owns a plot of land with boundaries that form right triangle ACD with right angle C, $AC = 9$ yards, and $AD = 15$ yards. He purchases an adjacent plot of land that extends $\overline{CD}$ to B with $AB = 41$ yards, as shown in the accompanying figure.

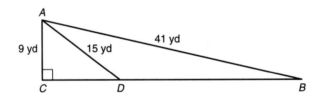

If the additional plot of land costs $217 per square yard, how many dollars did Mr. Carlin pay for this land?

9.2 CIRCUMFERENCE AND AREA OF A CIRCLE

KEY IDEAS

If you were able to locate and connect all points that were a distance of 5 inches from a given point, a *circle* having a *radius* of 5 inches would be formed. The distance around the circle is called the *circumference* of the circle. The region enclosed by the circle represents the *area* of the circle.

The size of a circle depends on the length of its radius. The longer the radius, the greater the circumference and the area of the circle.

Some Parts of a Circle

A **circle** (Figure 9.1) is the set of all points at a given distance from a fixed point. The fixed point is called the center of the circle. The center of a circle is named by a single capital letter. If the center of a circle is named by the letter *O*, then the circle is referred to as *circle O*.

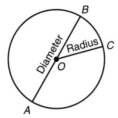

Figure 9.1   Circle

Any segment drawn from the center of the circle to a point on the circle is called a **radius** (plural: *radii*) of the circle. All radii of the same circle have the same length. A **diameter** is a line segment that passes through the center of a circle and whose endpoints are points on the circle. The length of a diameter is twice the length of a radius.

Any curved section of the circle is called an **arc** of the circle. A diameter divides a circle into two congruent arcs, each of which is called a **semicircle**. In Figure 9.1, $\overset{\frown}{AB}$ (read as "arc *AB*") is a semicircle. Also, $\overset{\frown}{ACB}$ (read as "arc *ACB*") is a semicircle.

Circumference of a Circle

When a circle's **circumference** (that is, the distance around the circle), is divided by its diameter, the number obtained is always the same regardless of the size of the circle. This constant value is represented by the Greek letter π, which is read as "pi."

$$\frac{\text{Circumference}}{\text{Diameter}} = \frac{C}{D} = 3.1415926\ldots = \pi(\text{pi})$$

Since the length of the diameter *D* is twice the length of the radius *r*, the circumference *C* may also be expressed in the following equivalent ways:

$$C = \pi D \quad \text{and} \quad C = 2\pi r$$

The quantity pi is an irrational number that takes the form of a never-ending, nonrepeating decimal number. When an estimate of the value of the circumference is needed, π may be replaced by a rational approximation, such as 3.14 or $\frac{22}{7}$. An exact value of the circumference is represented whenever the circumference is written in terms of π.

Area of a Circle

To find the area, A, of a circle, multiply π by the square of the radius, r. Thus:

$$A = \pi r^2$$

Example 1

What is the circumference of a circle whose area, in square units, is 49π?

   Solution: **14π**

- The area of the circle is $49\pi = \pi r^2$, so $r^2 = 49$ and $r = \sqrt{49} = 7$.
- Since $r = 7$, circumference $= 2\pi r = 2\pi (7) = 14\pi$.

Example 2

A point P on a bicycle wheel is touching the ground. The next time point P hits the ground, the wheel has rolled 3.5 feet in a straight line along the ground. What is the radius of the wheel correct to the *nearest tenth of an inch*?

   Solution: **6.7**

Since the distance the wheel rolls is equal to the circumference of the circle, $3.5 \text{ ft} = 2\pi r$, so

$$
\begin{aligned}
r &= \frac{3.5 \text{ ft}}{2\pi} \\
&= \frac{3.5 \times 12 \text{ in}}{2\pi} \\
&= \frac{42 \text{ in}}{2\pi} \approx 6.7 \text{ in}
\end{aligned}
$$

Example 3

The circumference of a circle is $\dfrac{6}{\pi}$.

(a) What is the diameter of the circle in terms of π?
(b) What is the area of the circle in terms of π?

Solution: (a) $\frac{6}{\pi^2}$

The circumference of the circle is

$$\frac{6}{\pi} = \pi \times \text{diameter}$$

so the diameter is $\frac{6}{\pi^2}$.

(b) $\frac{9}{\pi^3}$ **square units**

Since the diameter of the circle is $\frac{6}{\pi^2}$, the radius of the circle is $\frac{3}{\pi^2}$. Hence, the area of the circle is

$$\pi r^2 = \pi \left(\frac{3}{\pi^2} \right)^2$$

$$= \pi \left(\frac{9}{\pi^4} \right)$$

$$= \frac{9}{\pi^3} \text{ square units}$$

Check Your Understanding of Section 9.2

A. Multiple Choice

1. What is the circumference of a circle whose area is 16π square units?
   (1) 4π (2) 8π (3) 16π (4) 64π

2. What is the area of a circle whose circumference is 10π centimeters?
   (1) 5π cm^2 (2) 20π cm^2 (3) 25π cm^2 (4) 100π cm^2

3. What is the radius of a circle whose circumference is 7?
   (1) $\dfrac{3.5}{\pi}$ (2) $\dfrac{\sqrt{3.5}}{\pi}$ (3) $\dfrac{7}{\pi}$ (4) $\dfrac{\sqrt{7}}{\pi}$

4. If the area, in square units, of a circle increases from π to 4π, by what percent does the length of the radius increase?
   (1) 100   (2) 200   (3) 300   (4) 400

5. In the accompanying diagram of circle O, radius $\overline{OB}$ is perpendicular to diameter $\overline{AOC}$ and $OC = 8$. What is the number of square units in the area of $\triangle ABC$?

    (1) 32　　　　　(2) 32π　　　　　(3) 64　　　　　(4) 64π

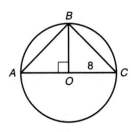

6. In the accompanying figure, circles A, B, and C touch each other at exactly one point. If the area, in square units, of each circle is 9π, what is the perimeter of $\triangle ABC$?

    (1) 12　　　　　(2) 15　　　　　(3) 180　　　　　(4) 24

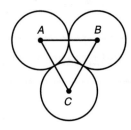

7. In the accompanying figure of square $ABCD$, circle O touches each of the sides of $ABCD$ at one point. If the area of square $ABCD$ is 64 square inches, what is the number of square units in the area of circle O?

    (1) 4π　　　　　(2) 8π　　　　　(3) 16π　　　　　(4) 18π

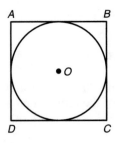

8. Each time the pedals of a certain bicycle go through one complete circular rotation the tires rotate three times. If the tires are 24 inches in diameter, what is the minimum number of complete rotations of pedals needed for the bicycle to travel at least 1 mile?
   (1) 841   (2) 281   (3) 561   (4) 70.1

B. In each case, show how you arrived at your answer by clearly indicating all of the necessary steps, formula substitutions, diagrams, graphs, charts, etc.

9. In the accompanying diagram of semicircle O, $\overline{AC} \perp \overline{BC}$, diameter $AB = 20$, and $AC = 12$.

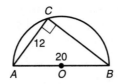

(a) What is the number of square units in the area of semicircle O correct to the *nearest tenth*?
(b) The area of right triangle ACB is what percent, correct to the *nearest tenth*, of the area of semicircle O?

10. In six complete revolutions, a bicycle wheel rolls 27 feet. What is the number of inches, correct to the *nearest tenth*, in the radius of the bicycle wheel?

11. The ratio of the areas of two circles is 9 to 16. If the length of the radius of the larger circle is 20 centimeters, what is the length of the radius of the smaller circle?

12. To measure the number of miles in a hiking trail, a worker uses a device with a 2-foot-diameter wheel that counts the number of revolutions the wheel makes. If the device reads 0 revolution at the beginning of the trail and 2300 revolutions at the end of the trail, how many miles long, correct to the *nearest tenth of a mile*, is the trail?

9.3 AREAS OF OVERLAPPING FIGURES

KEY IDEAS

When one geometric figure lies inside the other, the area of the region between the two figures is the difference between the areas of the two figures.

Finding Areas Indirectly

When finding the area of a region between two or more overlapping figures, it may be necessary to change your point of view. Rather than trying to find the area of the region directly, you may need to find it indirectly by subtracting the areas of the other figures.

Example 1

Circles with the same center are called **concentric circles**. Find in terms of π the number of square units in the area of the shaded region between the two concentric circles with radii of 5 and 8, as shown in the accompanying figure.

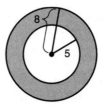

Solution: **39π**

The shaded region represents the area that the concentric circles do *not* have in common. To find this area, *subtract* the area of the inner circle from the area of the outer circle.

$$\text{Area of outer circle } = \pi(8^2) = 64\pi$$
$$- \text{ Area of inner circle } = \pi(5^2) = 25\pi$$
$$\overline{\text{Area of shaded region} \qquad = 39\pi}$$

Example 2

In the accompanying diagram, the length of the rectangle is 8 inches and the width is 6 inches. Find in terms of π the number of square inches in the area of the shaded region.

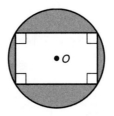

Solution: **25π − 48**

Each diagonal of the rectangle is a diameter of the circle and the hypotenuse of a right triangle whose legs measure 6 in and 8 in.

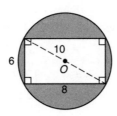

- Because 6-8-10 is a Pythagorean triple, the diameter of the circle is 10 in, so its radius is 5 in. The area of the circle is $\pi \times 5^2 = 25\pi$ in.^{2}
- The area of the rectangle is $6 \times 8 = 48$ in.^{2}
- The area of the shaded region is $25\pi - 48$ in.^{2}

Example 3

In the accompanying diagram, *ABCD* is a square. Find in terms of π the area of the shaded region.

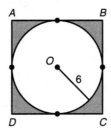

Solution: **144 – 36π**

The diameter of the inscribed circle is 12, so the length of a side of the circumscribed square is also 12.

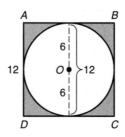

Area of shaded region = area of square – area of circle

$$= \quad s^2 \quad -\pi r^2$$

$$= \quad 12^2 \quad -\pi(6)^2$$

$$= \quad 144 \quad -36\pi$$

Check Your Understanding of Section 9.3

In each case, show how you arrived at your answer by clearly indicating all of the necessary steps, formula substitutions, diagrams, graphs, charts, etc.

1–4. In each case, find the area of the shaded region. Assume that ABCD is a square or rectangle.

1.

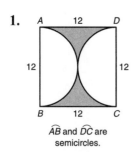

$\overarc{AB}$ and $\overarc{DC}$ are semicircles.

2.

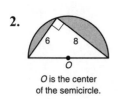

O is the center of the semicircle.

3.

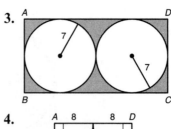

4.

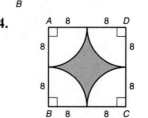

The four quarter-circles have centers at the vertices of the square.

5–6. In each of the following diagrams, each dot is one unit from an adjacent horizontal or vertical dot. Find the number of square units in the area of ABCD.

5.

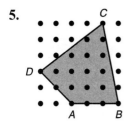

6.

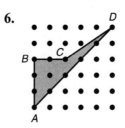

7. Ms. Vine wants to carpet her living room floor. The living room is a square that measures 20 feet on each side. She wants to leave the entranceway uncovered by carpeting a quarter of a circle as shown in the accompanying diagram.

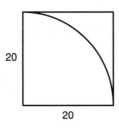

(a) Find, correct to the *nearest square foot*, the area that will be carpeted.

(b) Find, correct to the *nearest tenth*, the percent of the living room floor that will remain uncarpeted.

8. The target shown in the accompanying diagram consists of three circles with the same center. The lengths of the radii of the three circles are 2 inches, 5 inches, and 8 inches.

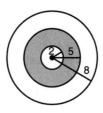

(a) What is the area of the shaded region to the *nearest tenth of a square inch*?

(b) To the *nearest percent*, what percent of the target in *not* shaded?

9. The shaded circular field with an inner radius of 25 feet and an outer radius of 40 feet, shown in the accompanying diagram, is to be planted at a cost of $2.15 per square foot. What is the cost, to the *nearest dollar*, of planting the shaded circular field?

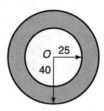

10. The accompanying diagram shows three congruent circles in a row. $\overline{AD}$ and $\overline{BC}$ are the diameters of the first and third circles, $\overline{AB}$ and $\overline{CD}$ are drawn to form rectangle $ABCD$, the radius of each circle is 5, and the distance between adjacent circles is 3.

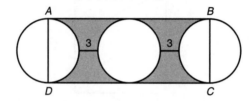

(a) Find the number of square units in the area of the shaded region to the *nearest integer*.
(b) To the *nearest percent*, what percent of the area of rectangle $ABCD$ is shaded?

9.4   VOLUMES OF CIRCULAR SOLIDS

KEY IDEAS

Some familiar solids have circular bases. Finding the volumes of these solids involves finding the areas of the circular bases.

Volume of a Cylinder and Volume of a Cone

A **cylinder** (see Figure 9.2) has two congruent circular bases, and a **cone** (see Figure 9.3) has one circular base.

The volume of a cylinder is the area of a circular base times its height.

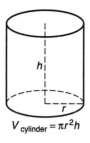

$V_{\text{cylinder}} = \pi r^2 h$

Figure 9.2 Cylinder

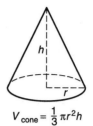

$V_{\text{cone}} = \frac{1}{3} \pi r^2 h$

Figure 9.3 Cone

The volume of a cone is $\frac{1}{3}$ times the area of its circular base times its height.

Example 1

The radius of a base of a cylinder is 5, and the height of the cylinder is 6. What is the number of cubic units in the volume of this cylinder?

Solution: **150π**

$$V = \pi r^2 h$$
$$= \pi \times (5^2) \times 6$$
$$= \pi \times (25) \times (6)$$
$$= 150\pi \text{ cubic units}$$

Example 2

The *slant height*, ℓ, of a cone is a segment that connects the vertex to a point on the circular base. What is the volume of a cone having a radius of 6 and a slant height of 10?

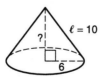

$\ell = 10$

?

6

Solution: **96π cubic units**

The slant height is the hypotenuse of a right triangle whose legs are a radius and the altitude of the cone. The lengths of this right triangle form a 6-8-10 right triangle, where 8 represents the height of the cone. Therefore:

261

$$V = \frac{1}{3}\pi r^2 h$$

$$= \frac{1}{3} \times \pi \times (6^2) \times 8$$

$$= \frac{1}{\cancel{3}} \times \pi \times (\overset{12}{\cancel{36}}) \times 8$$

$$= 96\pi \text{ cubic units}$$

Example 3

The circumference of the base of a cone is 8π centimeters. If the volume of the cone is 16π cubic centimeters, what is the height?

Solution: **3 cm**

Since $C = \pi D = 8\pi$, the diameter is 8 cm; therefore, the radius of the base is 4 cm. Then:

$$V = \frac{1}{3}\pi r^2 h$$

$$16\pi = \frac{1}{3} \times \pi \times (4^2) \times h$$

$$48\pi = 16\pi h$$

$$\frac{48\pi}{16\pi} = h$$

$$3 = h$$

Volume of a Sphere

A sphere (Figure 9.4) is a solid that represents the set of all points in space that are at a given distance from a fixed point. The volume of the sphere can be determined using the formula

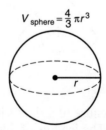

$$V_{sphere} = \frac{4}{3}\pi r^3$$

Figure 9.4   Sphere

Example 4

Find the volume of a sphere whose diameter is 6 centimeters.

Solution: **36π cm³**

$$\text{Radius} = \frac{1}{2}(6) = 3$$

$$V = \frac{4}{3} \times \pi \times (3)^3$$

$$= \frac{4}{3} \times \pi \times (27)$$

$$= 4 \times \pi \times (9)$$

$$= 36\pi \text{ cm}^3$$

Check Your Understanding of Section 9.4

A. *Multiple Choice*

1. If the diameter of a cylinder is tripled, then the volume of the cylinder is multiplied by
   (1) 9   (2) 6   (3) 3   (4) 7

2. If the circumference of the circular base of a cone is doubled and the height is also doubled, then the volume of the cone is multiplied by what number?
   (1) 2   (2) 4   (3) 6   (4) 8

3. If the radius of a cylinder is doubled and the height is tripled, then the volume of the cylinder is multiplied by
   (1) 6   (2) 9   (3) 12   (4) 36

B. *In each case, show how you arrived at your answer by clearly indicating all of the necessary steps, formula substitutions, diagrams, graphs, charts, etc.*

4 and 5. What is the volume in terms of π of the cone in each of the accompanying diagrams?

4.

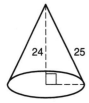

5.

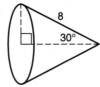

6. The length of the radius of a sphere is π centimeters. What is the volume of the sphere correct to the *nearest cubic centimeter*?

7. If the radius of the base of a cone is multiplied by 2 while its height is multiplied by 3, by what number is its volume multiplied?

8. A ball in the shape of a sphere has a volume of 36π cubic inches. The ball is placed in a box that has the shape of a cube. When the top of the box is closed, the shortest distance between the ball and the top of the box is 1 inch. What is the number of cubic inches in the volume of the box?

9. The circumference of a cylinder is 10π centimeters and its height is 3.5 centimeters. What is the volume, in terms of π, of the cylinder?

10. A cylinder and a cone have congruent bases and equal heights. How many times greater is the volume of the cylinder than the volume of the cone?

11. The radius of a sphere is 6 centimeters, and the radius of a cylinder is 4 centimeters. If these solids have equal volumes, what is the number of centimeters in the height of the cylinder?

12. The radii of two spheres are in the ratio of 3 to 5. If the volume of the smaller sphere is 270π cubic inches, what is the number of cubic inches, in terms of π, in the volume of the larger sphere?

13. An isosceles right triangle whose hypotenuse measures 12 centimeters is revolved in space about one of its legs. What is the volume, correct to the *nearest cubic centimeter*, of the resulting cone?

14. A container having the shape of a cylinder with a radius of 4 inches and a height of 18 inches is filled with water.
(a) What is the number of inches in the radius of a container having the shape of a sphere that can hold the same amount of water?
(b) If water leaks out of the cylinder at a constant rate of 1.5π cubic inches per minute, how many minutes will elapse before the cylinder is exactly half full?

15. A cylindrical container is manufactured so that its radius is one-half of its height. The container must have a volume of at least 250 cubic inches, and the height must be a whole number. What is the *least* number of inches the height can be?

16. A rectangular container with the dimensions 10 inches by 15 inches by 20 inches is to be filled with water, using a cylindrical cup whose radius is 2 inches and whose height is 5 inches. What is the maximum number of full cups of water that can be poured into the container without the water overflowing the container?

CHAPTER 10

SIMILAR TRIANGLES AND TRIGONOMETRIC RATIOS

10.1 COMPARING SIDES OF SIMILAR POLYGONS

===== KEY IDEAS =====

When a photograph is enlarged, the original and enlarged figures are *similar* since they have exactly the same shape. Two *polygons* with the same number of sides are **similar** if:

- corresponding angles are congruent and
- ratios of the lengths of corresponding sides are equal.

  Two *triangles* are **similar** if two angles of one triangle are congruent to two angles of the other triangle.

Congruent Versus Similar Polygons

Congruent polygons have the same shape *and* the same size, while similar figures have the same shape but may have different sizes. Triangles I and II in Figure 10.1 are similar; the symbol ~ means "is similar to."

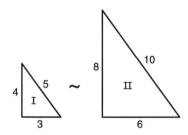

Figure 10.1 Similar Triangles

Since triangles I and II are similar, the lengths of corresponding sides have the same ratio and, as a result, are "in proportion":

$$\frac{\text{Side in } \triangle I}{\text{Corresponding side in } \triangle II} = \frac{3}{6} = \frac{4}{8} = \frac{5}{10} = \frac{1}{2} \text{ or } 1:2$$

Example 1

At a certain time during the day, light falls so that a pole 10 feet in height casts a shadow 15 feet in length on level ground. At the same time, a man casts a shadow that is 9 feet in length. How tall is the man?

Solution: **6 ft**

The shadows cast by the pole and the man can be represented as legs of right triangles in which the hypotenuses represent the rays of light.

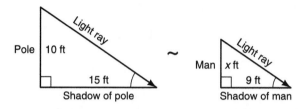

Assume that in each right triangle the ray of light makes the same angle with the ground. Since two angles of the first right triangle are congruent to two angles of the second right triangle, the two right triangles are similar and, as a result, the lengths of their corresponding sides are in proportion. Hence:

$$\frac{\text{Height of pole}}{\text{Height of man}} = \frac{\text{shadow of pole}}{\text{shadow of man}}$$

$$\frac{10}{x} = \frac{15}{9}$$

$$15x = 90$$

$$x = \frac{90}{15}$$

$$= 6$$

Example 2

In the accompanying figure, angles D and B are right angles. If $BC = 80$ meters, $ED = 15$ meters, and $BD = 171$ meters, what is the length of $\overline{AB}$?

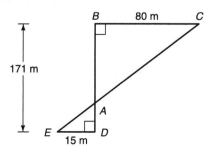

Solution: **144 m**

- Since right angles *B* and *D* are congruent and vertical angles *EAD* and *CAB* are congruent, $\triangle ADE \sim \triangle ABC$.
- Because lengths of corresponding sides of similar triangles are in proportion:

$$\frac{AB}{BC} = \frac{AD}{DE}$$

- If $x = AB$, then $AD = 171 - x$:

$$\frac{x}{80} = \frac{171 - x}{15}$$
$$15x = 80(171 - x)$$
$$= 13,680 - 80x$$
$$15x + 80x = 13,680$$
$$x = \frac{13,680}{95}$$
$$= 144$$

Perimeters of Similar Polygons

In Figure 10.1, the perimeter of $\triangle I$ is $3 + 4 + 5 = 12$ and the perimeter of $\triangle II$ is $6 + 8 + 10 = 24$. Since $\frac{12}{24} = \frac{1}{2}$ or $1 : 2$, the perimeters of triangles I and II have the same ratio as the ratio of the lengths of any pair of corresponding sides.

Example 3

Quadrilateral *ABCD* is similar to quadrilateral *RSTW*. The lengths of the sides of quadrilateral *ABCD* are 6, 9, 12, and 18. If the length of the longest side of quadrilateral *RSTW* is 24, find the length of its shortest side.

Solution: **8**

- Let $x =$ length of the shortest side of quadrilateral *RSTW*. Then:

$$\frac{\text{Shortest side of quad } ABCD}{\text{Shortest side of quad } RSTW} = \frac{\text{longest side of quad } ABCD}{\text{longest side of quad } RSTW}$$

$$\frac{6}{x} = \frac{18}{24}$$

- Write $\frac{18}{24}$ in lowest terms: $\dfrac{6}{x} = \dfrac{3}{4}$

- Cross-multiply: $3x = 24$

$$x = \frac{24}{3}$$

$$= 8$$

Example 4

The lengths of the five sides of a pentagon are 1, 3, 5, 7, and 12. If the length of the longest side of a similar pentagon is 18, find the perimeter of the larger pentagon.

Solution: **42**

Let x = perimeter of the larger pentagon. Then:

$$\frac{\text{Longest side of smaller pentagon}}{\text{Longest side of larger pentagon}} = \frac{\text{perimeter of smaller pentagon}}{\text{perimeter of larger pentagon}}$$

$$\frac{12}{18} = \frac{1+3+5+7+12}{x}$$

$$\frac{2}{3} = \frac{28}{x}$$

$$2x = 84$$

$$x = \frac{84}{2}$$

$$= 42$$

MATH FACTS

If you know that two angles of one triangle are congruent to two angles of another triangle, you can conclude that the two triangles are similar. When two triangles are similar, the ratio of the lengths of a pair of corresponding sides is the same as the ratio of any other pair of corresponding linear measurements of the two triangles. Therefore:

- Lengths of corresponding sides are in proportion.
- The ratio of the lengths of any pair of corresponding sides is the same as the ratio of corresponding perimeters, altitudes, or medians.

Check Your Understanding of Section 10.1

A. *Multiple Choice*

1. The length of the shortest side of a triangle is 12, and the length of the shortest side of a similar triangle is 5. If the longest side of the larger triangle is 15, what is the longest side of the smaller triangle?
   (1) 2.4 (2) 6.25 (3) 24 (4) 36

2. The ratio of the perimeters of two similar pentagons is $3:4$. If the length of the shortest side of the smaller pentagon is 48, what is the length of the shortest side of the larger pentagon?
   (1) 24 (2) 36 (3) 64 (4) 72

3. A person 5 feet tall is standing near a tree 30 feet high. If the length of the person's shadow is 3 feet, what is the length of the shadow of the tree?
   (1) 12 ft (2) 18 ft (3) 24 ft (4) 50 ft

B. *In each case, show how you arrived at your answer by clearly indicating all of the necessary steps, formula substitutions, diagrams, graphs, charts, etc.*

4. In the accompanying diagram of $\triangle ABC$, $\overline{DE} \parallel \overline{AB}$, $DE = 8$, $CD = 12$, and $DA = 3$. What is the length of $\overline{AB}$?

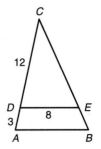

5. In the accompanying diagram, $\overline{AB} \perp \overline{BE}$ and $\overline{DE} \perp \overline{BE}$.
   (a) Explain why $\triangle ABC \sim \triangle DEC$.
   (b) If $AB = 8$, $AC = 10$, and $DC = 25$, find DE.

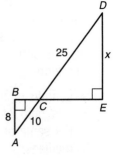

6 and 7. In the accompanying diagram of △RST, $\overline{EF}$ ∥ $\overline{RT}$.

6. (a) Explain why △*ESF* ~ △*RST*.
   (b) If *SE* = 8, *ER* = 6, and *FT* = 15, find *SF*.

7. If *ER* = 5, *EF* = 8, and *RT* is 2 more than *SE*, find *SE*.

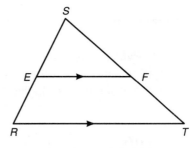

Exercises and 6 and 7

8. In the accompanying diagram of △*ABC*, $\overline{DE}$ ∥ $\overline{BC}$, *AD* = 8, *AB* = 12, and *EC* = 5. What is the length of $\overline{AE}$?

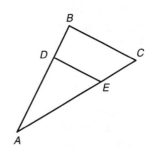

9. In the accompanying diagram, $\overline{AB}$ ⊥ $\overline{BE}$ and $\overline{DE}$ ⊥ $\overline{BE}$, *AC* = 8, *DC* = 12, and *BE* = 15. What is *BC*?

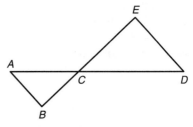

10. In the accompanying diagram of △*PRT*, $\overline{KG}$ ∥ $\overline{PR}$. If *TP* = 20, *KP* = 4, and *GR* = 7, what is the length of *TG*?

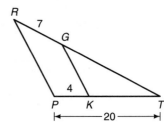

11. In the accompanying diagram of right triangle *ACB*, ∠*C* is the right angle. Point *D* is on leg $\overline{AC}$, and point *E* is on hypotenuse $\overline{AB}$ with $\overline{AD}$ ⊥ $\overline{DE}$. If *BC* = 20, *CD* is 3 less than *AD*, and *DE* is 3 more than *AD*, what is the length of $\overline{AD}$?

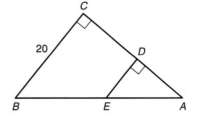

10.2 RIGHT TRIANGLE TRIGONOMETRY

KEY IDEAS

The *Pythagorean theorem* relates the measures of the *three* sides of a right triangle. A *trigonometric ratio* relates the measures of *two* sides and *one* of the acute angles of a right triangle.

The Three Basic Trigonometric Ratios

In Figure 10.2, $\triangle ABC$, $\triangle ADE$, and $\triangle AFG$ each contain a right angle and have $\angle A$ in common. As a result, these triangles are similar to each other and the lengths of the following sides are in proportion:

$$\frac{BC}{AC} = \frac{DE}{AE} = \frac{FG}{AG} = \frac{\text{length of leg opposite } \angle A}{\text{length of hypotenuse}}$$

The proportion states that in a right triangle the ratio of the length of the leg opposite an acute angle to the length of the hypotenuse is a constant. This constant is called the **sine** of the angle. In a right triangle, there are two other trigonometric ratios that you should know, **cosine** and **tangent**.

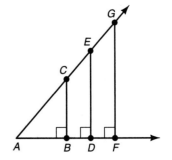

Figure 10.2 Similar Triangles

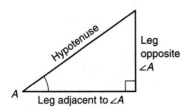

Figure 10.3 Parts of a Right Triangle

Figure 10.3 identifies the parts of a right triangle referred to in the following table:

$\text{Sin } A = \dfrac{\text{Opposite leg}}{\text{Hypotenuse}}$	$\text{Cos } A = \dfrac{\text{Adjacent leg}}{\text{Hypotenuse}}$	$\text{Tan } A = \dfrac{\text{Opposite leg}}{\text{Adjacent leg}}$
S O H	**C A H**	**T O A**

Notice that the first letters of the three key words in each of the definitions of the three trigonometric ratios spell *SOH-CAH-TOA*. Remembering *SOH-CAH-TOA* can help you recall the definitions of sine, cosine, and tangent.

Trigonometric Relationships

In the accompanying diagram of right triangle *ABC*, since leg *a* is opposite $\angle A$ and adjacent to $\angle B$:

$$\sin A = \frac{a}{c} \quad \text{and} \quad \cos B = \frac{a}{c}$$

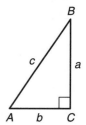

Since $\sin A$ and $\cos B$ are both equal to $\frac{a}{c}$, $\sin A = \cos B$. If $\angle A = 60°$, then $\angle B = 90° - 60° = 30°$ so $\sin 60° = \cos 30°$. In general, *the sine of an angle is numerically equal to the cosine of the complement of that angle.* For example:

- $\sin 25° = \cos(90 - 25)° = \cos 65°$.

- If $\sin(2x)° = \cos(3x)°$, then $2x + 3x = 90$ so $5x = 90$ and $x = \dfrac{90}{5} = 18$.

Example 1

In right $\triangle ABC$, shown in the accompanying diagram, $\angle C$ is the right angle, $AC = 12$, and $BC = 5$. Find the values of:
(a) $\sin A$
(b) $\cos A$
(c) $\tan A$

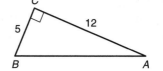

Solution: Since $\triangle ABC$ is a 5-12-*13* right triangle, $AB = 13$.

(a) **0.3846** (b) **0.9231** (c) **0.4167**

$\sin A$ $\cos A$ $\tan A$

$$= \frac{\text{leg opposite } \angle A}{\text{hypotenuse}}$$
$$= \frac{5}{13}$$
$$\approx 0.3846$$

$$= \frac{\text{leg adjacent } \angle A}{\text{hypotenuse}}$$
$$= \frac{12}{13}$$
$$\approx 0.9231$$

$$= \frac{\text{leg opposite } \angle A}{\text{leg adjacent } \angle A}$$
$$= \frac{5}{12}$$
$$\approx 0.4167$$

Using a Scientific Calculator

When you first turn on your calculator, the unit of angle measurement is automatically set to DEGree mode. Many calculators with show DEG in small type in the display window. If necessary, press the $\boxed{\text{DRG}}$ or $\boxed{\text{MODE}}$ key, which will allow you to choose degrees as the unit of angle measurement. Since not all calculators look or work the same, as you read through this section you may have to modify the instructions or read the instruction manual for your calculator.

To evaluate sin 57°:

- Enter 57.
- Press the $\boxed{\text{sin}}$ key. (If you needed to find cos 57° or tan 57°, you would press the $\boxed{\text{cos}}$ or $\boxed{\text{tan}}$ key.)
- Round off the decimal number that appears in the display window. Correct to *four decimal places*, sin 57° = 0.8387.

Some calculators require that you first press the $\boxed{\text{sin}}$ key, enter 57, and then press the $\boxed{=}$ key.

Sometimes it is necessary to find the measure of an angle when you know the value of the sine, cosine, or tangent of that angle. For example, if you know that sin $\angle A$ = 0.8387, you can use the calculator to perform the inverse operation of working back to find the degree measure of $\angle A$ as follows:

- Enter 0.8387.
- Press the $\boxed{\text{INV}}$ key. (Your calculator may label this button $\boxed{\text{2nd}}$ or $\boxed{\text{SHIFT}}$.)
- Press the $\boxed{\text{sin}}$ key. Since you have already pressed the $\boxed{\text{INV}}$ key, the calculator automatically makes this key have the $[\sin]^{-1}$ inverse function, which is printed above it.
- Round off the display value of 57.0030. . . .

If this procedure does not work for your calculator, try this key sequence:

$$\boxed{\text{SHIFT}} \rightarrow \boxed{\text{sin}} \rightarrow \boxed{.}\,\boxed{8}\,\boxed{3}\,\boxed{8}\,\boxed{7} \rightarrow \boxed{=}$$

Solving Triangles

Trigonometric ratios can be used to find the measure of an unknown part of a right triangle when:

- the lengths of two sides are known, and you need to find the measure of an acute angle, or
- the measures of a side and an acute angle are known, and you need to find the length of another side.

Example 2

For right triangle *JLK* shown in the accompanying diagram, find the value of *x* correct to the *nearest tenth*.

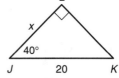

Solution: **15.3**

Decide which trigonometric ratio to use, and then write the corresponding equation. Since the length of the hypotenuse is given and the length of the leg *adjacent* to ∠*J* needs to be determined, use the cosine ratio:

$$\cos J = \frac{\text{leg adjacent to } \angle J}{\text{hypotenuse}}$$

$$\cos 40° = \frac{x}{20}$$

$$0.7660 = \frac{x}{20}$$

$$x = 20(0.7660)$$

$$= 15.32 \approx 15.3 \text{ to the } nearest\ tenth$$

Example 3

A plane takes off from a runway, and climbs while maintaining a constant angle with the ground. When the plane has traveled 1000 meters, its altitude is 290 meters. Find, correct to the *nearest degree*, the angle at which the plane has risen with respect to the horizontal ground.

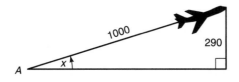

Solution: **17**

To find the value of $\angle x$, first determine the appropriate trigonometric ratio. The sine ratio relates the three quantities under consideration:

$$\sin x = \frac{\text{leg opposite } \angle A}{\text{hypotenuse}}$$
$$= \frac{290}{1000}$$
$$= 0.2900$$

Use a scientific calculator to find that $\angle x = 17°$.

Indirect Measurement and Trigonometry

A trigonometric ratio may be used to calculate the measure of a part of a right triangle that may be difficult, if not impossible, to obtain by direct measurement.

Example 4

To determine the distance across a river, a surveyor marked two points on one river-bank, H and F, 65 meters apart. She also marked one point, K, on the opposite bank such that $\overline{KH} \perp \overline{HF}$, as shown in the accompanying figure. If $m\angle HKF = 54$, what is the width, to the *nearest tenth of a meter*, of the river?

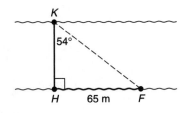

Solution: **47.2**

To find KH, the width of the river, use the tangent ratio:

$$\tan \angle HKF = \frac{\text{leg opposite } \angle HKF}{\text{leg adjacent } \angle HKF}$$
$$\tan 54° = \frac{HF}{KH}$$
$$1.3764 = \frac{65}{KH}$$
$$1.3764\,KH = 65$$
$$KH = 65 \div 1.3764 \approx 47.22$$

Check Your Understanding of Section 10.2

A. *Multiple Choice*

1. In right triangle ACB, $\angle C$ is the right angle. If $\tan A = \dfrac{4}{3}$, then $\sin B$ is equal to

    (1) $\dfrac{3}{5}$ (2) $\dfrac{4}{5}$ (3) $\dfrac{5}{3}$ (4) $\dfrac{5}{4}$

2. In the accompanying diagram of right triangle $\triangle ACB$, $\angle C$ is the right angle, m$\angle A = 63$, and $AB = 10$. If BC is represented by x, which equation can be used to find x?

    (1) $\sin 63° = \dfrac{x}{10}$ (3) $x = 10\cos 63°$

    (2) $\tan 63° = \dfrac{x}{10}$ (4) $x = \dfrac{\tan 27°}{10}$

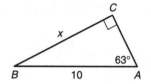

3. In the accompanying diagram of right triangle ABC, $AB = 4$, and $BC = 9$. What is the measure of $\angle A$ correct to the *nearest degree*?

    (1) 24   (3) 55
    (2) 35   (4) 66

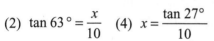

4. In right triangle ACB, hypotenuse $AB = 13$ and leg $BC = 5$. What is $\tan A$?

    (1) $\dfrac{5}{13}$ (2) $\dfrac{13}{5}$ (3) $\dfrac{12}{5}$ (4) $\dfrac{5}{12}$

5. If $\cos(3x°) = \sin 72°$, what is the value of x?

    (1) 6   (2) 18   (3) 24   (4) 30

B. *In each case, show how you arrived at your answer by clearly indicating all of the necessary steps, formula substitutions, diagrams, graphs, charts, etc.*

6. In right triangle ACB, $\angle C$ is the right angle, $BC = 6$, and $AB = 10$. What is the value of $\tan A + \tan B$?

7. If $\sin(3x - 19)° = \cos(x + 31)°$, what is the value of x?

8. As shown in the accompanying diagram, a kite is flying at the end of a 200-meter straight string. If the string makes an angle of 68° with the level ground, how high, to the *nearest meter*, is the kite?

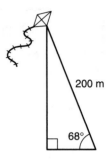

200 m

68°

9. To figure out the approximate distance across Lily Lake, a surveyor marked off right triangle ABC, so that $BC = 150$ feet and m∠$ACB = 43$, as shown in the accompanying diagram. What is the approximate distance, correct to the *nearest foot*, across the lake?

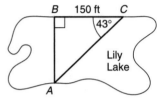

B   150 ft   C
43°
Lily Lake
A

10. In quadrilateral $ABCD$, shown in the accompanying diagram, $\overline{AB} \perp \overline{BC}$, $\overline{AD} \perp \overline{CD}$, $AB = 4$, $BC = 3$, and $AD = 2$.
    (a) What are m∠A and m∠C correct to the *nearest degree*?
    (b) What is the perimeter of quadrilateral $ABCD$ correct to the *nearest tenth*?

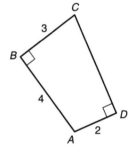

C
3
B
4
D
2
A

11. A plane is climbing at an angle of 11° with the ground. Find, to the *nearest foot*, the horizontal-ground distance the plane has traveled when it has attained an altitude of 400 feet.

12. A 20-foot ladder is leaning against the side of a building. The foot of the ladder is 5 feet from the base of the building. Find, to the *nearest degree*, the angle that the ladder makes with the ground.

13. At the Slippery Ski Resort, the beginner's slope is inclined at an angle of 12° while the advanced slope is inclined at an angle of 26°. Rudy skis 1000 meters down the advanced slope, while Valerie skis the same distance on the beginner's slope. To the *nearest tenth of a meter*, how much longer was the horizontal distance that Valerie covered?

14. A triangular access ramp to an office building is constructed so that it rises $1\frac{1}{2}$ feet for every 8 feet of horizontal ground distance.
    (a) Find, to the *nearest degree*, the angle of incline of the access ramp.
    (b) Find, to the *nearest foot*, the length of the access ramp at the point at which the height of the ramp is 4 feet.

10.3 SOLVING PROBLEMS USING TRIGONOMETRY

KEY IDEAS

The need to use trigonometry may arise in real-life settings as well as in geometric situations.

Angles of Elevation and Depression

The angles formed by an observer's line of vision and a horizontal line are sometimes referred to by special names. The **angle of elevation** represents the angle through which an observer must *raise* his or her line of sight with respect to a horizontal line in order to see an object. For example, if to see a bird in flight John must raise his line of sight 35° with respect to the horizontal ground, then the angle of elevation of the bird is 35°.

If in order to view an object the observer must *lower* his or her line of sight with respect to a horizontal line, the angle thus formed is called the **angle of depression**. For example, if a pilot of an airplane in flight must lower her line of sight 23° to spot a landmark on the ground, then the angle of depression of the landmark is 23°. As Figure 10.4 illustrates:

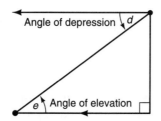

Figure 10.4   Angles of Elevation and Depression

- The angle of elevation e is an angle of a right triangle that has the line of sight as its hypotenuse, while the angle of depression d falls outside this triangle.
- The angle of elevation and the angle of depression are numerically equal ($e = d$) since they are alternate interior angles formed by parallel (horizontal) lines and a transversal (the line of sight).

Example 1

A man standing 30 feet from a flagpole observes the angle of elevation of its top to be 48°. Find the height of the flagpole, correct to the *nearest tenth of a foot.*

279

Solution: **33.3**

- Draw a right triangle, and label it with the given information as shown in the accompanying diagram.

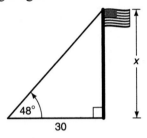

30

- Decide which trigonometric ratio to use, and then write the corresponding equation:

$$\tan 48° = \frac{x(\text{leg opposite angle})}{30(\text{leg adjacent to angle})}$$

- Replace tan 48° by its value, obtained by using a scientific calculator, and then solve for x:

$$1.1106 = \frac{x}{30}$$
$$x = 30(1.1106)$$
$$= 33.318$$
$$\approx 33.3 \, (\text{correct to the } \textit{nearest tenth of a foot})$$

Example 2

An airplane pilot observes the angle of depression of a point on a landing field to be 28°. If the plane's altitude at this moment is 900 meters, find the distance from the pilot to the observed point on the landing field, correct to the *nearest meter*.

Solution: **1917**

- Draw a right triangle, and label it with the given information, as shown in the accompanying diagram. Use the fact that the angle of elevation and the angle of depression are numerically equal.

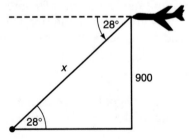

280

- Decide which trigonometric ratio to use, and then write the corresponding equation:

$$\sin 28° = \frac{900(\text{leg opposite angle})}{x(\text{hypotenuse})}$$

- Note that in this case forming a ratio in which the variable is in the denominator is unavoidable. Replace sin 28° by its value obtained by using a scientific calculator, and then solve for *x*.

$$0.4695 = \frac{900}{x}$$

$$x = \frac{900}{0.4695} = 1916.93$$

$$= 1917 \text{ (correct to the } nearest\ meter)$$

Solving Geometry Problems

Sometimes you need to use a trigonometric ratio to find the measure of an unknown part of a geometric figure that contains a right triangle.

Example 3

The vertex angle of an isosceles triangle measures 70°, and the length of the base is 30 inches.
(a) What is the length of the altitude of the triangle correct to the *nearest tenth of an inch*?
(b) What is the area of the triangle correct to the *nearest square inch*?

   Solution: (a) **21.4**

Draw a diagram of an isosceles triangle. Form a right triangle by drawing an altitude from the vertex to the base, as shown in the accompanying diagram.

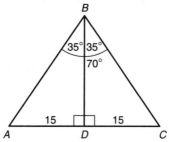

- Since the altitude drawn to the base of an isosceles triangle bisects the vertex angle and bisects the base, $AD = \frac{1}{2}(30) = 15$ and $m\angle ABD = \frac{1}{2}(70) = 35$.

- To find the length of the altitude, $\overline{BD}$, use the tangent ratio in right triangle ADB:

$$\tan 35° = \frac{\text{leg opposite } \angle ABD}{\text{leg adjacent } \angle ABD}$$

$$= \frac{AD}{BD}$$

$$0.7002 = \frac{15}{BD}$$

$$0.7002 \times BD = 15$$

$$BD = \frac{15}{0.7002}$$

$$= 21.42 \approx 21.4$$

(b) **321**

The area A of a triangle is equal to one-half the product of its base, b, and height, h. Hence:

$$A = \frac{1}{2} \times AC \times BD$$

$$= \frac{1}{2} \times 30 \times 21.4$$

$$= 321$$

Example 4

In rhombus $ABCD$, diagonal $AC = 40$ centimeters, and $m\angle DAB = 72$.

(a) What is the length of diagonal $\overline{DB}$ correct to the *nearest tenth of a centimeter?*

(b) What is the area of rhombus $ABCD$ correct to the *nearest square centimeter?*

Solution: (a) **29.1**

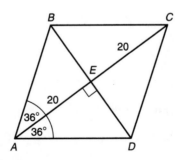

As shown in the accompanying diagram:

- Since the diagonals of a rhombus bisect each other,

$$AE = \frac{1}{2} \times AC = \frac{1}{2} \times 40 = 20$$

- Each diagonal of a rhombus bisects the angles at the vertices they connect. Hence:

$$m\angle EAD = \frac{1}{2} m\angle DAB = \frac{1}{2} \times 72 = 36$$

- Since the diagonals of a rhombus intersect at right angles, $\angle AED$ is a right angle. In right triangle AED, use the tangent ratio to find DE.

$$\tan 36° = \frac{DE(\text{leg opposite } \angle EAD)}{AE(\text{leg adjacent } \angle EAD)}$$

$$0.7265 = \frac{DE}{20}$$

$$0.7265 \times 20 = DE$$

$$14.53 = DE$$

Because the diagonals of a rhombus bisect each other,

$$DB = 2 \times DE = 2 \times 14.53 = 29.06 \approx 29.1$$

(b) **582**

The area of a rhombus is equal to one-half of the product of the lengths of its diagonals. Hence:

$$\text{Area of rhombus } ABCD = \frac{1}{2} \times AC \times DB$$

$$= \frac{1}{2} \times 40 \times 29.1$$

$$= 20 \times 29.1$$

$$= 582$$

Check Your Understanding of Section 10.3

A. Multiple Choice

1. In the accompanying diagram of isosceles triangle $\triangle ABC$, $AB = 20$, m$\angle A = 68$, and $\overline{CD}$ is the altitude to side $\overline{AB}$. What is the length of $\overline{CD}$ to the *nearest tenth?*
   (1) 24.8   (2) 9.3   (3) 3.7   (4) 4.0

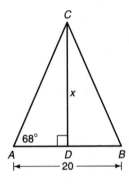

2. In the accompanying diagram of trapezoid *DEFG*, altitude $\overline{EH}$ is drawn, $\overline{FG} \perp \overline{DG}$, $EF = 9$, $FG = 8$, and $GD = 15$. What is the measure of $\angle D$ to the *nearest degree?*
   (1) 37   (2) 53   (3) 60   (4) 80

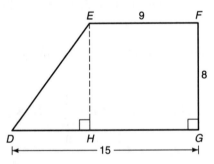

B. In each case, show how you arrived at your answer by clearly indicating all of the necessary steps, formula substitutions, diagrams, graphs, charts, etc.

3. At noon, a tree having a height of 10 feet casts a shadow 15 feet in length. Find, to the *nearest degree*, the angle of elevation of the Sun at this time.

4. Find, to the *nearest tenth of a foot*, the height of a building that casts a shadow 80 feet long when the angle of elevation of the Sun is 42°.

5. A man observes the angle of depression from the top of a cliff overlooking the ocean to a ship to be 37°. If at this moment the ship is 1000 meters from the foot of the cliff, find, to the *nearest meter*, the height of the cliff.

6. Find, to the *nearest degree*, the angle of elevation of the Sun when a man 6 feet tall casts a shadow $4\frac{1}{2}$ feet in length.

7. When the altitude of a plane is 800 meters, the pilot spots a target on the ground at a distance of 1200 meters. At what angle of depression, correct to the *nearest degree*, does the pilot observe the target?

8. At an angle of depression of 42°, an airplane pilot is able to view a target on the ground that is at a distance of 1000 meters from the pilot. Find, to the *nearest meter*, the altitude of the plane.

9. Robert is holding his kite string 4 feet above the ground, as shown in the accompanying diagram. When 125 feet of kite string is let out and the angle of elevation to the kite is 55°, what is the height, h, of the kite, to the *nearest foot*?

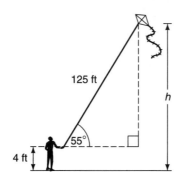

10. If the length of a rectangle is 8 and its width is 6, find, correct to the *nearest degree*, the measure of the angle a diagonal makes with the longer side.

11. The lengths of two adjacent sides of a parallelogram are 8 centimeters and 15 centimeters, and the measure of the angle between the two sides is 50°. What is the area of the parallelogram, correct to the *nearest square centimeter*?

12. The vertex angle of an isosceles triangle measures 76°, and the length of base is 24 inches. What is the perimeter of the triangle correct to the *nearest tenth of an inch*?

13. The measure of the vertex angle of an isosceles triangle is 78°. If the length of an altitude drawn to the base is 10, find:
(a) the length of the base, correct to the *nearest tenth*
(b) the area of the triangle, correct to the *nearest square unit*

14. In the accompanying diagram of square *MARY*, $\overline{ME}$ is drawn to side $\overline{YR}$, m∠*EMY* = 40, and *YE* = 12.
(a) Find the area of quadrilateral *MARE* correct to the *nearest square unit*.
(b) If diagonal $\overline{AY}$ is drawn, will $\overline{AY}$ be perpendicular to $\overline{ME}$? Give a reason for your answer.

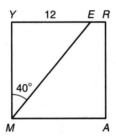

15. In the accompanying diagram of rectangle *ABCD*, diagonal $\overline{AC}$ is drawn, *DE* = 8, $\overline{DE} \perp \overline{AC}$, and m∠*DAC* = 55. What is the area of *ABCD* correct to the *nearest square unit*?

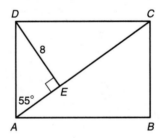

16. A circus acrobat is on a tightrope that places her line of vision 90 feet directly above the ground. When she looks down at a spot on the ground, her angle of depression is 39°. She continues to walk along the tightrope and a minute later stops to look down at the same spot at an angle of depression of 54°. How far, correct to the *nearest foot*, did she walk in the minute?

<table>
<tr><td>

**Unit
Five**

</td><td>

GRAPHS AND TRANSFORMATIONS

</td></tr>
</table>

CHAPTER 11

LINEAR EQUATIONS AND THEIR GRAPHS

11.1 COORDINATES AND AREA

===== **KEY IDEAS** =====

In order to be able to refer to the location of points in a plane, we can create a two-dimensional coordinate system by drawing two perpendicular number lines that intersect at their 0 points, called the **origin**. The location of a point is expressed by writing an ordered pair of numbers that indicate the horizontal and vertical distances and the direction of the point from the origin.

Locating Points in the Coordinate Plane

When a horizontal number line (called the *x*-**axis**) intersects a vertical number line (the *y*-**axis**), a two-dimensional coordinate system is created. The two axes divide the plane into four quadrants that are numbered, using roman numerals, in the counterclockwise direction, as shown in Figure 11.1.

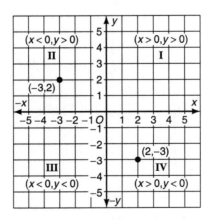

Figure 11.1   The Coordinate Plane

A point is located by writing an ordered pair of numbers of the form (*x*-coordinate, *y*-coordinate).

- The *x*-coordinate gives the horizontal distance and direction of the point from the origin. When *x* is *positive*, the point lies to the *right* of the origin; when *x* is *negative*, the point lies to the *left* of the origin. For example, point $(-3, 2)$ is located in Quadrant II since $x = -3$ and $y = 2$.
- The *y*-coordinate gives the vertical distance and direction of the point from the origin. When *y* is *positive*, the point lies *above* the origin; when *y* is *negative*, the point lies *below* the origin. For example, point $(2, -3)$ is located in Quadrant IV since $x = 2$ and $y = -3$.

Finding Area Using Coordinates

If one side of a triangle or special quadrilateral is parallel to a coordinate axis, the area of the figure can be determined by drawing an altitude to that side and then using the appropriate area formula.

Example 1

Graph a parallelogram whose vertices are $A(2, 2)$, $B(5, 6)$, $C(13, 6)$, and $D(10, 2)$, and then find its area.

Solution: **32 square units**

In the accompanying graph, altitude $\overline{BH}$ has been drawn to base $\overline{AD}$. Count boxes to find the lengths of these segments.

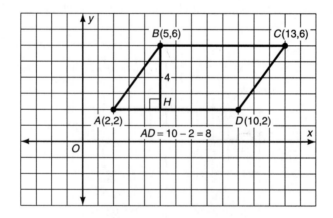

$$\text{Area of } ABCD = b \times h$$
$$= AD \times BH$$
$$= 8 \times 4$$
$$= 32 \text{ square units}$$

Finding Area Indirectly

When the area of a figure cannot be found directly, it may be possible to find the area indirectly.

Example 2

Find the area of the quadrilateral whose vertices are $A(-2, 2)$, $B(2, 5)$, $C(8, 1)$, and $D(-1, -2)$.

Solution: **36.5 square units**

Circumscribe rectangle $WXYZ$ about quadrilateral $ABCD$ by drawing intersecting horizontal and vertical segments through the vertices of the quadrilateral as shown in the accompanying diagram.

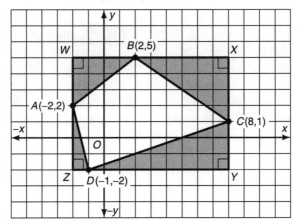

The area of quadrilateral $ABCD$ is calculated *indirectly* as follows:

- Find the area of the rectangle.

$$\text{Area of rect } WXYZ = ZY \times YX = 10 \times 7 = 70$$

- Find the sum of the areas of the right triangles in the four corners of the rectangle. Keep in mind that the area of a right triangle is equal to one-half the product of the lengths of the legs of the triangle.

$$\text{Sum} = \text{Area of rt } \triangle BWA = \frac{1}{2}(BW) \times (WA) = \frac{1}{2}(4) \times (3) = 6$$

$$+ \text{ Area of rt } \triangle BXC = \frac{1}{2}(BX) \times (XC) = \frac{1}{2}(6) \times (4) = 12$$

$$+ \text{ Area of rt } \triangle DYC = \frac{1}{2}(DY) \times (YC) = \frac{1}{2}(9) \times (3) = 13.5$$

$$+ \text{ Area of rt } \triangle DZA = \frac{1}{2}(DZ) \times (ZA) = \frac{1}{2}(1) \times (4) = 2$$

$$\text{Sum of } \triangle \text{ areas} = 33.5$$

sum of the areas of the right triangles from the area of the

quad $ABCD$ = area of rect $WXYZ$
$\quad\quad\quad$ – sum of areas of right triangles
$\quad\quad\quad$ = 70 – 33.5
$\quad\quad\quad$ = 36.5 square units

Check You Understanding of Section 11.1

A. Multiple Choice

1. The vertices of $\triangle ABC$ are $A(-4, 0)$, $B(2, 4)$, and $C(4, 0)$. What is the area, in square units, of $\triangle ABC$?
   (1) 8 $\quad\quad\quad$ (2) 16 $\quad\quad\quad$ (3) 32 $\quad\quad\quad$ (4) 64

2. The coordinates of the vertices of rectangle $ABCD$ are $A(2, 2)$, $B(2, 6)$, $C(8, 6)$, and $D(8, 2)$. The area, in square units, of rectangle $ABCD$ is
   (1) 16 $\quad\quad\quad$ (2) 24 $\quad\quad\quad$ (3) 36 $\quad\quad\quad$ (4) 48

3. The vertices of $\triangle ABC$ are $A(0, 0)$, $B(0, k)$, and $C(k, 0)$. The area of this triangle can be expressed in square units as
   (1) $\dfrac{k^2}{2}$ $\quad\quad$ (2) $\dfrac{k^2}{4}$ $\quad\quad$ (3) k^2 $\quad\quad$ (4) $2k$

4. If $x = 2$ and $y = -3$, in which quadrant is $P(-x, -y)$ located?
   (1) I $\quad\quad\quad$ (2) II $\quad\quad\quad$ (3) III $\quad\quad\quad$ (4) IV

5. In which quadrant is $P(x, y)$ located if $xy > 0$ and $x + y < 0$?
   (1) I $\quad\quad\quad$ (2) II $\quad\quad\quad$ (3) III $\quad\quad\quad$ (4) IV

B. In each case, show how you arrived at your answer by clearly indicating all of the necessary steps, formula substitutions, diagrams, graphs, charts, etc.

6. Find the area, in square units, of the triangle whose vertices are $A(0, 5)$, $B(6, 0)$, and $C(0, 0)$.

7. Find the area, in square units, of the triangle whose vertices are $A(2, 2)$, $B(2, 7)$, and $C(5, 2)$.

8. Find the area, in square units, of the parallelogram whose vertices are (2, 1), (7, 1), (9, 5), and (4, 5).

9. If the area of the triangle whose vertices are $A(1, 4)$, $B(1, 1)$, and $C(x, 1)$ is 6 square units, what is the value of x?

10. In the accompanying figure of right triangle OAP, the coordinates of P are $(6, k)$. Find the value for k if the area of $\triangle OAP$ is 12 square units.

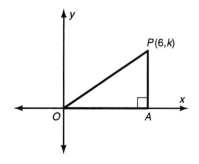

11. In the accompanying figure, if the area of square $ABCD$ is 36 square units, what is the area, in square units, of $\triangle APQ$?

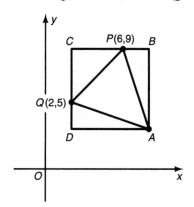

12. The rectangle whose vertices are $A(0, 0)$, $B(0, 5)$, $C(h, k)$, and $D(8, 0)$ lies in Quadrant I.
    (a) What are the values of h and k?
    (b) What is the area, in square units, of rectangle $ABCD$?

13. Find the area, in square units, of the trapezoid whose vertices are $T(-4, -4)$, $R(-1, 5)$, $A(6, 5)$, and $P(9, -4)$.

14. Find the area, in square units, of the quadrilateral whose vertices are $A(-4, -2)$, $B(0, 5)$, $C(9, 3)$, and $D(7, -4)$.

291

15. Pentagon *RSTUV* has coordinates *R*(1, 4), *S*(5, 0), *T*(3, –4), *U*(–1, –4), and *V*(–3, 0).
   (a) On graph paper, plot pentagon *RSTUV*, and draw a line of symmetry.
   (b) Find the area, in square units, of:
   (1) △*RVS*   (2) trapezoid *STUV*   (3) pentagon *RSTUV*

11.2 MIDPOINT AND DISTANCE FORMULAS

KEY IDEAS

The line segment that connects points $A(x_A, y_A)$ and $B(x_B, y_B)$ has its midpoint at $\left(\dfrac{x_A + x_B}{2}, \dfrac{x_A + x_B}{2} \right)$ and a length that is numerically equal to

$$\sqrt{(x_B - x_A)^2 + (y_B - y_A)^2}.$$

Using the Midpoint Formula

Figure 11.2 illustrates that the *x*- and *y*-coordinates of midpoint *M* of a line segment are equal to the averages of the corresponding coordinates of the endpoints of the segment.

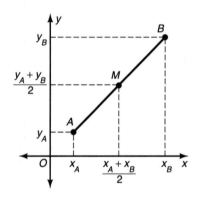
Figure 11.2 Midpoint Formula

Example 1

What are the coordinates of the center of a circle that has a diameter whose endpoints are (4, 9) and (–10, 1)?

Solution: **(–3, 5)**

Let $A(x_A, y_A) = A(4, 9)$ and $B(x_B, y_B) = B(-10, 1)$. Then, since the center of a circle is at the midpoint of any diameter of the circle:

and

$$\bar{x}_{\text{midpoint}} = \frac{x_A + x_B}{2} \qquad \bar{y}_{\text{midpoint}} = \frac{y_A + y_B}{2}$$

$$= \frac{4 + (-10)}{2} \qquad\qquad = \frac{9 + 1}{2}$$

$$= \frac{-6}{2} \qquad\qquad\qquad = \frac{10}{2}$$

$$= -3 \qquad\qquad\qquad\quad = 5$$

Example 2

The diagonals of parallelogram *ABCD* intersect at $(7, -1)$. If the coordinates of endpoint *A* of diagonal $\overline{AC}$ are $A(5, 4)$, what are the coordinates of endpoint *C*?

    Solution: **(9, −6)**

The diagonals of a parallelogram bisect each other. Hence, $(7, -1)$ is the midpoint of diagonal $\overline{AC}$, whose endpoints are $A(5, 4)$ and $C(x, y)$. According to the midpoint formula:

$$7 = \frac{5+x}{2} \quad \text{and} \quad -1 = \frac{4+y}{2}$$

Multiply each side of each equation by 2:

$$7 = \frac{5+x}{2} \quad \text{and} \quad -1 = \frac{4+y}{2}$$
$$2 \cdot 7 = 2\left(\frac{5+x}{2}\right) \quad \Big| \quad 2(-1) = 2\left(\frac{4+y}{2}\right)$$
$$14 = 5 + x \quad\quad\quad\quad\quad -2 = 4 + y$$
$$9 = x \quad\quad\quad\quad\quad\quad\quad -6 = y$$

Using the Distance Formula

Figure 11.3 illustrates that points $A(x_A, y_A)$ and $B(x_B, y_B)$ determine a right triangle in which the length of the horizontal side is $x_B - x_A$, the length of the vertical side is $y_B - y_A$, and the length of the hypotenuse is the distance between points *A* and *B*.

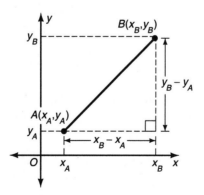

Figure 11.3   Distance Formula

According to the Pythagorean theorem:

$$AB = \sqrt{(x_B - x_A)^2 + (y_B - y_A)^2}$$

Since each difference under the radical sign is squared, the order in which the coordinates are subtracted does not matter.

Example 3

What is the distance between points $(6, -9)$ and $(-3, 4)$?

   Solution: $\sqrt{250}$

Let $A(x_A, y_A) = A(6, -9)$ and $B(x_B, y_B) = B(-3, 4)$. Use the distance formula:

$$AB = \sqrt{(x_B - x_A)^2 + (y_B - y_A)^2}$$
$$= \sqrt{(-3-6)^2 + (4-(-9))^2}$$
$$= \sqrt{(-9)^2 + (4+9)^2}$$
$$= \sqrt{81+169}$$
$$= \sqrt{250}$$

Example 4

Quadrilateral *FAME* has vertices $F(2, -2)$, $A(8, -1)$, $M(9, 3)$, and $E(3, 2)$.
(a) Show that quadrilateral *FAME* is a parallelogram.
(b) Show that *FAME* is or is *not* a rectangle.

   Solution: (a) **FAME is a parallelogram.**

A quadrilateral is a parallelogram if its diagonals have the same midpoint.

- The coordinates of the midpoint of diagonal $\overline{FM}$ are

$$\left(\frac{2+9}{2}, \frac{-2+3}{2}\right) = \left(\frac{11}{2}, \frac{1}{2}\right).$$

- The coordinates of the midpoint of diagonal $\overline{AE}$ are

$$\left(\frac{8+3}{2}, \frac{-1+2}{2}\right) = \left(\frac{11}{2}, \frac{1}{2}\right).$$

- Since the diagonals of quadrilateral *FAME* have the same midpoint, the diagonals bisect each other and, as a result, *FAME* is a parallelogram.

   (b) **FAME is *not* a rectangle.**

If the diagonals of a parallelogram have the same length, the parallelogram is a rectangle. Use the distance formula to find the lengths of diagonals $\overline{FM}$ and $\overline{AE}$:

$$FM = \sqrt{(9-2)^2 + (3-(-2))^2} = \sqrt{7^2 + 5^2} = \sqrt{74}$$
$$AE = \sqrt{(3-8)^2 + (2-(-1))^2} = \sqrt{(-5)^2 + 3^2} = \sqrt{34}$$

Since $FM \neq AE$, parallelogram *FAME* is *not* a rectangle.

To determine whether a quadrilateral is a parallelogram, find and then compare the midpoints of its two diagonals. A quadrilateral is a parallelogram only if its diagonals have the same midpoint. A parallelogram is a rectangle only if its diagonals have the same length.

Check Your Understanding of Section 11.2

A. Multiple Choice

1. If the endpoints of a diameter of a circle are $(2, -1)$ and $(4, 0)$, what are the coordinates of the center of the circle?

   (1) $(6, -1)$ (2) $\left(3, -\dfrac{1}{2}\right)$ (3) $\left(3, \dfrac{1}{2}\right)$ (4) $(2, -1)$

2. If the coordinates of A are $(-3, 2)$ and the coordinates of the midpoint of $\overline{AB}$ are $(-1, 5)$, what are the coordinates of B?

   (1) $(1, 10)$ (2) $(1, 8)$ (3) $(0, 7)$ (4) $(-5, 8)$

3. The length of the line segment connecting the points whose coordinates are $(3, -1)$ and $(6, 5)$ is

   (1) $\sqrt{45}$ (2) 5 (3) 3 (4) $\sqrt{97}$

4. Line segment $\overline{AB}$ has midpoints M. If the coordinates of A are $(2, 3)$ and the coordinates of M are $(-1, 0)$, what are the coordinates of B?

   (1) $(1, 3)$ (2) $\left(\dfrac{1}{2}, \dfrac{3}{2}\right)$ (3) $(-4, -3)$ (4) $(-4, 6)$

5. The coordinates of the endpoints of the base of an isosceles triangle are $(2, 1)$ and $(8, 1)$. The coordinates of the vertex of this triangle may be

   (1) $(1, 5)$ (2) $(2, 5)$ (3) $(2, -6)$ (4) $(5, -6)$

6. Which point is closest to the origin?

   (1) $(5, 12)$ (2) $(-6, -8)$ (3) $(10, 4)$ (4) $(0, 11)$

7. The distance between points $(4a, 3b)$ and $(3a, 2b)$ is

   (1) $a^2 + b^2$ (2) $\sqrt{a^2 + b^2}$ (3) $a + b$ (4) $\sqrt{a+b}$

B. In each case, show how you arrived at your answer by clearly indicating all of the necessary steps, formula substitutions, diagrams, graphs, charts, etc.

8. Find the area of a circle if the endpoints of a diameter are $(-1, 7)$ and $(9, -17)$.

9. The vertices of a rhombus are $A(0, -4)$, $B(8, -3)$, $C(4, 4)$, and $D(-4, 3)$.
   (a) Find, in simplest radical form, the length of each diagonal of the rhombus.
   (b) Find, to the nearest degree, the measure of angle CAD.

10. The vertices of parallelogram $ABCD$ are $A(-3, 1)$, $B(2, 6)$, $C(x, y)$, and $D(4, 0)$. What are the values of x and y?

11. Quadrilateral $STAR$ has coordinates $S(0, 0)$, $T(6, -1)$, $A(4, 2)$, and $R(-2, 3)$. Show that quadrilateral $STAR$ is or is *not* a parallelogram.

12. The coordinates of the vertices of $\triangle ABC$ are $A(-6, 8)$, $B(4, 0)$, and $C(0, 0)$. Find the number of square units in the area of $\triangle ABC$.

13. The vertices of right triangle RAG with hypotenuse $\overline{AG}$ are $R(-2, 4)$, $A(7, 4)$, and $G(x, y)$. If $m\angle RAG = 45$, what are the possible coordinates of vertex G?

14. The coordinates of the vertices of quadrilateral $ABCD$ are $A(2, 9)$, $B(11, 4)$, $C(6, -1)$, and $D(-5, 4)$.
   (a) Using graph paper, graph quadrilateral $ABCD$.
   (b) Determine whether diagonals $\overline{AC}$ and $\overline{BD}$ bisect each other. Give a reason for your answer.

11.3 SLOPE OF A LINE

KEY IDEAS

If you think of a line in the coordinate plane as a hill, then the slope of the line is a number that represents its steepness. "Walking down" the hill, from left to right, corresponds to a negative slope. "Walking up" the hill, also from left to right, corresponds to a positive slope. If two lines have positive slopes, the steeper line has the greater slope.

Meaning of Slope

A line that is "slanted" and thus is not parallel to either coordinate axis is called an **oblique line**. In moving from one point to another on an oblique line, there is a change in vertical distance, called the *rise*, and also a change in horizontal distance, called the *run*. The slope or steepness of a line is computed by forming the ratio $\frac{\text{rise}}{\text{run}}$, as illustrated in Figure 11.4. The letter *m* is used to represent slope.

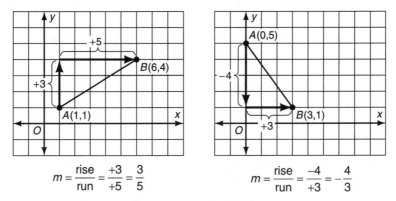

$$m = \frac{\text{rise}}{\text{run}} = \frac{+3}{+5} = \frac{3}{5}$$

$$m = \frac{\text{rise}}{\text{run}} = \frac{-4}{+3} = -\frac{4}{3}$$

Figure 11.4 Finding the Slope of $\overline{AB}$

Slope Formula

The rise between two points in the coordinate plane is measured by the difference in their *y*-coordinates, and the run is measured by the difference in their *x*-coordinates, as shown in Figure 11.5.

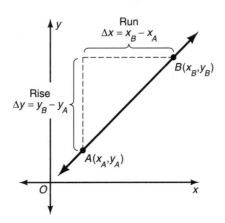

Figure 11.5 Slope Formula

The slope of an *oblique* line that contains points $A(x_A, y_A)$ and $B(x_B, y_B)$ is expressed as:

$$\text{Slope, } m = \frac{\text{rise}}{\text{run}} = \frac{\text{difference in } y\text{-values}}{\text{difference in } x\text{-values}} = \frac{y_B - y_A}{x_B - x_A}$$

297

Sometimes the symbols Δy (read as "delta y") and Δx (read as "delta x") are used to represent a change in coordinates, where $\Delta y = y_B - y_A$ and $\Delta x = x_B - x_A$. Since the order in which two numbers are subtracted matters, the order in which the y-coordinates are subtracted in the numerator of the slope fraction must be the same as the order in which the x-coordinates are subtracted in the denominator. Thus, it would *not* be correct to write $\dfrac{y_B - y_A}{x_A - x_B}$.

Example 1

Determine the slope of the line that contains points $A(-2, 3)$ and $B(1, 7)$.

   Solution: $\frac{4}{3}$

Let $A(x_A, y_A) = (-2, 3)$ and $B(x_B, y_B) = (1, 7)$. Then substitute -2 for x_A, 3 for y_A, 1 for x_B, and 7 for y_B in the slope formula:

$$m = \frac{y_B - y_A}{x_B - x_A} = \frac{7-3}{1-(-2)} = \frac{4}{1+2} = \frac{4}{3}$$

Positive Versus Negative Slope

The slope of an oblique line may be either a positive or a negative number.

- If, as x increases, a line rises, then its slope, m, is a positive number.
- If, as x increases, a line falls, then its slope, m, is a negative number.

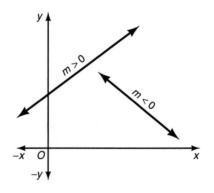

Example 2

Multiple Choice:
What is the slope of line ℓ, shown in the accompanying diagram?

(1) $-\dfrac{5}{3}$ (3) $\dfrac{5}{3}$

(2) $-\dfrac{3}{5}$ (4) $\dfrac{3}{5}$

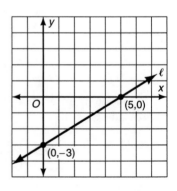

Solution: **(4)**

Since, as x increases, the line rises from left to right, the slope of the line is positive. Hence, eliminate choices (1) and (2). You can figure out the slope either by using the graph to form the slope ratio $\frac{\text{rise}}{\text{run}}$, or by using the slope formula.

Method 1: Use the graph. Complete a right triangle by starting at $(0, -3)$ and moving 3 units vertically in the positive direction (up), so the rise = +3, as shown in the accompanying figure. Then move 5 units horizontally in the positive direction (to the right) so the run = +5. Thus,

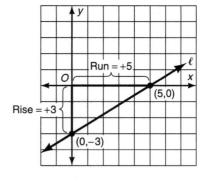

$$m = \frac{\text{rise}}{\text{run}} = \frac{\Delta y}{\Delta x} = \frac{3}{5}$$

Method 2: Use the slope formula. Let $(x_A, y_A) = (0, -3)$ and $(x_B, y_B) = (5, 0)$. Then

$$m = \frac{y_B - y_A}{x_B - x_A} = \frac{0 - (-3)}{5 - 0} = \frac{3}{5}$$

Slopes of Horizontal and Vertical Lines

All points on a *horizontal* or *flat line* must have the same y-coordinate, so $\Delta y = 0$, meaning that the slope m of the line is 0 since $m = \dfrac{\Delta y}{\Delta x} = \dfrac{0}{\Delta x} = 0$.

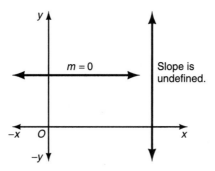

All points on a *vertical line* must have the same x-coordinate, so $\Delta x = 0$, making the slope fraction undefined since division by 0 is not allowed. Hence, the slope m of a vertical line is not defined.

Slopes of Parallel Lines

Two nonvertical lines will intersect only if the lines have *different* slopes.

- If two nonvertical lines have the *same* slope, the lines are parallel.
- Conversely, if two nonvertical lines are parallel, the lines have the *same* slope.

Thus, you can tell whether nonvertical lines are parallel by comparing their slopes.

Slopes of Perpendicular Lines

Pairs of numbers such as $\frac{3}{4}$ and $-\frac{4}{3}$ are negative reciprocals because their product is -1.

- If two nonvertical lines have slopes that are negative reciprocals, then the lines are perpendicular.
- Conversely, if two nonvertical lines are perpendicular, then the slopes of the lines are negative reciprocals.

Thus, you can tell whether two nonvertical lines are perpendicular by comparing their slopes. For example, if $\frac{3}{4}$ and $-\frac{4}{3}$ represent the slopes of two lines, then these lines are perpendicular.

Example 3

The coordinates of the vertices of $\triangle PQR$ are $P(-1,-1)$, $Q(1,-2)$ and $R(3,2)$.

(a) Find the slopes of the three sides of $\triangle PQR$.
(b) Is $\triangle PQR$ a special type of triangle? Explain your answer.

Solution: (a) **Slope of $\overline{PQ} = \dfrac{-1}{2}$, of $\overline{PR} = \dfrac{3}{4}$, of $\overline{QR} = \dfrac{2}{1}$**

Use the slope formula to find the slope of each of the three sides.

$$\text{Slope of } \overline{PQ} = \frac{-2-(-1)}{1-(-1)} = \frac{-2+1}{1+1} = \frac{-1}{2}$$

$$\text{Slope of } \overline{PR} = \frac{2-(-1)}{3-(-1)} = \frac{2+1}{3+1} = \frac{3}{4}$$

$$\text{Slope of } \overline{QR} = \frac{2-(-2)}{3-1} = \frac{2+2}{2} = \frac{4}{2} = \frac{2}{1}$$

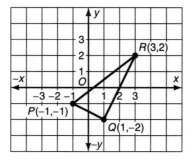

(b) *PQR* **is a right triangle.**

Compare the slopes of the three sides. Since the slopes of $\overline{PQ}$ and $\overline{QR}$ are negative reciprocals, $\overline{PQ} \perp \overline{QR}$, making $\triangle PQR$ a right triangle with right angle Q.

Collinear Points and Slope

Collinear points are points that lie on the same line. Since the slope of a line is a constant, it doesn't matter which two points on a line are used to calculate its slope. Hence, if points A, B, and C lie on the same line, the slopes calculated using any two of these points are equal.

Example 4

Line ℓ contains points (4, 0) and (0, 6). Show that point (−51, 40) does or does *not* lie on line ℓ.

   Solution: **Point (−51, 40) does *not* lie on line ℓ.**

Three points lie on the same line if the slope of the line determined by the first and the second points is the same as the slope of the line determined by the second and the third points.

- If $A(x_A, y_A) = (4, 0)$ and $B(x_B, y_B) = (0, 6)$, then the slope of the line determined by the first and the second points, that is, line ℓ, is:

$$m = \frac{y_B - y_A}{x_B - x_A} = \frac{6-0}{0-4} = -\frac{3}{2}$$

- To find the slope of the line containing (0, 6) and the third point, (−51, 40), use the slope formula with $A(x_A, y_A) = (0, 6)$ and $B(x_B, y_B) = (-51, 40)$:

$$m = \frac{y_B - y_A}{x_B - x_A} = \frac{40-6}{-51-0} = \frac{34}{-51} = -\frac{2}{3}$$

- Compare slopes. The slope of line ℓ is $-\dfrac{3}{2}$. The slope of the line containing the second and third points, $(0, 6)$ and $(-51, 40)$, is $-\dfrac{2}{3}$. Since the two slopes are not equal, point $(-51, 40)$ does *not* lie on line ℓ.

Check Your Understanding of Section 11.3

A. Multiple Choice

1. What is the slope of a line through points $(-4, 2)$ and $(6, 8)$?

   (1) $-\dfrac{3}{5}$ (2) $\dfrac{3}{5}$ (3) $\dfrac{5}{3}$ (4) $-\dfrac{5}{3}$

2. Which pair of points determine a line that is parallel to the *y*-axis?
   (1) $(1, 1)$ and $(2, 3)$ (3) $(2, 3)$ and $(2, 5)$
   (2) $(1, 1)$ and $(3, 3)$ (4) $(2, 5)$ and $(4, 5)$

3. Which pair of points determine a line that is parallel to the *x*-axis?
   (1) $(1, 3)$ and $(-2, 3)$ (3) $(1, 3)$ and $(1, -1)$
   (2) $(1, -1)$ and $(-1, 1)$ (4) $(1, 1)$ and $(-3, -3)$

4. If points $(3, 2)$ and $(x, -5)$ lie on a line whose slope is $-\dfrac{7}{2}$, then what is the value of *x*?

   (1) 5   (2) 6   (3) $\dfrac{15}{7}$ (4) 4

5. Given points $A(0, 0)$, $B(3, 2)$, and $C(-2, 3)$, which statement is true?
   (1) $\overline{AB} \parallel \overline{AC}$ (3) $AB > BC$
   (2) $\overline{AB} \perp \overline{AC}$ (4) $\overline{BC} \perp \overline{CA}$

6. The point whose coordinates are $(4, -2)$, lies on a line whose slope is $\dfrac{3}{2}$. The coordinates of another point on this line could be

   (1) $(1, 0)$ (2) $(2, 1)$ (3) $(6, 1)$ (4) $(7, 0)$

7. In the accompanying diagram, if the slope of line ℓ is $\dfrac{1}{4}$, what are coordinates (a, b) of point P?

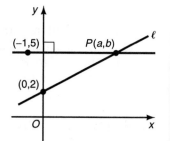

(1) $(12, 5)$ (3) $(8, 5)$
(2) $(16, 5)$ (4) $(5, -12)$

B. *In each case, show how you arrived at your answer by clearly indicating all of the necessary steps, formula substitutions, diagrams, graphs, charts, etc.*

8. If $C(x, 14)$ lies on the same line as $A(6, -1)$ and $B(2, 5)$, what is the value of x?

9. In the accompanying diagram, lines AB and CD intersect at E.
 (a) Find the slope of $\overleftrightarrow{AB}$ and the slope of $\overleftrightarrow{CD}$.
 (b) Is $\angle AED$ a right angle? Give a reason for your answer.

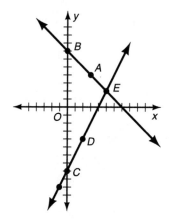

10. The vertices of $\triangle PQR$ are $P(1, 2)$, $Q(-3, 6)$, and $R(4, 8)$. A line through Q is parallel to $\overline{PR}$ and passes through $(k, 14)$. What is the value of k?

11. The vertices of $\triangle ABC$ are $A(-2, 1)$, $B(4, 8)$, and $C(14, -3)$.
 (a) If $M(h, 0)$ lies on $\overline{AC}$, what is the value of h?
 (b) Show by means of slope that $\overline{BM} \perp \overline{AC}$.
 (c) Find, to the *nearest degree*, the measure of angle A.

12. The coordinates of the vertices of quadrilateral *TEAM* are *T*(−2, 3), *E*(1, 0), *A*(7, 6), and *M*(0, 5).
    (a) Show that quadrilateral *TEAM* is or is *not* a trapezoid.
    (b) Points *E*, *F*(*h*, *k*), and *A* are on the same line. If *T*, *E*, *F*, and *M* are the vertices of a square, what are the values of *h* and *k*?

11.4 EQUATIONS OF LINES

KEY IDEAS

An equation of a line gives the general relationship between the *x*- and the *y*-coordinates of any point on the line. When an equation of a line is written in the form $y = mx + b$, the coefficient of the *x*-term, **m**, represents the slope of the line. The constant term, **b**, is the *y*-coordinate of the point at which the line crosses the *y*-axis.

Meaning of an Equation of a Line

A linear equation in two variables, such as $x + y = 5$, has an infinite number of solutions, including (1, 4), (2, 3), and (−3, 8). The graph of that equation visually displays the set of *all* possible solutions, as shown in Figure 11.6.

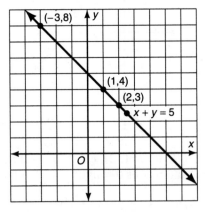

Figure 11.6   The Graph of $x + y = 5$

- If an ordered pair of numbers (*a*, *b*) satisfies the equation $x + y = 5$, then point (*a*, *b*) lies on the graph in Figure 11.6. Thus, since 4.9 + 0.1 = 5, you can predict that point (4.9, 0.1) lies on the line $x + y = 5$.
- If point (*a*, *b*) lies on the line in Figure 11.6, then the coordinates of that point satisfy the equation $x + y = 5$.

304

An equation of a line may be written in more than one way. For example, the equation $x + y = 5$ may also be written as $y = -x + 5$.

Example 1

Multiple Choice:
The graph of $3x - y = 6$ contains which of the following points?
(1) $(0, 2)$ (2) $(5, 9)$ (3) $(4, -6)$ (4) $(1, 9)$

Solution: **(2)**

For each choice, plug the x- and the y-values into the equation $3x - y = 6$ until you find the coordinates that work. For choice (2), $x = 5$ and $y = 9$:

$$3x - y = 6$$
$$3(5) - 9 \overset{?}{=} 6$$
$$15 - 9 \overset{?}{=} 6$$
$$6 \overset{\checkmark}{=} 6$$

Example 2

The line whose equation is $y = 3x - 1$ contains point $A(2, k)$. What is the value of k?

Solution: **5**

Since point $A(2, k)$ lies on the line, its coordinates must satisfy the equation of the line. To find k, replace x by 2 and y by k in $y = 3x - 1$, making $k = 3(2) - 1 = 5$.

Example 3

The line whose equation is $kx + 3y = 13$ passes through point $(8, -1)$. What is the value of k?

Solution: **2**

Since the line contains point $(8, -1)$, replace x with 8 and y with -1 in $kx + 3y = 13$. Then

$$8k + 3(-1) = 13$$
$$8k = 16$$
$$k = \frac{16}{8} = 2$$

Equation of a Line: $y = mx + b$

The equation of the line in Figure 11.7 is $y = 2x + 4$. You can verify that the slope of the line is 2 and that the y-coordinate of the point at which the line crosses the y-axis, called the **y-intercept**, is 4. When an equation of a line is written in the form $y = mx + b$:

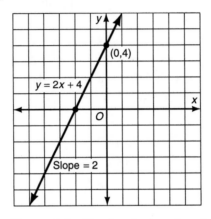

Figure 11.7 The Graph of $y = 2x + 4$

- m, the coefficient of the x-term, is the slope of the line.
- b is the y-intercept of the line.

If an equation is not written in the form $y = mx + b$ and you need to know the slope or the y-intercept, solve the equation for y. For example, if the equation of line ℓ is $y + 3x = 6$, then:

- The equation $y + 3x = 6$ can be rewritten as $y = -3x + 6$, so the slope of line ℓ is -3 and the y-intercept is 6.
- If another line, p, is parallel to line ℓ, the slope of line p is also -3 since parallel lines have equal slopes.
- If another line, q, is perpendicular to line ℓ, the slope of line q is $\frac{1}{3}$ since the product of the slopes of perpendicular lines is -1.

Example 4

Multiple Choice:
What is the slope of the line whose equation is $4x - 3y - 8 = 0$?

(1) $\dfrac{3}{4}$ (2) $-\dfrac{3}{4}$ (3) $\dfrac{4}{3}$ (4) -4

Solution: **(3)**

Since the given equation is not in $y = mx + b$ form, first solve for y in terms of x. The equation $4x - 3y - 8 = 0$ can be written as $-3y = -4x + 8$, so

$$y = \frac{-4x}{-3} + \frac{8}{-3} \quad \text{or} \quad y = \frac{4}{3}x - \frac{8}{3}$$

Since $m = \frac{4}{3}$ and $b = -\frac{8}{3}$, the slope of the line is $\frac{4}{3}$, which is choice (3).

Example 5

If line ℓ is perpendicular to the line whose equation is $2y - 3x = 6$, what is the slope of line ℓ?

Solution: $-\frac{2}{3}$

Put equation $2y - 3x = 6$ into slope-intercept form by solving for y in terms of x:

$$2y = 3x + 6$$
$$\frac{2y}{2} = \frac{3}{2}x + \frac{6}{2}$$
$$y = \frac{3}{2}x + 3$$

For $y = \frac{3}{2}x + 3$, $m = \frac{3}{2}$ and $b = 3$. The slopes of perpendicular lines are negative reciprocals. Then, since the negative reciprocal of $\frac{3}{2}$ is $-\frac{2}{3}$, the slope of line ℓ is $-\frac{2}{3}$.

Writing an Equation of a Line

To write an equation of a line, use the given information to find the slope, m, of the line and its y-intercept, b. Then substitute these values in the equation $y = mx + b$.

Example 6

Write an equation of the line that is parallel to the line $y + 3x = 5$ and passes through point $(1, 4)$.

Solution: **$y = -3x + 7$**

- First find the slope of the desired line. Since the equation $y + 3x = 5$ can be rewritten as $y = -3x + 5$, the slope of this line is -3. Parallel lines have the same slope. Therefore, for the desired line, $m = -3$ and $y = -3x + b$.

- Next, find the value of b. Since this line passes through point $(1, 4)$, find b by substituting $x = 1$ and $y = 4$ into $y = -x + b$; then $4 = -3(1) + b$, so $7 = b$.
- Since $m = -3$ and $b = 7$, an equation of the desired line is $y = -3x + 7$.

Example 7

Write an equation of a line that contains points $(1, -7)$ and $(3, 5)$.

Solution: $\boldsymbol{y = 6x - 13}$

Use the two points to find the slope and y-intercept of the line:

- Calculate the slope: $m = \dfrac{5 - (-7)}{3 - 1} = \dfrac{12}{2} = 6$
- Let $m = 6$; then: $y = 6x + b$
- Use $(x, y) = (3, 5)$ to find b: $5 = 6(3) + b$, so $b = -13$

Since $m = 6$ and $b = -13$, an equation of the line is $y = 6x - 13$.

TIP

- You can use the statistics features of a graphing calculator to find an equation of the line that contains two given points. Enter the two x-values in list L1 and the corresponding y-values in list L2. Then press the $\boxed{\text{STAT}}$ key and highlight the $\boxed{\text{CALC}}$ menu. Select option **4:Lin Reg(ax + b)**, and then press the $\boxed{\text{ENTER}}$ key, which will display the coefficients of the equation of the line that contains the two points.
- If you need to tell the slope of a line from its equation, it may be necessary to solve the equation for y so that it has the form $y = mx + b$.
- If you know the slope, m, of a line and the coordinates of any point $P(h, k)$ on that line, you can quickly write an equation of the line by substituting these values into the **point-slope** equation form, $y - h = m(x - k)$. In Example 7, $m = 6$ and $(h, k) = (3, 5)$. Substituting 3 for h and 5 for k gives $y - 5 = 6(x - 3)$. Solving for y leads to $y = 6x - 13$.

Equations of Vertical and Horizontal Lines

In Figure 11.8 (a), each point on the vertical line $x = a$ has an x-coordinate of a. Similarly, in Figure 11.8 (a), each point on the horizontal line $y = b$ has a y-coordinate of b.

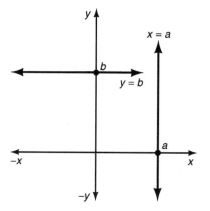

Figure 11.8 (a) Equations of Vertical and Horizontal Lines

Figure 11.8 (b) Vertical and Horizontal Lines Through (–3, 5)

In the coordinate plane, a horizontal line is parallel to the x-axis and a vertical line is parallel to the y-axis. For example, if a line passes through point $(-3, 5)$ and is:

- parallel to the x-axis, then an equation of this horizontal line is $y = 5$ as shown in Figure 11.8 (b).
- parallel to the y-axis, then an equation of this vertical line is $x = -3$ as shown in Figure 11.8 (b).

Check Your Understanding of Section 11.4

A. Multiple Choice

1. What is an equation of a line that is parallel to the y-axis and contains point $(-3, 1)$?
(1) $x = -3$ (2) $x = 1$ (3) $y = -3$ (4) $y = 1$

2. Which point does *not* lie on the graph of $3x - y = 7$?
(1) $(2, -1)$ (2) $(3, 2)$ (3) $(-1, 4)$ (4) $(1, -4)$

309

3. What is an equation of a line that is parallel to the x-axis and contains point $(4, -2)$?
   (1) $x = 4$ (2) $x = -2$ (3) $y = 2$ (4) $y = -2$

4. If point $(k, 2)$ is on the line whose equation is $2x + 3y = 4$, what is the value of k?

   (1) 1   (2) 0   (3) -1 (4) $\dfrac{1}{2}$

5. Which is an equation of a line that is parallel to the y-axis and 2 units to the right of it?
   (1) $x = 2$ (2) $x = -2$ (3) $y = 2$ (4) $y = -2$

6. Which is an equation of the line that passes through point $(0, 2)$ and has a slope of 4?
   (1) $x = 2y - 4$ (3) $4x + y = 2$
   (2) $y = 2x + 4$ (4) $y - 2 = 4x$

7. Which is an equation of the line that is parallel to $y = 2x - 8$ and passes through point $(0, -3)$?

   (1) $y = 2x + 3$ (3) $y = -\dfrac{1}{2}x + 3$

   (2) $y = 2x - 3$ (4) $y = -\dfrac{1}{2}x - 3$

8. Which is an equation of the line that passes through points $(1, 3)$ and $(-1, 1)$?
   (1) $x = 1$ (3) $y = x + 2$
   (2) $y = 2x + 1$ (4) $y = 3$

9. Which is an equation of the line that is parallel to $y = 3x - 5$ and has the same y-intercept as $y = -2x + 7$?
   (1) $y = 3x - 2$ (3) $y = 3x + 7$
   (2) $y = -2x - 5$ (4) $y = -2x + 7$

10. The graph of $x - 3y = 6$ is parallel to the graph of which equation?
    (1) $y = -3x + 7$ (3) $y = 3x - 8$

    (2) $y = -\dfrac{1}{3}x + 5$ (4) $y = \dfrac{1}{3}x + 8$

11. The graph of which equation is perpendicular to the graph of $2y = x + 3$?

    (1) $y = -\dfrac{1}{2}x + 5$ (3) $y = 2x + 5$

    (2) $y = \dfrac{1}{2}x - 3$ (4) $y = -2x + 3$

12. Which statement is true about the line in the accompanying graph?
(1) The slope is negative when x is negative.
(2) The slope is positive and the y-intercept is positive.
(3) The slope is negative and the y-intercept is negative.
(4) The slope is positive and the y-intercept is negative.

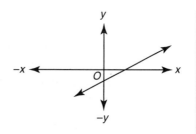

B. In each case, show how you arrived at your answer by clearly indicating all of the necessary steps, formula substitutions, diagrams, graphs, charts, etc.

13. If the lines whose equations are $3x + y - 8 = 0$ and $-2x + by + 9 = 0$ are perpendicular, what is the value of b?

14. For the accompanying graph:
(a) Determine and then write an equation of line ℓ in slope-intercept form.
(b) Determine and then write an equation of line m in slope-intercept form.

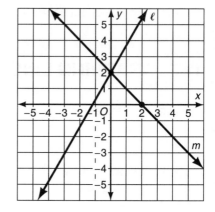

15. Write an equation of the perpendicular bisector of the segment whose endpoints are $(-2, -3)$ and $(2, -5)$.

16. Molly charges a fixed amount for babysitting plus an additional charge for each hour that she works. Molly's total fee for babysitting for 3 hours one evening was $39, and her total fee for babysitting 5 hours on another evening was $57.
(a) Molly's next babysitting job is for h hours. Find an equation that represents the number of dollars, D, expressed in terms of h, that Molly will earn for babysitting for h hours.
(b) What is Molly's hourly rate for babysitting?

17. Given points $A(4, 6)$ and $B(0, -2)$:
(a) Write an equation of $\overleftrightarrow{AB}$.
(b) Write an equation of the line that is parallel to $\overleftrightarrow{AB}$ and contains point $(2, -5)$.
(c) Write an equation of the perpendicular bisector of $\overline{AB}$.

18. Kevin knows that $-40°C = -40°F$ and $20°C = 68°F$, and that the conversion relationship between Celsius and Fahrenheit temperatures is linear.
    (a) What is an equation of the line that contains all paired conversion temperatures of the form (°C, °F)?
    (b) Use the equation determined in part (a) to find the number of degrees Fahrenheit that is equivalent to 35 degrees Celsius.

19. Determine the distance between $A(4, 1)$ and $B(-2, -7)$. Write an equation of the perpendicular bisector of $\overline{AB}$.

20. Brittany graphed the line represented by the equation $y + 6 = 2x$.
    (a) Write an equation of a line that is parallel to the given line.
    (b) Write an equation of a line that is perpendicular to the given line and contains point (0, 3).
    (c) Write an equation of a line that is identical to the given line but has different numerical coefficients.

11.5 GRAPHING LINEAR EQUATIONS

KEY IDEAS

A line may be graphed from its equation by:

- finding at least two points that satisfy the equation, plotting these points on graph paper, and then drawing a line that contains these points, or
- using a graphing calculator.

Graphing by Finding Particular Points

Here are two methods that you can use to find the points needed to graph a line using pencil, ruler, and graph paper.

Method 1: Three-point table

To graph the equation $y - 2x = 3$:

- Solve for y in terms of x: $y = 2x + 3$.
- Pick three easy numbers for x, and find the corresponding values for y:

x	$y = 2x + 3$	(x, y)
-2	$y = 2(-2) + 3 = -1$	(-2, -1)
0	$y = 2(0) + 3 = 3$	(0, 3)
2	$y = 2(2) + 3 = 7$	(2, 7)

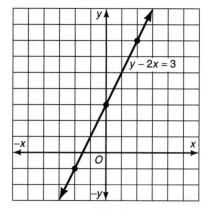

Figure 11.9 Graph of $y - 2x = 3$

- Graph the three points in the table, as shown in Figure 11.9. Make sure that the axes are labeled and that you chose an appropriate scale. Use a straightedge to draw the line, and then label the line with its equation. Although two points determine a line, the third point serves as a check point. If a line cannot be drawn that contains all three points, you need to check your work.

<u>Method 2: Slope-intercept</u>

To graph the equation $y - 2x = 3$:

- Solve for y in terms of x:

$$y = 2x + 3.$$

Then locate the y-intercept, as shown in Figure 11.10.

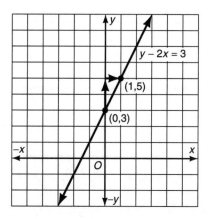

Figure 11.10 Graphing $y - 2x = 3$ Using Slope-Intercept Method

- Using the slope of the line, find another point on the line. The slope of $y = 2x + 3$ is $2 = \dfrac{2(=\text{rise})}{1(=\text{run})}$. Starting at $(0, 3)$, move up 2 units and then move 1 unit to the right. This point is $(1, 5)$.
- Draw a line that contains $(0, 3)$ and $(1, 5)$, and label the line with its equation.

Graphing Using a Calculator

If you have the Texas Instruments TI-83 graphing calculator, the equation $y - 2x = 3$ can be graphed as follows:

- Using paper and pencil, first solve for y in terms of x:

$$y = 2x + 3.$$

- Press the $\boxed{Y=}$ key.

- At the blinking square cursor on the line that begins $\backslash Y_1 =$, enter the right side of the equation $y = 2x + 3$ by pressing these keys:

$$\boxed{2}\ \boxed{x,\ T,\ \theta,\ n}\ \boxed{+}\ \boxed{3}$$

The blinking cursor should now be at the end of a line that reads $\backslash Y_1 = 2X + 3$. Notice that pressing $\boxed{x,\ T,\ \theta,\ n}$ makes variable X appear in the display window.
- Set the size of the viewing rectangle by pressing the $\boxed{\text{ZOOM}}$ key. Press $\boxed{4}$ from the $\boxed{\text{ZOOM}}$ menu to make the graph appear in a "friendly" window in which the cursor moves in convenient steps of 0.1. Since the graph of the line, shown in Figure 11.11, looks as though it is cut off at 3, its y-intercept, the size of the viewing rectangle needs to be changed.

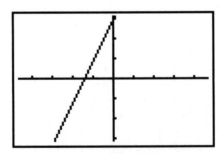

Figure 11.11 Graph of $y = 2x + 3$

- Change the size of the viewing rectangle to improve the fit of the line within the rectangle. Press the $\boxed{\text{WINDOW}}$ key. The values of Xmin and Xmax determine the left and right edges of the viewing rectangle, and the values of Ymin and Ymax determine the bottom and top edges of the viewing rectangle. To expand the *y*-range of values:

1. Press the down arrow key until the cursor is on the line that begins Ymin =. Change the Ymin value to −6 by pressing

$$\boxed{\text{CLEAR}}\;\boxed{(-)}\;\boxed{6}$$

2. Move the cursor to the next line by pressing the down arrow key. Increase the current Ymax value to 6 by pressing the $\boxed{\text{CLEAR}}$ key followed by the $\boxed{6}$ key. The window settings should now look like those shown in Figure 11.12.

```
WINDOW
 Xmin=-4.7
 Xmax=4.7
 Xscl=1
 Ymin=-6
 Ymax=6█
 Yscl=1
 Xres=1
```

Figure 11.12   Adjusted Window Values

- Press the $\boxed{\text{GRAPH}}$ key so that you can see the graph in the [−4.7, +4.7] by [−6, +6] viewing rectangle, as shown in Figure 11.13.

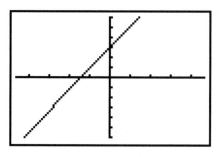

Figure 11.13   Graph of *y* − 2*x* = 3 in [−4.7, +4.7] by [−6, +6] Window.

⎛ **TIPS** ⎞

- The $\boxed{(-)}$ key means "take the opposite of" and the $\boxed{-}$ key means "subtraction."

  For example, to graph $y = -3x$, enter $-3x$ in the $\boxed{Y =}$ editor by pressing the keys

$$\boxed{(-)}\ \boxed{3}\ \boxed{x,\ T,\ \theta,\ n}$$

  To graph $y = x - 3$, enter $x - 3$ in the $\boxed{Y =}$ editor by pressing the keys

$$\boxed{x,\ T,\ \theta,\ n}\ \boxed{-}\ \boxed{3}$$

- The notation [Xmin, Xmax] and [Ymin, Ymax] represents the size of the viewing rectangle.
- After setting the screen size, by entering $\boxed{4}$ from the $\boxed{\text{ZOOM}}$ menu, you may find the graph needs to be in a larger viewing rectangle. For example, when graphing a slanted line, you may need to size the viewing rectangle so that the x-intercept falls within the interval from Xmin to Xmax and the y-intercept falls within the interval from Ymin to Ymax.
- You can create other "friendly" windows with more tick marks on the x-axis by pressing the $\boxed{\text{WINDOW}}$ key and then multiplying the preset values for Xmin and Xmax by the same whole number. For example, when Xmin $= -4.7 \times 2$ and Xmax $= 4.7 \times 2$, the x-axis will be drawn with 9 tick marks on either side of the origin and, as the cursor moves along the graph, the x-coordinates of points will change in steps of 0.2.
- Not all graphing calculators work the same or have the same arrangement of keys. If you are using a different calculator, you may need to read the manual for your calculator.

Check Your Understanding of Section 11.5

In each case, show how you arrived at your answer by clearly indicating all of the necessary steps, formula substitutions, diagrams, graphs, charts, etc.

1–6. *(a) Using graph paper, draw the graph of each equation.*
   (b) Use a graphing calculator to confirm that your graph is correct.

1. $y + 1 = 3x$

2. $x - 2y = 8$

3. $3y - 2x = 6$

4. $y = -\dfrac{1}{2}x + 4$

5. $2y + 5x = 10$

6. $\dfrac{y}{2} - x = 3$

7. Find the area, in square units, of the region bounded by these three lines:
   (1) $y = x + 2$ (2) $y = 1$ (3) $x = 7$

8. Find the number of square units in the area of the triangular region formed in the first quadrant by the lines $y = x$, and $y = 4$ and the y-axis.

9. Find the number of square units in the area of the triangular region formed in the first quadrant by the intersection of the lines $y = x$ and $x + y = 6$ and the x-axis.

10. Find the number of square units in the area of the trapezoid formed in the first quadrant by the intersection of the lines $y = x$, $y = 4$, and $x = 8$ and the x-axis.

11.6 DIRECT VARIATION

KEY IDEAS

When two variables are related so that the ratio of their values always remains the same, the two variables are said to be in **direct variation**. If y varies directly as x, the graph that describes this relationship is a line through the origin whose slope is called the *constant of variation*.

Expressing Direct Variation as an Equation

The equation $\dfrac{y}{x} = 2$ states that y "varies directly as" x since the ratio of y to x remains constant. The number 2 is the constant of variation. The equation $\dfrac{y}{x} = 2$ can also be written in the equivalent form, $y = 2x$.

MATH FACTS

EQUATION OF DIRECT VARIATION

If y varies directly as x, then $y = kx$, where k is some fixed, nonzero number called the **constant of variation**.

Algebraic Interpretation of Direct Variation

For an equation of the form $y = kx$, multiplying x by some fixed amount also multiplies y by the same fixed amount. For example, since the perimeter P of a square varies directly as the length of a side s of the square, $P = 4s$. If the length of a side of a square is doubled, the perimeter of the square is also doubled. If the length of a side of a square is tripled, the perimeter of the square is also tripled, and so forth.

Geometric Interpretation of Direct Variation

The equation $y = kx$ is a special case of the linear equation $y = mx + b$, with $b = 0$. Thus, a line through the origin always represents a direct variation between y and x. The slope of this line is the constant of variation.

Example 1

Multiple Choice:
(a) Which table of values shows that y varies directly as x?

(1)

x	2	4	6
y	5	7	9

(3)

x	3	5	7
y	9	15	21

(2)

x	2	4	6
y	6	3	2

(4)

x	3	5	7
y	2	4	6

(b) Write an equation that expresses the direct variation.

Solution: (a) **(3)**

The table of values that shows that y varies directly as x is choice (3), in which the ratio of y to x remains the same for each ordered pair:

$$\frac{y}{x} = \frac{9}{3} = \frac{15}{5} = \frac{21}{7} = 3 = k$$

(b) $y = 3x$

Since the constant of variation is 3, the equation $y = 3x$ expresses this direct variation.

Example 2

If y varies directly as x, and $y = 8$ when $x = 12$, write an equation that expresses this variation.

318

Solution: $y = \dfrac{2}{3}x$

Find k. Since y varies directly as x:

$$y = kx$$
$$8 = k \times 12$$
$$\frac{8}{12} = \frac{12k}{12}$$
$$\frac{2}{3} = k$$

Hence, $y = \dfrac{2}{3}x$.

Example 3

If y varies directly as x, and $y = 24$ when $x = 16$, find y when $x = 12$.

Solution: **18**

<u>Method 1:</u> First find k and then an equation of variation. Let $y = 24$ and $x = 16$:

$$y = kx$$
$$24 = k \times 16$$
$$\frac{24}{16} = \frac{16k}{16}$$
$$\frac{3}{2} = k$$

Hence, an equation of direct variation is $y = \dfrac{3}{2}x$. If $x = 12$, then

$$y = \frac{3}{2} \times 12$$
$$= 18$$

<u>Method 2:</u> Since the ratios of corresponding values of y to x are always equal, form and then solve the related proportion. Thus:

$$\frac{y}{x} = \frac{24}{16} = \frac{y}{12}$$

Equate the cross-products:

$$16y = 24 \times 12$$
$$y = \frac{288}{16}$$
$$= 18$$

Example 4

If 20 identical coins weigh 42 grams, how many coins weigh 105 grams?

Solution: **50**

Since the coins are identical, you can assume that their weight varies directly as their number. If x represents the number of coins that weigh 105, then

$$\frac{\text{weight}}{\text{number}} = \frac{42}{20} = \frac{105}{x}$$

Equate the cross-products:

$$42x = 20 \times 105$$
$$\frac{42x}{42} = \frac{2100}{42}$$
$$x = 50$$

Check Your Understanding of Section 11.6

A. *Multiple Choice*

1. If y varies directly as x and $y = 8$ when $x = 3$, what is the value of y when $x = 9$?

   (1) $\dfrac{8}{3}$ (2) $\dfrac{27}{8}$ (3) 14 (4) 24

2. In which table does y *not* vary directly as x?

   (1)
x	2	4	6
y	3	6	9

   (3)
x	4	8	12
y	2	3	4

   (2)
x	3	6	8
y	4.5	9	12

   (4)
x	10	15	20
y	6	9	12

B. *In each case, show how you arrived at your answer by clearly indicating all of the necessary steps, formula substitutions, diagrams, graphs, charts, etc.*

3. A recipe for four servings requires $\frac{2}{3}$ cup of sugar. How many cups of sugar are needed if the same recipe is used to prepare nine servings?

4. On a certain map 1.25 inches represents 150 miles. If two cities are 1 foot apart on this map, what is their distance in miles?

5. If four identical pens cost $5.80, how many pens of this same kind can be purchased for $13.05?

6. If 38.8 Hong Kong dollars can be exchanged for 5 U.S. dollars, then 12 U.S. dollars can be exchanged for what number of Hong Kong dollars?

11.7 USING GRAPHS AS MATHEMATICAL MODELS

KEY IDEAS

A **mathematical model** is a mathematical representation of a real-world situation. When the variables in a linear equation that has the form $y = mx + b$ stand for real-world quantities, the graph of the linear equation becomes a *mathematical model*.

Using a Line as a Model

When a weight is attached to a spring, as shown in Figure 11.14, the number of centimeters in the total length, L, of the elongated spring depends on the number of grams, w, in the weight. The graph in Figure 11.15 shows the relationship between L and w for a particular spring.

You can learn these facts from the graph:

• The length of the spring with no weight attached is 15 centimeters since $L = 15$ when $w = 0$.

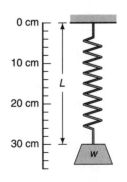

Figure 11.14

321

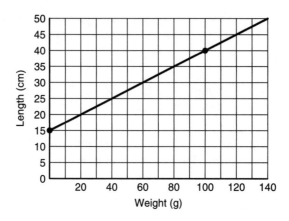

Figure 11.15 Graph of Spring Length Versus Weight

- When a weight of 100 grams is hung from the spring, the total length of the elongated spring is 40 centimeters since, when $w = 100$, $L = 40$.
- The number of centimeters the spring stretches per gram of weight that is hung from it is 0.25 since the slope of the line is 0.25:

$$\text{Slope} = \frac{\text{rise}}{\text{run}} = \frac{40 - 15 \,\text{cm}}{100 - 0 \,\text{g}} = \frac{25 \,\text{cm}}{100 \,\text{g}} = 0.25 \frac{\text{cm}}{\text{g}}$$

- An equation for this spring is $L = 0.25w + 15$ since the slope of the line is 0.25 and the y-intercept of the line is 15.

| **MATH FACTS** |

In a linear graphical model, the slope of the line represents the rate of change of the quantity measured along the y-axis per unit of the quantity measured along the x-axis.

Example 1

The accompanying graph represents the distances traveled by car A and car B at the end of 6 hours.

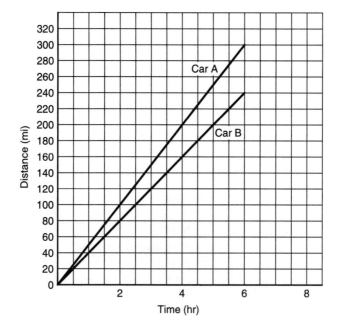

(a) Car A is traveling how many miles per hour faster than car B?
(b) Write an equation of the line for each car.
(c) If the cars continue to travel at the same rates of speed, how many more miles will car A have traveled at the end of 8 hours than car B?

Solution: (a) **10**

Find the average rate of speed of each car by finding the slope of each line.

- From 0 to 6 hr, car A traveled 300 mi, so its average rate of speed is

$$\frac{\text{change in distance}}{\text{change in time}} = \frac{300 - 0 \text{ mi}}{6 - 0 \text{ hr}} = 50 \text{ mph}$$

- From 0 to 6 hr, car B traveled 240 mi, so its average rate of speed is

$$\frac{\text{change in distance}}{\text{change in time}} = \frac{240 - 0 \text{ mi}}{6 - 0 \text{ hr}} = 40 \text{ mph}$$

Car A is going $50 - 40 = 10$ mph faster than car B.

(b) **Car A: $y = 50x$, car b: $y = 40x$**

The general form of an equation of a line that passes through the origin is $y = kx$, where k is the slope of the line. Hence, an equation of the line for car A is $y = 50x$ and an equation of the line for car B is $y = 40x$.

(c) **80**

At the end of 8 hr, car A will have traveled $y = 50(8) = 400$ mi and car B will have traveled $y = 40(8) = 320$ mi. Hence, at the end of 8 hr car A will have traveled 80 mi more than car B.

You could have also reasoned that, since car A is traveling 10 mph faster than car B, at the end of 8 hr car A will have traveled 8 hr × 10 mph = 80 mi more than car B.

Scatter Plot

A **scatter plot** is a graph of points whose x-coordinates are from one set of measurements and whose y-coordinates are from another set. A scatter plot can be used to help determine whether a line can be used to describe the relationship, if any, between the two sets of measurements.

Suppose two different tests were designed to measure students' understanding of a topic. Both tests were given to ten students with the following results:

Test x	75	78	88	92	95	67	58	72	74	81
Test y	81	73	85	88	89	73	66	75	70	78

A scatter plot for these test scores is shown in Figure 11.16.

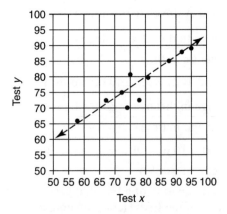

Figure 11.16 Scatter Plot

Since it is possible to draw a line about which most of the data points appear to be clustered, the relationship between the two sets of test scores is *approximately* linear. Therefore, the scores on test y can be estimated from the scores on test x by using an equation that has the form $y = ax + b$. The statistical technique for obtaining the numerical values of the constants a and b for this equation is studied in later math courses.

Check Your Understanding of Section 11.7

In each case, show how you arrived at your answer by clearly indicating all of the necessary steps, formula substitutions, diagrams, graphs, charts, etc.

1. When Carol works as a babysitter she charges a fixed amount of $7 plus an additional $6 per hour for each hour that she works.
   (a) Write an equation that shows the relationship between the amount, A, Carol earns for each babysitting job and the number of hours, h, she works on that job.
   (b) Graph the equation written in part (a).
   (c) From the graph drawn in part (b), determine how much Carol earns when she baby-sits for $4\frac{1}{2}$ hours.

2. Mary and Martin start at the same time from point A, shown on the accompanying grid. Mary walks at a constant average rate of 3.5 miles per hour from point A to point R to point S to point C. Martin walks at a constant average rate of 3 miles per hour directly from point A to point C on $\overline{AC}$.
   (a) Which person reaches point C first?
   (b) How many minutes later does the second person reach point C?

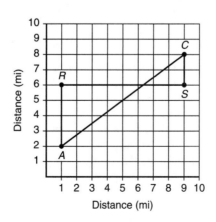

3. During a gym class, Allan jogged and Bill walked at constant rates around a circular $\frac{1}{4}$-mile track. Their times in minutes and distances in miles are shown in the accompanying graph.

(a) At the end of 30 minutes, how many more times had Allan completed the track than Bill?

(b) What was Allan's average rate of jogging in miles per hour?

(c) How much faster did Allan jog than Bill walked?

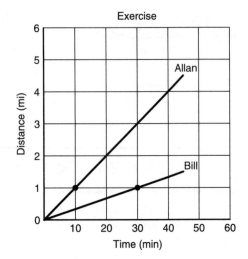

4. Two health clubs offer different membership plans. The accompanying graph represents the total costs of belonging to club A and club B for 1 year. The yearly cost for each club includes an initial membership fee plus a monthly charge.

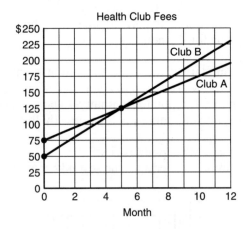

(a) By what amount does the membership fee for club A exceed the membership fee for club B?

(b) If Jack's membership in club A begins in January and Jill's membership in club B begins in the same month, in what month will the total cost of club membership be the same for Jack and Jill?

(c) What is the *monthly* charge for club B?

(d) What is an equation for the line that represents the total charge for membership in club A?

5. Two heating systems, electric and oil, are being considered for a new apartment complex. The accompanying graph compares the total costs of using these electric and oil systems to heat the apartment complex.

(a) For which heating system does the cost increase at the greater rate? Explain your answer.

(b) After 10 years, which system is projected to be more expensive and by how many thousands of dollars?

(c) Write an equation that represents the total cost of oil heat, C, in Y years, where C is expressed in terms of Y.

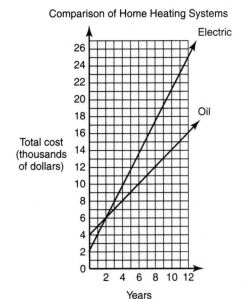

Comparison of Home Heating Systems

Electric

Oil

Total cost (thousands of dollars)

Years

CHAPTER 12

POINTS THAT SATISFY CONDITIONS

12.1 OBSERVING SYMMETRY

KEY IDEAS

There are many examples of *symmetry* in nature. People's faces, leaves, and butterflies have real or imaginary "lines of symmetry" that divide the figures into two parts that are mirror images. If a geometric shape has line symmetry, it can be "folded" along the line of symmetry so that the two parts coincide. A figure can also be symmetric with respect to a point.

Line Symmetry

A figure has **line symmetry** if a line can be drawn that divides the figure into two parts that are mirror images. The line of symmetry may be a horizontal line, a vertical line, or neither.

The examples in Figure 12.1 have horizontal line symmetry.

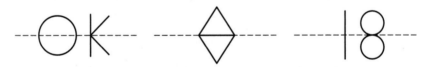

Figure 12.1   Horizontal Line Symmetry

The examples in Figure 12.2 have vertical line symmetry.

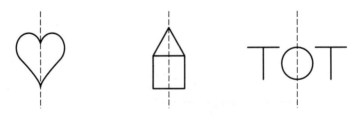

Figure 12.2   Vertical Line Symmetry

328

A figure may have a line of symmetry that is neither horizontal nor vertical, as shown in Figure 12.3. Figure 12.4 illustrates that a figure may have both a horizontal and a vertical line of symmetry.

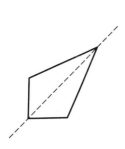

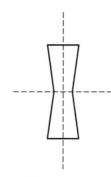

Figure 12.3 Neither Horizontal nor Vertical Line Symmetry

Figure 12.4 Both Horizontal and Vertical Line Symmetry

As shown in Figure 12.5, a figure may have more than one line of symmetry or may have no line of symmetry.

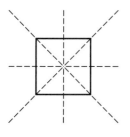

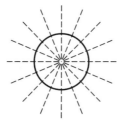

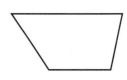

Four lines of symmetry

An infinite number of lines of symmetry

No line of symmetry

Figure 12.5 Figures with More Than One or No Line of Symmetry

Point Symmetry

A figure has **point symmetry** if it is possible to locate a point, Q, such that, if any line is drawn through point Q and intersects the figure in another point, P, that line will also intersect the figure in a different point, call it P', so that $PQ = QP'$. The letter **S** has point symmetry, as shown in Figure 12.6. Since the letter **S** has point symmetry about point Q, turning this figure 180° about point Q does not change how the figure looks.

Figure 12.6 Point Symmetry

Check Your Understanding of Section 12.1

Multiple Choice

1. If a rectangle is not a square, what is the greatest number of lines of symmetry that can be drawn?
   (1) 1   (2) 2   (3) 3   (4) 4

2. Which figure has one and only one line of symmetry?
   (1) rhombus   (3) square
   (2) circle   (4) isosceles triangle

3. Which type of symmetry, if any, does a square have?
   (1) line symmetry, only   (3) both line and point symmetry
   (2) point symmetry, only   (4) no symmetry

4. What is the total number of lines of symmetry for the letter **W**?
   (1) 1   (2) 2   (3) 0   (4) 4

5. What is the total number of lines of symmetry for an equilateral triangle?
   (1) 1   (2) 2   (3) 3   (4) 4

6. Which letter has vertical line symmetry?
   (1) **T**   (2) **P**   (3) **E**   (4) **S**

7. Which letter has only horizontal line symmetry?
   (1) **A**   (2) **D**   (3) **H**   (4) **F**

8. Which letter has *both* line and point symmetry?
   (1) **A**   (2) **S**   (3) **Z**   (4) **H**

9. Which number has horizontal and vertical line symmetry?
   (1) **414**   (2) **383**   (3) **818**   (4) **l00**

10. Which figures, if any, have *both* point symmetry and line symmetry?
    (1) *A* and *C*, only
    (2) *B* and *C*, only
    (3) none of the figures
    (4) all of the figures

A   B   C

330

12.2 TYPES OF TRANSFORMATIONS

$\equiv\!\!\equiv$ **KEY IDEAS** $\equiv\!\!\equiv$

The process of "moving" each point of a figure according to some given rule is called a **transformation**. Each point of the new figure corresponds to exactly one point of the original figure and is called the **image** of that point. The image of a point, *P*, under a transformation is usually denoted by *P′*, read as "*P* prime." There are three basic types of transformations that produce images congruent to the original figure:

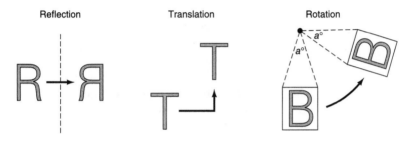

A **dilation** changes the size of a figure without affecting its shape.

Line Reflections

A **reflection** of a point or a figure in a line may be thought of as the mirror image of the point or figure, with the line serving as the reflecting surface. The image of point *A* after a reflection in line *ℓ* is point *A′*, which is located on the opposite side of line *ℓ* and at the same distance from line *ℓ* as point *A*. This means that line *ℓ* is the perpendicular bisector of the line segment that connects *A* and *A′*, as shown in Figure 12.7.

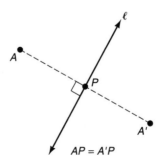

Figure 12.7 Reflecting a Point in a Line

To reflect $\overline{AB}$ in line ℓ, reflect the endpoints of line segment AB in line ℓ. Then connect the images of the two endpoints. Figure 12.8 shows the reflection of $\triangle ABC$ in line ℓ.

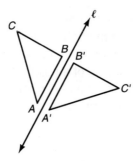

Figure 12.8 Reflection of $\triangle ABC$ in a Line

Translations

A **translation** "slides" a figure in such a way that each point of the figure is moved the same distance in the same direction, as shown in Figure 12.9. Under a translation, a figure and its image are congruent.

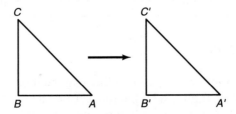

Figure 12.9 Translation of $\triangle ABC$

Rotations

A **rotation** "turns" a figure a fixed number of degrees about some given point called the *center of rotation* while preserving distance. Figure 12.10 shows the rotation of $\triangle ABC\ x°$ about point O. The images of points A, B, and C are determined so that corresponding sides of the figure and its image have the same lengths and

$$m\angle AOA' = m\angle BOB' = m\angle COC' = x$$

Under a rotation, a figure and its image are congruent.

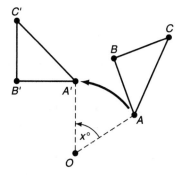

Figure 12.10 Counterclockwise Rotation of △*ABC x°* about Point *O*

Dilations

A **dilation** is a transformation that shrinks or enlarges a figure by a fixed amount called the *scale factor* or *constant of dilation*. A dilation preserves angle measure, so the shape of the figure does not change. Therefore, a dilation produces a figure that is *similar* to the original figure. For example, the image of a square under a dilation with a scale factor of 3 is another square whose sides are three times as long as the sides of the original square.

Example

Multiple Choice:
Under what type of transformation, shown in the accompanying figure, is △*AB'C'* the image of △*ABC*?

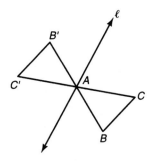

(1) dilation
(2) translation
(3) rotation about point *A*
(4) reflection in line ℓ

Solution: **(3)**

Consider each choice in turn:

- Choice (1): A dilation changes the size of the original figure. Since △*ABC* and △*AB'C'* are the same size, the transformation is *not* a dilation.
- Choice (2): Since △*AB'C'* cannot be obtained by "sliding" △*ABC* in the horizontal ("sideways") or vertical (up and down) direction, or in both directions, the transformation is *not* a translation.
- Choice (3): A rotation about a fixed point "turns" a figure about that point. Since angles *BAB'* and *CAC'* are straight angles, △*AB'C'* is the image of △*ABC* after a rotation of 180° about point *A*.
- Choice (4): Since line ℓ is not the perpendicular bisector of $\overline{BB'}$ and $\overline{CC'}$, points *B'* and *C'* are not the images of points *B* and *C*, respectively, after a reflection in line ℓ.

Check Your Understanding of Section 12.2

Multiple Choice

1. Which type of transformation produces a figure similar to the original figure?
   (1) reflection   (2) dilation   (3) translation   (4) rotation

2. Which property is *not* preserved under a line reflection?
   (1) collinearity of points   (3) congruence of angles
   (2) congruence of line segments   (4) orientation

3. In the accompanying diagram, △*A'B'C'* is the image of △*ABC*. Which type of transformation is shown in the diagram?
   (1) line reflection   (3) translation
   (2) rotation   (4) dilation

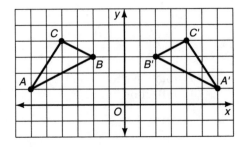

4. If △*OAB*, shown in the accompanying diagram, is rotated clockwise 90° about point *O*, which figure represents the image of this rotation?

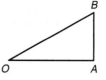

(1)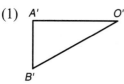

(2)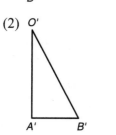

(3)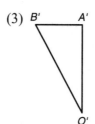

(4)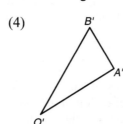

5. In the accompanying diagram, △*R'S'T'* is the image of △*RST*. Which type of transformation is shown in the diagram?
(1) dilation   (3) rotation
(2) reflection   (4) translation

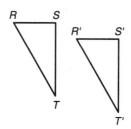

6. In the accompanying diagram of right triangle *ACB* with the right angle at *C*, line ℓ is drawn through *C* and is parallel to $\overline{AB}$. If △*ABC* is reflected in line ℓ, forming the image △*A'B'C'*, which statement is *not* true?
(1) *C* and *C'* are the same point.
(2) m∠*ABC* = m∠*A'B'C'*
(3) The area of △*A'B'C'* is twice the area of △*ABC*.
(4) Line ℓ is equidistant from *A* and *A'*.

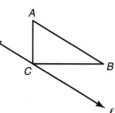

7. In the diagram below, figure B is the image of figure A under which transformation?

(1) line reflection
(2) rotation
(3) translation
(4) dilation

8. The area of a circle is 16π square inches. After the circle is dilated, the circumference of the new circle is 16π inches. What is the scale factor?

(1) 1   (2) 2   (3) $\dfrac{1}{4}$ (4) $\dfrac{1}{2}$

9. The three vertices of $\triangle ABC$ are located in Quadrant II. The image of $\triangle ABC$ after a reflection in the x-axis is $\triangle A'B'C'$. The image of $\triangle A'B'C'$ after a reflection in the y-axis is $\triangle A''B''C''$. In which quadrant is $\triangle A''B''C''$ located?
(1) I   (2) II   (3) III   (4) IV

10. In the accompanying diagram, K is the image of A after a translation. Under the same translation, which point is the image of J?

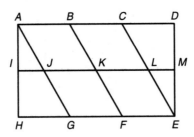

(1) B   (2) C   (3) E   (4) F

12.3 TRANSFORMATIONS IN THE COORDINATE PLANE

KEY IDEAS

There are special rules for determining the coordinates of the images of points of a figure that are reflected, translated, or dilated in the coordinate plane.

Reflection of a Point in the Origin

The image of $P(a, b)$ after a reflection in the origin is $P'(-a, -b)$, as shown in Figure 12.11. A reflection of a point in the origin is equivalent to a rotation of that point 180° about the origin.

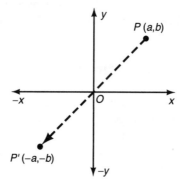

Figure 12.11   Reflection in the Origin

Reflections in a Coordinate Axis

To reflect a *point* in a coordinate axis, flip it over the axis so that the image is the same distance from the reflecting line as the original point, as shown in Figure 12.12. In general:

- The reflection of $P(a, b)$ in the x-axis is $P'(a, -b)$.
- The reflection of $P(a, b)$ in the y-axis is $P'(-a, b)$.

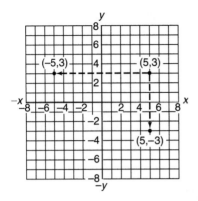

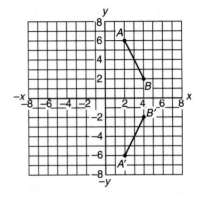

Figure 12.12 Reflecting a Point   **Figure 12.13** Reflecting a Line Segment

To reflect a *line segment* in a coordinate axis, flip it over the reflecting line by reflecting each endpoint in that line. If the endpoints of $\overline{AB}$ are $A(2, 6)$ and $B(4, 2)$, then, after a reflection of $\overline{AB}$ in the x-axis, the image is $\overline{A'B'}$ with endpoints $A'(2, -6)$ and $B'(4, -2)$, as shown in Figure 12.13.

Translations in the Coordinate Plane

Sliding a point $P(x, y)$ horizontally h units and then vertically k units places the image at $P'(x + h, y + k)$. Figure 12.14 illustrates a translation in which both h and k stand for positive numbers. The signs of h and k have the same meaning as the signs of x and y. For example, $P'(x + 1, y - 2)$ is a translation of $P(x, y)$. Since h is $+1$ and k is -2, point P is shifted 1 unit to the right and then 2 units down.

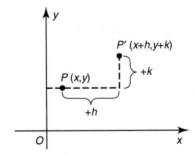

Figure 12.14 Translation of a Point

Example 1

The coordinates of the vertices of $\triangle ABC$ are $A(2, -3)$, $B(0, 4)$ and $C(-1, 5)$. If the image of point A under a translation is point $A'(0, 0)$, find the images of points B and C under this translation.

Solution: $\boldsymbol{B'(-2, 7)}$, $\boldsymbol{C'(-3, 8)}$

In general, after a translation of h units in the horizontal direction and k units in the vertical direction, the image of $P(x, y)$ is $P'(x + h, y + k)$. Since

$$A(2, -3) \to A'(2+h, -3+k) = A'(0, 0)$$

it follows that

$$2 + h = 0 \quad \text{and} \quad h = -2,$$
$$-3 + k = 0 \quad \text{and} \quad k = 3$$

Therefore:

$$B(0, 4) \to B'(0 + [-2], 4 + 3) = B'(-2, 7),$$
$$C(-1, 5) \to C'(-1 + [-2], 5 + 3) = C'(-3, 8)$$

Dilations in the Coordinate Plane

The image of $P(x, y)$ under a dilation with respect to the origin is $P'(cx, cy)$, where c is the constant of dilation, so $c \neq 0$. Figure 12.15 illustrates a dilation in which $c > 1$ so that $OP' > OP$.

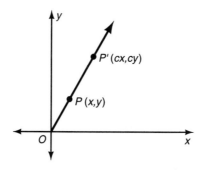

Figure 12.15   Dilation of a Point

Example 2

After a dilation with respect to the origin, the image of $A(2, 3)$ is $A'(4, 6)$. What are the coordinates of the point that is the image of $B(1, 5)$ after the same dilation?

Solution: $B'(2, 10)$

- Determine the constant of dilation. The constant of dilation is 2 since

$$A(2, 3) \rightarrow A'(\underline{2} \times 2, \underline{2} \times 3) = A'(4, 6)$$

- Under the same dilation, the x- and y-coordinates of point B are also multiplied by 2:

$$B(1, 5) \rightarrow B'(\underline{2} \times 1, \underline{2} \times 5) = B'(2, 10)$$

Transformations of Figures Using Coordinates

A transformation can be applied to a figure such as a triangle by applying the transformation to each point of the figure. For example, if the coordinates of $\triangle ABC$ are $A(0, 3)$, $B(4, -1)$, and $C(6, 4)$ then the coordinates of the vertices of $\triangle A'B'C'$, the image of $\triangle ABC$ after a dilation of 2, are $A'(0, 6)$, $B'(8, -2)$, and $C'(12, 8)$ since:

$$A(0, 3) \rightarrow A'(0 \times 2, 3 \times 2) = A'(0, 6)$$
$$B(4, -1) \rightarrow B'(4 \times 2, -1 \times 2) = B'(8, -2)$$
$$C(6, 4) \rightarrow C'(6 \times 2, 4 \times 2) = C'(12, 8)$$

Example 3

Graph $\triangle ABC$ with coordinates $A(1, 3)$, $B(5, 7)$, and $C(8, -3)$. On the same set of axes graph $\triangle A'B'C'$, the reflection of $\triangle ABC$ in the y-axis.

Solution: See the accompanying graph.

After graphing $\triangle ABC$, reflect points A, B, and C in the y-axis, as shown in the accompanying figure. Then connect the image points A', B', and C' with line segments.

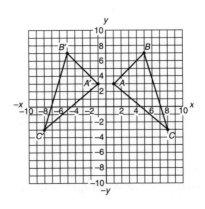

340

Check Your Understanding of Section 12.3

A. *Multiple Choice*

1. A translation moves $A(2, 3)$ onto $A'(4, 8)$. What are the coordinates of the point that is the image of $B(4, 6)$ under the same translation?
   (1) $(12, 18)$ (2) $(6, 8)$ (3) $(8, 12)$ (4) $(6, 11)$

2. Under the translation that maps point $(3, -4)$ onto its image, $(1, 0)$, what is the image of any point (x, y)?
   (1) $(x + 2, y + 4)$ (3) $(x + 2, y - 4)$
   (2) $(x - 2, y - 4)$ (4) $(x - 2, y + 4)$

3. After a reflection in the origin, the image of $A(3, -2)$ is
   (1) $A'(-2, 3)$ (2) $A'(-3, 2)$ (3) $A'(-3, -2)$ (4) $A'(3, 2)$

4. Which transformation represents a dilation?
   (1) $A(-3, 5) \rightarrow A'(-6, 10)$ (3) $A(-3, 5) \rightarrow A'(1, 9)$
   (2) $A(-3, 5) \rightarrow A'(5, -3)$ (4) $A(-3, 5) \rightarrow A'(-3, -5)$

5. If a point in Quadrant IV is reflected in the y-axis, its image will lie in Quadrant
   (1) I   (2) II   (3) III   (4) IV

6. A translation maps $A(1, 2)$ onto $A'(-1, 3)$. What are the coordinates of the image of the origin under the same translation?
   (1) $(0, 0)$ (2) $(2, -1)$ (3) $(-2, 1)$ (4) $(-1, 2)$

7. In which quadrant does the image of $(4, -7)$ lie after the translation that shifts (x, y) to $(x - 6, y + 3)$?
   (1) I   (2) II   (3) III   (4) IV

B. *In each case, show how you arrived at your answer by clearly indicating all of the necessary steps, formula substitutions, diagrams, graphs, charts, etc.*

8. What are the coordinates of the point that is the image of $Z(4, -1)$ after a reflection in point $(0, 3)$?

9. What are the coordinates of the point that is the image of $N(5, -3)$ under a reflection in the y-axis?

10. A translation maps $A(-3, 4)$ onto $A'(2, -6)$. Find the coordinates of the point that is the image of $B(-4, 0)$ under the same translation.

11. Using graph paper, graph $\triangle RST$ with coordinates $R(-2, 1)$, $S(2, 6)$ and $T(8, -4)$. On the same set of axes, graph $\triangle R'S'T'$, the image of $\triangle RST$ after a reflection in the x-axis.

12. The vertices of $\triangle ABC$ are $A(-4, 7)$, $B(3, -2)$, and $C(8, -2)$. After a translation that maps (x, y) onto $(x + h, y + k)$, $\triangle A'B'C'$ is the image of $\triangle ABC$.
    (a) If $\triangle A'B'C'$ lies completely in Quadrant I, what are the smallest possible integer values of h and k?
    (b) How many square units are in the area of $\triangle ABC$?
    (c) If $\triangle A''B''C''$ is the image of $\triangle ABC$ after a dilation using a scale factor of 2 with respect to the origin, how many square units are in the area of $\triangle A''B''C''$?

13. Using graph paper, graph $\triangle BIG$ with coordinates $B(-2, -1)$, $I(1, 4)$, and $G(5, -1)$.
    (a) On the same set of axes, graph $\triangle B'I'G'$, the image of $\triangle BIG$ after a dilation of scale factor 2.
    (b) What is the ratio of the area of $\triangle B'I'G'$ to the area of $\triangle BIG$?

14. Using graph paper, graph $\triangle ABC$ with coordinates $A(-1, -3)$, $B(6, 2)$ and $C(8, -5)$. On the same set of axes, graph:
    (a) $\triangle A'B'C'$, the image of $\triangle ABC$ after a reflection in the x-axis.
    (b) $\triangle A''B''C''$, the image of $\triangle A'B'C'$ after a reflection in the y-axis.

15. The coordinates of the endpoints of $\overline{AB}$ are $A(2, 8)$ and $B(6, -2)$. Shari claims that, when $\overline{AB}$ is reflected in the x-axis and its image, $\overline{A'B'}$, is dilated by $\frac{1}{2}$, the resulting image, $\overline{A''B''}$, is the same as when $\overline{AB}$ is first dilated by $\frac{1}{2}$ and then reflected in the x-axis. Is Shari correct? Justify your answer.

12.4 SIMPLE LOCUS

A **locus** may be thought of as a *path* consisting of the set of all points, and only those points, that satisfy one or more conditions. The plural of *locus* is *loci*.

An Example of Finding a Simple Locus

To find the *locus* of points that are 3 inches from point K:

- Make a diagram. Keep drawing points 3 inches from point K until you discover a pattern.

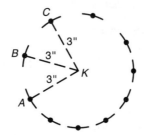

- Connect the points with a broken line or curve, as shown in the accompanying figure.
- Write a sentence that describes what you have discovered. For this example, write, "The locus of points that are 3 inches from point K is a circle with K as its center and a radius of 3 inches."

Five Basic Loci

The accompanying table summarizes the five basic loci that you should know.

THE FIVE BASIC LOCI

Locus	Diagram
1. **Condition:** Locus at a fixed distance d from a given point P. **Locus:** A circle with radius d and center P.	1.
2. **Condition:** Locus at a fixed distance d from a line ℓ. **Locus:** Two lines parallel to line ℓ and each at distance d from line ℓ.	2.
3. **Condition:** Locus equidistant from two points A and B. **Locus:** The perpendicular bisector of the segments whose endpoints are A and B.	3.
4. **Condition:** Locus equidistant from two parallel lines. **Locus:** A line parallel to the pair of lines and mid-way between them.	4.
5. **Condition:** Locus equidistant from two intersecting lines. **Locus:** The bisectors of each pair of vertical angles formed by the lines.	5.

Example

Find the locus of points:
(a) equidistant from two concentric circles having radii of lengths 3 and 7 centimeters, respectively
(b) equidistant from two parallel lines that are 8 inches apart

344

Solution: (a) **A concentric circle with radius 5 cm**

The locus of points equidistant from two concentric circles is another circle having the same center as the original circles and a radius length equal to 5 cm, which is the average of the lengths of the radii of the two given circles:

$$\frac{3 \text{ cm} + 7 \text{ cm}}{2} = 5 \text{ cm}$$

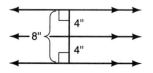

(b) **A line midway between the parallel lines**

The locus is a line that is parallel to the given lines and is at a distance of 4 in from each of them.

Check Your Understanding of Section 12.4

In each case, show how you arrived at your answer by clearly indicating all of the necessary steps, formula substitutions, diagrams, graphs, charts, etc.

1–5. Describe each locus.

1. The locus of points 1 centimeter from a circle whose radius is 8 centimeters.

2. The locus of points 3 centimeters from a given line.

3. The locus of points that are equidistant from sides $\overline{AB}$ and $\overline{AC}$ of $\triangle ABC$.

4. The locus of points equidistant from two circles having the same center and radii of 7 centimeters and 11 centimeters.

5. The locus of the center of a coin 1 inch in diameter that rolls along a flat table in a straight line.

6–8. Write an equation of the line (or lines) that satisfies each of the following conditions:

6. All points 5 units from the line $x = 3$.

7. All points 2 units from the line $y = -1$.

8. All points equidistant from the lines $y = 3x - 1$ and $y = 3x + 9$.

12.5 SATISFYING MORE THAN ONE LOCUS CONDITION

KEY IDEAS

To find the number of points that satisfy more than one locus condition, use the same diagram to describe each locus condition. Then locate the points, if any, at which all of the loci intersect.

An Example of More Than One Locus Condition

Suppose point P is on line ℓ. To find the number of points that are 2 inches from line ℓ (first locus condition) and also 2 inches from point P (second locus condition), proceed as follows:

- Draw a diagram of the first locus. The locus of points that are 2 inches from line ℓ, as shown in the accompanying diagram, is a pair of parallel lines on either side of line ℓ, and at a distance of 2 inches from line ℓ.

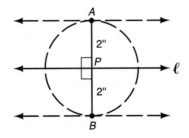

- Using the same diagram, draw the second locus. The locus of points 2 inches from point P is a circle that has P as its center and a radius of 2 inches.
- Count the number of points at which the two loci intersect. Since the two loci intersect at points A and B, there are two points that satisfy both of the given locus conditions.

Example 1

Tom's backyard has two grown trees that are 40 feet apart, as shown in the accompanying diagram. Tom wants to plant new trees that are 30 feet from each grown tree. In how many different locations could Tom plant the new trees?

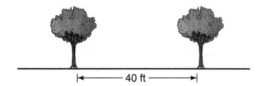

Solution: **2**

The locus of points 30 ft from each grown tree is a circle having that tree as its center and a radius of 30 ft, as shown in the accompanying diagram. The two points at which the two circles intersect represent the two possible locations for the new trees.

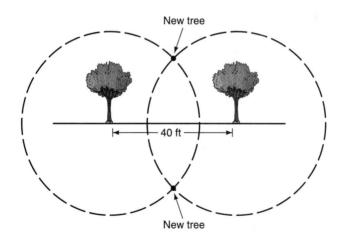

Example 2

Two parallel lines are 8 inches apart. Point P is located on one of the lines. What is the number of points that are equidistant from the parallel lines and also 4 inches from P?

Solution: **1**

Since the circle has a radius of 4 in, it is tangent to the line that is midway between the original pair of parallel lines. The loci intersect at point A, the only point that satisfies both conditions.

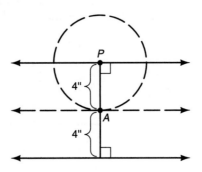

Example 3

In Example 2 what is the number of points that are equidistant from the parallel lines and also 3 inches from P?

Solution: **0**

Since the circle has a radius of 3 in, it does *not* intersect the line that is midway between the original pair of parallel lines. Since the loci do *not* intersect, no point satisfies both conditions.

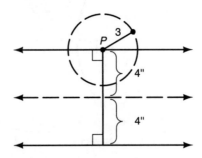

Example 4

How many points are 3 units from the origin and 2 units from the *y*-axis?

Solution: **4**

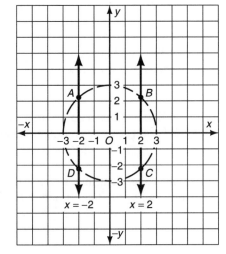

- *Locus condition 1:* All points 3 units from the origin. The desired locus is a circle with the origin as its center and a radius of 3 units. Note in the accompanying diagram that the circle intersects each coordinate axis at 3 and −3.
- *Locus condition 2:* All points 2 units from the *y*-axis. The desired locus is a pair of parallel lines; one line is 2 units to the right of the *y*-axis ($x = 2$), and the other line is 2 units to the left of the *y*-axis ($x = -2$). See the accompanying diagram.
- Since the loci intersect at points *A*, *B*, *C*, and *D*, there are 4 points that satisfy both conditions.

Check Your Understanding of Section 12.5

A. *Multiple Choice*

1. How many points are equidistant from points *A* and *B* and also 4 inches from $\overline{AB}$?
   (1) 1   (2) 2   (3) 3   (4) 4

2. How many points are equidistant from two intersecting lines and also 3 inches from the point of intersection of the lines?
   (1) 1   (2) 2   (3) 3   (4) 4

3. How many points are 4 units from the origin and also 4 units from the *x*-axis?
   (1) 1   (2) 2   (3) 3   (4) 4

4. How many points are equidistant from points *P*(2, 1) and *Q*(2, 5) and also 3 units from the origin?
   (1) 1   (2) 2   (3) 3   (4) 4

5. What is an equation of the locus of points equidistant from points $A(3, 1)$ and $B(7, 1)$?
   (1) $x = 5$ (2) $y = 5$ (3) $y = x + 5$ (4) $x^2 + y^2 = 5$

6. The distance between points P and Q is 9 inches. How many points are equidistant from P and Q and also 4 inches from Q?
   (1) 1   (2) 2   (3) 0   (4) 4

7. The distance between points P and Q is 9 inches. How many points are equidistant from P and Q and also 5 inches from P?
   (1) 1   (2) 2   (3) 0   (4) 4

8. The distance between points P and Q is 8 inches. How many points are equidistant from P and Q and also 4 inches from P?
   (1) 1   (2) 2   (3) 0   (4) 4

9. Point A is 4 centimeters from line k. How many points are 1 centimeter from line k and 3 centimeters from A?
   (1) 1   (2) 2   (3) 0   (4) 4

10. What is an equation of the locus of all points whose ordinates exceed twice their abscissas by 3?
    (1) $x = 2y + 3$ (2) $x + 3 = 2y$ (3) $y = 2x + 3$ (4) $y + 3 = 2x$

B. *In each case, show how you arrived at your answer by clearly indicating all of the necessary steps, formula substitutions, diagrams, graphs, charts, etc.*

11. A tree is located 30 feet east of a fence that runs north to south. Kelly tells her brother Billy that their dog buried Billy's hat a distance of 15 feet from the fence and also 20 feet from the tree.
    (a) Draw a sketch to show where Billy should dig to find his hat.
    (b) How many locations for the hat are possible?

12. Point P is x inches from line ℓ. If there are exactly 3 points that are 2 inches from line ℓ and also 6 inches from P, what is the value of x?

13. Point P is located on $\overleftrightarrow{AB}$.
    (a) Describe the locus of points that are:
        (1) 3 units from $\overleftrightarrow{AB}$ (2) 5 units from point P
    (b) How many points satisfy both conditions in part (a)?

14. Lines AB and CD are parallel to each other and 6 inches apart. Point P is located between the two parallel lines and 1 inch from $\overleftrightarrow{AB}$.
   (a) Describe the locus of points that are:
       (1) equidistant from $\overleftrightarrow{AB}$ and $\overleftrightarrow{CD}$ (2) 2 inches from point P
   (b) How many points satisfy both conditions in part (a)?

15. (a) Describe completely the locus of points 2 units from the line whose equation is $x = 3$.
   (b) Describe completely the locus of points n units from point $P(3, 2)$.
   (c) What is the total number of points that satisfy the conditions in parts (a) and (b) simultaneously for $n = 2$?

CHAPTER 13

SYSTEMS OF LINEAR EQUATIONS AND INEQUALITIES

13.1 SOLVING LINEAR SYSTEMS GRAPHICALLY

=== **KEY IDEAS** ===

Solving a system of equations or inequalities means finding the set of all ordered pairs of numbers that make each equation or inequality true at the same time. A system of equations can be solved either graphically or algebraically.

Solving a System of Linear Equations Using Graph Paper

To solve the system

$$2x + y = 5$$
$$y - x = -4$$

graph each equation on graph paper using the same set of axes. Write $2x + y = 5$ as $y = -2x + 5$, and $y - x = -4$ as $y = x - 4$, so that each equation can be graphed using the slope-intercept method. As shown in Figure 13.1, the solution of the system of equations is $(3, -1)$, the point at which the two lines intersect.

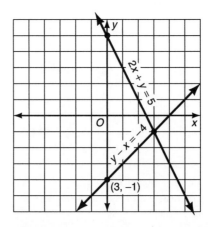

Figure 13.1 Solution of a System of Linear Equations

You can check algebraically that $(3, -1)$ is the solution point by verifying that $x = 3$ and $y = -1$ make both of the *original* equations true at the same time:

$$\begin{array}{c|c}
2x + y = 5 & y - x = -4 \\
\hline
2(3) + (-1) \overset{?}{=} 5 & -1 - (3) \overset{?}{=} -4 \\
6 - 1 \overset{?}{=} 5 & -4 \overset{\checkmark}{=} -4 \\
5 \overset{\checkmark}{=} 5 &
\end{array}$$

Solving a System of Linear Equations Using a Graphing Calculator

If you have the Texas Instruments TI-83 graphing calculator, you can solve a system of linear equations such as $2x + y = 5$ and $y - x = -4$ by entering the equations in the $\boxed{Y=}$ editor and then using the special features of the graphing calculator to find the coordinates of the point of intersection of the graphs.

Method 1: Use the TRACE function.

- Using paper and pen, solve each equation for y in terms of x. The result is $y = -2x + 5$ and $y = x - 4$.
- Press the $\boxed{Y=}$ key. On the line that begins $\backslash Y_1 =$, enter the right side of the equation $y = -2x + 5$. Then press the down arrow key to get to the line that begins $\backslash Y_2 =$. On this line enter the right side of the equation $y = x - 4$.
- Press the $\boxed{\text{ZOOM}}$ key, followed by the $\boxed{4}$ key. Then press the $\boxed{\text{GRAPH}}$ key.
- Find the coordinates of the point of intersection of the two lines by pressing the $\boxed{\text{TRACE}}$ key. When the $\boxed{\text{TRACE}}$ key is pressed, the TRACE cursor appears on the graph of the first equation in the $\boxed{Y=}$ menu. As you press the right or left arrow key, the TRACE cursor moves along the line $y = -2x + 5$. The coordinates of the TRACE cursor appear at the bottom of the screen and are updated as the cursor moves along the graph. When the TRACE cursor is on the point of intersection of the two lines, the coordinates displayed are $x = 3$, $y = -1$, as shown in Figure 13.2.

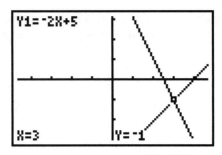

Figure 13.2 Using the ⌈TRACE⌉ Key

- Verify that the coordinates of the point of intersection are correct by using the up and down arrow keys to toggle between the two lines at the point of intersection. When toggling between the two lines, the equation displayed alternates between $Y_1 = -2X + 5$ and $Y_2 = X - 4$ while the trace coordinates stay fixed at $(3, -1)$. This confirms that $(3, -1)$ is a point on both curves and, as a result, is the solution to the system of equations that was graphed.

Method 2: Use the **intersect** function in the CALCULATE menu.

After graphing the two equations, find the coordinates of the point of intersection of the two lines by selecting **intersect** in the CALCULATE menu.

- To view the CALCULATE menu, press the ⌈2nd⌉ key followed by the ⌈TRACE⌉ key, which has the CALC label above it. Press 5 to select the **intersect** function. The bottom left corner of the window now displays the prompt **First curve?**
- Press the up or down arrow key to move the cursor to the first graph, if necessary. Then press the ⌈ENTER⌉ key. The bottom left corner of the window now displays the prompt **Second curve?**
- Press the up or down arrow key to move the cursor to the second graph, if necessary. Then press the ⌈ENTER⌉ key. The bottom left corner of the window now displays the prompt **Guess?**
- Press the right or left arrow key to move the cursor to a "guess" point that is close to the point of intersection. Then press the ⌈ENTER⌉ key. The cursor will move, if necessary, to the point of intersection. The bottom left corner of the window now displays **Intersection** with the x- and the y-coordinates of the point of intersection appearing below it.

Solving a System of Linear Inequalities Graphically

A line whose equation has the form $y = mx + b$ divides the coordinate plane into two regions, as shown in Figure 13.3.

- The region *above* the line represents the solution set of $y > mx + b$.
- The region *below* the line represents the solution set of $y < mx + b$.

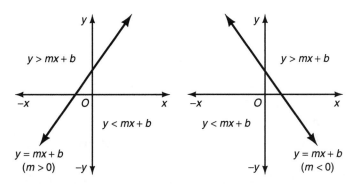

Figure 13.3 Solution of a Linear Inequality

To find the set of all ordered pairs that satisfy two linear inequalities simultaneously, graph each inequality on the same set of axes. Then determine the region in which the two solution sets overlap. For example, to solve the system of inequalities

$$y > 3x - 4$$

$$x + 2y \leq 6$$

graphically, proceed as follows:

- To graph $y > 3x - 4$, first graph the boundary line, $y = 3x - 4$. Since the solution set of $y > 3x - 4$ does *not* include the points on the boundary line, draw the graph of $y = 3x - 4$ as a *broken* line, as shown in Figure 13.4. Also, since the solution set of $y > 3x - 4$ lies *above* the line $y = 3x - 4$, shade in this region. To check that the points that lie in the shaded region make $y > 3x - 4$ a true statement, pick a test point from this region, say $(0, 0)$, and substitute the coordinates of the test point into the inequality. Since $0 > 3(0) - 4$ simplifies to $0 > -4$, which is a true statement, the region on the side of the boundary line that contains $(0, 0)$ represents the solution set of $y > 3x - 4$.

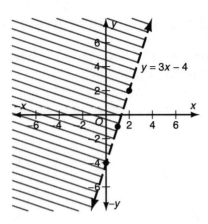

Figure 13.4 Graph of $y > 3x - 4$

- To graph $x + 2y \leq 6$, first solve for y in terms of x: $y \leq -\dfrac{1}{2}x + 3$. Then graph the boundary line, $y = -\dfrac{1}{2}x + 3$. Since the solution set of $y \leq -\dfrac{1}{2}x + 3$ includes the points on the boundary line, draw the graph of $y = -\dfrac{1}{2}x + 3$ as a *solid* line, as shown in the Figure 13.5. Since the solution set of $y \leq -\dfrac{1}{2}x + 3$ lies *below* the line $y = -\dfrac{1}{2}x + 3$, shade in the region. Check that you have shaded in the correct side of the boundary line by using (0, 0) as a test point. Since $0 \leq -\dfrac{1}{2}(0) + 3$ simplifies to $0 \leq 3$, which is a true statement, the region on the side of the boundary line $y = -\dfrac{1}{2}x + 3$ that includes (0, 0) represents the solution set of y $y \leq -\dfrac{1}{2}x + 3$.

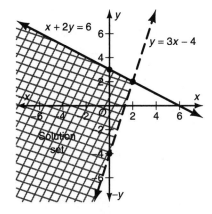

Figure 13.5

- Label the cross-hatched region in which the solution sets of the two inequalities overlap as the solution set of the system of inequalities.

If you have the TI-83 graphing calculator, you can easily confirm your solution by entering the equations of the boundary lines, $y = 3x - 4$ and $y = -\dfrac{1}{2}x + 3$, in the $\boxed{Y=}$ editor and then shading in the appropriate sides of the boundary lines as follows:

- In the $\boxed{Y=}$ editor, move the cursor over the backward slash at the left of Y_1. Then press the $\boxed{ENTER}$ key until a darkened right triangle appears *above* the slash. This shades the points *above* (>) the graph of $y = 3x - 4$.

- In the $\boxed{Y=}$ editor, move the cursor over the backward slash at the left of Y_2. Then press the $\boxed{ENTER}$ key until a darkened right triangle appears *below* the slash. This shades the points *below* (<) the graph of $y = -\dfrac{1}{2}x + 3$

- Press the $\boxed{GRAPH}$ key.

Check Your Understanding of Section 13.1

A. Multiple Choice

1. Which ordered pair is in the solution set of $x < 6 - y$?
   (1) $(0, 6)$ (2) $(6, 0)$ (3) $(0, 5)$ (4) $(7, 0)$

2. Which ordered pair is *not* a member of the solution set of $2x - 3y \geq 12$?
   (1) $(0, -4)$ (2) $(-3, -6)$ (3) $(10, 3)$ (4) $(6, 0)$

3. The accompanying diagram shows the graph of which inequality?

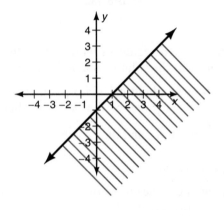

     (1) $y > x - 1$ (2) $y \geq x - 1$ (3) $y < x - 1$ (4) $y \leq x - 1$

B. In each case, show how you arrived at your answer by clearly indicating all of the necessary steps, formula substitutions, diagrams, graphs, charts, etc. If you use a graphing calculator, show each of the following: (1) a sketch of the viewing window; (2) scales indicated on the x- and y-axes; (3) clearly labeled x- and y-intercepts and point(s) of intersection.

4 and 5. Solve each system of equations graphically.

4. $2x + 2y = -10$

   $3x - \dfrac{1}{2}y = 6$

5. $2x = 8 - 5y$

   $x + y = 1$

6. (a) On the same set of coordinate axes, graph the following equations:
        (1) $x + y = 7$ (2) $3y - 2x = 6$ (3) $y = -2$

   (b) Find the area of the triangle formed by the lines graphed in part (a).

7. (a) On the same set of coordinate axes, graph the following system of inequalities:

$$3y \geq 2x - 6$$
$$x + y > 7$$

(b) Which ordered pair is in the solution set of the graph drawn in answer to part (a)?
(1) (3, 4)　　　(2) (8, 1)　　　(3) (2, 6)　　　(4) (−2, 7)

8. Solve the equation $1.5x - 5.7 = 4.8 - 2x$ by graphing two lines using a graphing calculator.

9. Marisa's music CD collection includes jazz CDs and classical CDs. She has, at most, 14 jazz CDs and, at most, 8 classical CDs. The combined number of jazz and classical CDs she has is not greater than 18. If x is the number of jazz CDs and y is the number of classical CDs, graph the region that contains the number of jazz and classical CDs in Marisa's music CD collection.

10. (a) On the same set of axes, graph the lines $y = -4$ and $2x + y = 6$.
(b) Find the area, in square units, of the trapezoid in Quadrant IV bounded by the x-axis, the y-axis, and the lines graphed in part (a).

11. (a) On the same set of axes, graph the lines $y = 2$, $y = 6$, $y = 2x + 12$, and $y = 2x - 12$.
(b) Find the area, in square units, of the parallelogram formed by these lines.

12. Two video rental stores offer two different membership plans:
Store A charges $15 for membership and $2 for each video that is rented.
Store B does not have a membership fee but charges $3.50 for each video that is rented.
(a) Write an equation that shows the total cost, C, of renting n videos from store A.
(b) Write an equation that shows the total cost, C, or renting n videos from store B.
(c) Graph the equations in part (a) and part (b) on the same set of axes.
(d) According to the graphs drawn in part (c), for what number of video rentals is the cost of membership in the two clubs the same?

13.2 SOLVING LINEAR SYSTEMS BY SUBSTITUTION

KEY IDEAS

To solve a system of linear equations algebraically, you must reduce the system to a single equation with only one variable. For example, if the original system of equations is

$$\left. \begin{array}{l} y = 2x \\ 3x + y = 10 \end{array} \right\} \text{System}$$

then *substituting* $2x$ for y in the second equation reduces the original system of two equations to the single equation $3x + 2x = 10$, which means that $5x = 10$, so $x = \dfrac{10}{5} = 2$.

To find the solution for y, substitute 2 for x in either of the two original equations. Since the first equation is simpler than the second equation, replace x with 2 in the equation $y = 2x$; then $y = 2(2) = 4$. The solution for the system is (2, 4).

Using the Substitution Method

Before using the substitution method, in one of the equations you may have to solve for one variable in terms of the other variable.

Example 1

Solve the following system of equations:

$$x + 2y = 7$$
$$y - 1 = 2x$$

Solution: **(1, 3)**
- Solve the second equation for y:

$$y = 2x + 1$$

- Eliminate y in the first equation by substituting $2x + 1$ for y:

$$x + \quad 2y = 7$$
$$x + 2(2x + 1) = 7$$
$$x + \quad 4x + 2 = 7$$
$$5x + 2 = 7$$
$$5x = 5$$
$$x = \frac{5}{5} = 1$$

- Find the value of y when $x = 1$ by substituting 1 for x in either of the two original equations:

$$x + 2y = 7$$
$$1 + 2y = 7$$
$$2y = 6$$
$$y = \frac{6}{2} = 3$$

- Check that the solution ($x = 1$, $y = 3$) works in each of the two original equations:

$$
\begin{array}{c|c}
x + 2y = 7 & y - 1 = 2x \\
1 + 2(3) \overset{?}{=} 7 & 3 - 1 \overset{?}{=} 2(1) \\
1 + 6 \overset{\checkmark}{=} 7 & 2 \overset{\checkmark}{=} 2
\end{array}
$$

TIP

You could also have solved this system of equations by solving the first equation for x: $x = 7 - 2y$, and then using that expression to eliminate x in the second equation.

Solving Word Problems

When a word problem involves two conditions, you may be able to translate each condition into an equation. The numbers that satisfy both equations at the same time represent the solution of the word problem.

Example 2

Smithtown Recreation Department ordered a total of 100 balls and bats for the summer baseball camp. Balls cost $4.50 each, and bats cost $20.00 each. The total purchase cost was $822. How many of each item were ordered?

Solution: **76 balls, 24 bats**

Method I: Create an algebraic model that uses two variables.

Let x = number of balls ordered, and
 y = number of bats ordered. Then:
   Condition 1:   The total number or balls and bats is 100.
   Hence:

$$x + y = 100 \qquad \text{[Condition 1]}$$

Condition 2:   The total purchase cost is $822 when one ball costs $4.50 and one bat costs $20.00. Hence:

$$\underbrace{\$4.50x}_{\text{Cost of } x \text{ balls}} + \underbrace{\$20.00y}_{\text{Cost of } y \text{ balls}} = \$822 \qquad \text{[Condition 2]}$$

- Solve the system of two equations in two unknowns:

$$\left.\begin{array}{c} x + y = 100 \\ \$4.50x + \$20.00y = \$822 \end{array}\right\} \text{System}$$

Since $x + y = 100$, $y = 100 - x$:

$$\$4.50x + \$20(100 - x) = \$822$$

- Remove the parentheses:

$$\$4.50x + \$2000 - \$20x = \$822$$

- Combine like terms:

$$-\$15.50x = \$822 - \$2000$$

- Solve for x:

$$x = \frac{-\$1178}{-\$15.50} = 76$$

- Find y:

$$y = 100 - x = 100 - 76 = 24$$

Method II: Create an algebraic model that uses one variable.

If x = number of balls, then the difference $100 - x$ must represent the number of bats, so $\$4.50x + \$20(100 - x) = \$822$. Thus,

$$\$4.50x + \$2000 - \$20x = \$822$$

$$\$1178 = \$15.50x$$

$$\frac{\$1178}{\$15.50} = x$$

$$76 = x$$

Method III: Guess, check, and revise.

Start with the same number, say 50, of balls and bats. If this guess leads to a cost that is too high, reduce the number of bats.

Guesses		Total Cost
Balls	Bats	
50	50	$(50 \times 4.50) + (50 \times 20)$ = 1225 ← Too high
70	30	$(70 \times 4.50) + (30 \times 20)$ = 915 ← Too high
80	20	$(80 \times 4.50) + (20 \times 20)$ = 760 ← Too low
75	25	$(75 \times 4.50) + (25 \times 20)$ = 837.5 ← Too high
76	**24**	$(76 \times 4.50) + (24 \times 20)$ = 822 ← This is the answer!

TIP

Method III can be helpful when you know that the solutions must be whole numbers.

Check Your Understanding of Section 13.2

A. Multiple Choice

1. What is the solution set for the following system of equations?

$$x = -y$$
$$x + 2y = 6$$

(1) $(-2, 2)$ (2) $(2, -2)$ (3) $(6, -6)$ (4) $(-6, 6)$

2. What is the solution for x in the following system of equations?

$$-y = 2x - 3$$
$$y = -x + 1$$

(1) $\dfrac{2}{3}$ (2) 2   (3) $\dfrac{4}{3}$ (4) 4

3. What is the solution for y in the following system of equations?

$$\frac{3}{4} = \frac{2}{x}$$
$$6x + y = 17$$

(1) 1   (2) −1   (3) 9   (4) −9

B. *In each case, show how you arrived at your answer by clearly indicating all of the necessary steps, formula substitutions, diagrams, graphs, charts, etc.*

4. Solve the following system of equations for y:

$$2x - y = 13$$
$$2x = y$$

5–10. Solve each system of equations algebraically, and check.

5. $y + 5x = 0$
$3x - 2y = 26$

6. $\dfrac{y}{4} - x = 0$
$2y = 3x + 7$

7. $\dfrac{a - 2}{3} = b$
$2a + 3b = -5$

8. $y - 1 = x$
$7y - 3x = -1$

9. $y - x = 1$
$7y - 11x = -5$

10. $0.3y - 0.2x = 2.1$
$0.75y = 2.25x$

11. Five pens of the same type cost the same as two notebooks of the same type. If one pen and two notebooks cost $4.20, what is the cost of one pen?

12. Two angles are complementary. The difference between four times the measure of the smaller angle and the measure of the larger angle is 10. What is the measure of the larger angle?

13. The denominator of a fraction exceeds its numerator by 7. If the numerator is increased by 3 and the denominator is decreased by 2, the new fraction equals $\frac{4}{5}$. Find the original fraction.

14. A club consisting of seniors and juniors has 15 members. After seven more seniors and three more juniors join the club, the ratio of juniors to seniors is 2 to 3. How many juniors are now in the club?

15. (a) If t represents the tens digit of a two-digit number and u represents the units digit, what is the two-digit number in terms of t and u?
(b) The units digit of a two-digit number is 1 more than the tens digit. The two-digit number is 2 less than five times the sum of its digits. What is the number?

16. The length of a side of an equilateral triangle exceeds twice the length of a side of a smaller equilateral triangle by 1. If the difference in the perimeters of the triangles is 15, what is the perimeter of the smaller equilateral triangle?

17. A store charges a fixed amount to fax one page and another amount to fax each additional page that is faxed to the same telephone number. If the cost of faxing 5 pages is $3.05 and the cost of faxing 13 pages is $6.65:
(a) What is the cost of faxing only one page?
(b) What is the cost of faxing each page after the first?

18. In $\triangle ABC$, the measure of $\angle B$ is twice the measure of $\angle A$. If the measure of $\angle A$ is subtracted from the measure of $\angle C$, the difference is 20. Find the measure of $\angle C$.

19. There were 100 more balcony tickets than main-floor tickets sold for a concert. The balcony tickets sold for $4, and the main-floor tickets sold for $12. The total amount of sales for both types of tickets was $3056.
(a) Write a system of two equations that describes the given conditions. Define the variables.
(b) Using the system of equations in part (a), find the number of balcony tickets that were sold.

20. The senior class at Northwest High School needed to raise money for the yearbook. A local sporting goods store donated hats and T-shirts. The number of T-shirts was three times the number of hats. The seniors charged $5 for each hat and $8 for each T-shirt. If the seniors sold every item and raised $435, what were the total number of hats and the total number of T-shirts that were sold?

21. In the accompanying diagram, $ABCD$ is a parallelogram with $m\angle B = 120$, $m\angle D = 2x + 5y$, and $m\angle A = 4x + y$. Find the values of x and y. Check your solution. [*Only an algebraic solution will be accepted.*]

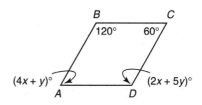

365

13.3 SOLVING LINEAR SYSTEMS BY ADDITION

KEY IDEAS

The following system of linear equations:

$$2x - 9y = 17$$
$$5x + 9y = 11$$

is difficult to solve using substitution, but can be easily solved by *adding* corresponding sides of the equations.

$$2x - 9y = 17$$
$$5x + 9y = 11$$
$$\overline{7x + 0 = 28}$$
$$x = \frac{28}{7} = 4$$

The new equation does not contain variable *y* since the coefficients of *y* in the original equations are additive inverses (opposites), so their sum is 0.

Using the Addition Method

Another way of solving a system of two first-degree equations in two variables is to eliminate one of the variables by adding corresponding sides of the two equations. Before adding the two equations, it may be necessary to:
- Write each equation in the form $Ax + By = C$.
- Multiply one or both equations by a number that will make the numerical coefficients of either the *x*-terms or the *y*-terms in the two equations have opposite values.

Example 1

Solve for *x* and *y*:

$$3y = 2x + 1$$
$$5y - 2x = 7$$

Solution: **(4, 3)**

- Put the first equation in the form $Ax + By = C$, and write it above the second equation with like terms aligned in the same columns:

$$3y - 2x = 1$$
$$5y - 2x = 7$$

- Multiply each term of the first (or the second) equation by -1 so that the coefficients of the x-terms will be additive inverses:

$$(-1)\{3y - 2x = 1\} \Rightarrow -3y + 2x = -1$$
$$5y - 2x = 7 \Rightarrow 5y - 2x = 7$$

- Add the two equations to eliminate the variable x:

$$\begin{array}{r} -3y + 2x = -1 \\ 5y - 2x = 7 \\ \hline 2y + 0 = 6 \end{array}$$

$$y = \frac{6}{2} = 3$$

- Find the corresponding value of x by substituting 3 for y in either of the two original equations:

$$5y - 2x = 7$$
$$5(3) - 2x = 7$$
$$15 - 2x = 7$$
$$-2x = -8$$
$$x = \frac{-8}{-2} = 4$$

Example 2

Solve for x and y:

$$3x + 4y = 9$$
$$5x + 6y = 13$$

Solution: **(−1, 3)**

Although both equations are already in the form $Ax + By = C$, the coefficients of the x-terms and the y-terms in the two equations are not opposites.

- To eliminate x, multiply the first equation by 5 and multiply the second equation by -3. The result is an equivalent system of two equations in which the coefficient of the x-term of the first equation is 15 and the coefficient of the x-term in the second equation is -15:

$$5\{3x+4y=9\} \Rightarrow \quad 15x+20y=45$$
$$-3\{5x+6y=13\} \Rightarrow -15x-18y=-39$$
$$2y=6$$
$$y=\frac{6}{2}=3$$

- To find x, substitute 3 for y in either of the two original equations:

$$3x+ 4y=9$$
$$3x+4(3)=9$$
$$3x+ 12=9$$
$$3x=-3$$
$$x=\frac{-3}{3}=-1$$

TIP

To eliminate y in the original system of equations, multiply the first equation by 3 and the second equation by -2. The result is two equivalent equations in which the coefficients of the y-terms are 12 and -12.

Solving Word Problems

Some word problems can be solved by writing either one equation with one unknown or two equations with two unknowns, or by using a guess-and-check method.

Example 3

Farmer Gray has only chickens and cows in his barnyard. If the animals in his barnyard have a total of 60 heads and 140 legs, how many chickens and how many cows are in the barnyard?

Solution: **50 chickens, 10 cows**

Method I: Create an algebraic model that uses two variables.

Let x = number of chickens, and
$\quad y$ = number of cows.

- Since there are 60 heads, the sum of the number of cows and number of chickens is 60. Hence, $x + y = 60$.
- A chicken has 2 legs, and a cow has 4 legs. Hence, $2x + 4y = 140$.
- Solve the system of equations by multiplying the first equation by -2 and then adding corresponding sides of the two equations, thereby eliminating x:

$$-2\{x + y = 60\} \Rightarrow -2x - 2y = -120$$
$$2x + 4y = 140 \Rightarrow -2x + 4y = 140$$
$$\overline{\, 2y = 20}$$
$$y = \frac{20}{2} = 10$$

- To solve for x, substitute 10 for y in the first equation: $x + 10 = 60$, so $x = 50$.

<u>Method II: Create an algebraic model that uses one variable.</u>

Let $\qquad x =$ number of chickens.
$\quad$ Then $60 - x =$ number of cows.
Hence:

$$2x + 4(60 - x) = 140$$
$$2x + 240 - 4x = 140$$
$$-2x + 240 = 140$$
$$-2x = -100$$
$$x = \frac{-100}{-2} = 50$$

Since there are 50 chickens, there are $60 - 50 = 10$ cows.

<u>Method III: Guess, check, and revise.</u>

Number of		Total Number of Legs
Chickens	Cows	
0	60	$(0 \times 2) + (60 \times 4) = 240 \quad \leftarrow \quad$ Too high
30	30	$(30 \times 2) + (30 \times 4) = 180 \quad \leftarrow \quad$ Too high
40	20	$(40 \times 2) + (20 \times 4) = 160 \quad \leftarrow \quad$ Too high
50	**10**	$(50 \times 2) + (10 \times 4) = 140 \quad \leftarrow \quad$ This is the answer!

Example 4

Carla and Steve went shopping at Office Plus. Carla bought 3 boxes of computer disks and 2 notebooks for a total cost of $15.00 excluding tax. Steve bought 2 boxes of the same computer disks and 5 of the same notebooks for a total cost of $18.25 excluding tax. Find the cost of 1 box of computer disks.

Solution: **$3.50**

To solve this problem algebraically, use two equations with two unknowns.

Let x = price of 1 box of computer disks, and
   y = price of 1 notebook.

- Write a system of two equations with two unknowns:

$$\text{Carla's purchase: } 3x + 2y = \$15.00$$
$$\text{Steve's purchase: } 2x + 5y = \$18.25$$

- Eliminate y. Multiply each member of the first equation by -5, and multiply each member of the second equation by 2 so that the coefficients of y are additive inverses. Then add corresponding sides of the two equations.

$$\begin{aligned}
-15x - 10y &= -\$75.00 \\
\underline{4x + 10y} &= \underline{\ \ \$36.50} \\
-11x\ \ \ \ \ \ &= -\$38.50
\end{aligned}$$

- Solve for x:

$$x = \frac{-\$38.50}{-11} = \$3.50$$

Check Your Understanding of Section 13.3

A. Multiple Choice

1. Which ordered pair is the solution to the following system of equations?

$$2x - y = 10$$
$$x + y = 2$$

   (1) $(4, -2)$ (2) $(4, 2)$ (3) $(2, -4)$ (4) $(-4, 2)$

2. At what point do the graphs of the equations $2x + y = 8$ and $x - y = 4$ intersect?

   (1) $(0, 4)$ (2) $(4, 0)$ (3) $(-4, 0)$ (4) $(5, -2)$

3. What is the solution for y in the following system of equations?

$$-2x + y = -3$$
$$x + y = 1$$

(1) $\dfrac{2}{3}$ (2) $-\dfrac{1}{3}$ (3) $\dfrac{4}{3}$ (4) $\dfrac{7}{3}$

B. In each case, show how you arrived at your answer by clearly indicating all of the necessary steps, formula substitutions, diagrams, graphs, charts, etc.

4. Solve the following system of equations for y:

$$2x + y = 10$$
$$3x - y = 15$$

5. Solve the following system of equations for x:

$$2x + y = 12$$
$$3y = 2x - 4$$

6–11. Solve each system of equations algebraically, and check.

6. $2x + 3y = -6$
$5x + 2y = 7$

7. $2x = 5y + 8$
$3x + 2y = 31$

8. $3y = 2x - 6$
$x + y = 8$

9. $2x + 3y = 17$
$3x - 2y = -0.5$

10. $0.4a + 1.5b = -1$
$1.2a - b = 8$

11. $\dfrac{2}{3}x + y = 13$
$-x + 2y = 5$

12. Three shirts and two neckties cost $69. At the same prices, two shirts and three neckties cost $61. What is the combined cost of one shirt and one necktie?

13. The difference between two positive numbers is 9. If four times the larger number is ten times the smaller number, what is the larger number?

14. The sophomore class at South High School raised $860 from the sale of tickets to a concert. Tickets sold for $2.50 if purchased in advance and for $4.00 if purchased at the door. If a total of 275 tickets was sold, how many tickets were sold at the door?

15. Three bags of potatoes and four cases of corn cost $40. Five bags of potatoes and two cases of corn cost $34. Find the cost of:
   (a) one bag of potatoes
   (b) one case of corn

16. In $\triangle DEF$, the measure of $\angle D$ is 1 more than twice the measure of $\angle E$, and the measure of $\angle F$ is 7 less than the measure of $\angle D$. Find the measure of the largest angle of the triangle.

17. In the accompanying diagram, lines $\overleftrightarrow{AB}$ and $\overleftrightarrow{CD}$ intersect at E, m$\angle AED$ = 110, m$\angle DEB = 3x + 2y$, m$\angle BEC = 9x + y$, and m$\angle CEA = 70$. Find the values of x and y. Check your answer.

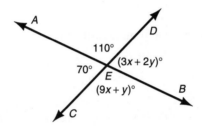

18. The accompanying table shows the number of grams of protein and the number of calories in a single serving of bran flakes cereal and milk. How many servings of (a) cereal and (b) milk are needed to get a total of 35 grams of protein and 470 calories?

Food	Protein	Calories
Cereal	5 g	90
Milk	8 g	80

19. Billy rented a sprayer and a generator. On his first job, he used each piece of equipment for 6 hours at a total cost of $90. On his second job, he used the sprayer for 4 hours and the generator for 8 hours at a total cost of $100.
   (a) Write a system of two equations that describes the given conditions. Define the variables.
   (b) Using the system of equations written in part (a), find the hourly cost of the generator.

20. Cedric and Zelda went shopping at Price Buster. Cedric bought two jumbo rolls of aluminum foil and three packages of AA batteries for a total cost of $21. Zelda bought five identical jumbo rolls of aluminum foil and two identical packages of AA batteries for a total cost of $25. Find the cost of (a) one jumbo roll of aluminum foil and (b) one package of AA batteries.

21. At the local video rental store, Jose rents two movies and three games for a total of $15.50. At the same time, Meg rents three movies and one game for a total of $12.05. How much money does Chris need to rent a combination of one game and one movie from the same video rental store?

22. The cost of a taxi ride is calculated by adding a fixed amount to a charge for each $\frac{1}{4}$ mile of the trip. If a 1-mile trip costs $4.00 and a $1\frac{1}{2}$-mile trip costs $5.50, what will be the cost of a 3-mile trip?

23. The cost of a long-distance telephone call is determined by a flat fee for the first 5 minutes and a fixed amount for each additional minute. Roseanne is charged $3.25 for a long-distance telephone call that lasts 15 minutes and $5.17 for a long-distance call that lasts 23 minutes. At the same rate, what is the cost of a long-distance telephone call that lasts 30 minutes?

CHAPTER 14

QUADRATIC EQUATIONS AND THEIR GRAPHS

14.1 GENERAL EQUATION OF A CIRCLE

=== **KEY IDEAS** ===

The graph of the equation $x^2 + y^2 = r^2$ is a circle whose center is at the origin and whose radius is r units, as shown on the diagram on the left.

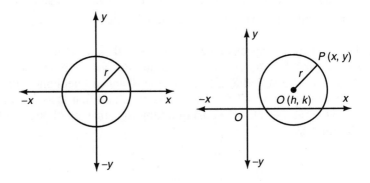

Translating this circle h units in the horizontal direction and k units in the vertical direction "moves" the center of the circle from (0, 0) to (h, k), as shown in the diagram on the right. As a result, the equation of the translated circle is $(x - h)^2 + (y - k)^2 = r^2$.

Working with Equations of Circles

- If you know the coordinates of the center, (h, k), of a circle and its radius, r, you can write an equation of the circle by substituting these values into the equation $(x - h)^2 + (y - k)^2 = r^2$. For example, the locus of points 5 units from point $(2, -1)$ is a circle with center at $(h, k) = (2, -1)$ and radius $r = 5$, so its equation is $(x - 2)^2 + (y - [-1])^2 = 5^2$ or, equivalently, $(x - 2)^2 + (y + 1)^2 = 25$.
- If you know an equation of a circle, you can determine the coordinates of its center and its radius by writing the equation in the form $(x - h)^2 + (y - k)^2 = r^2$. For example, if an equation of a circle is $(x - 3)^2 + (y + 4)^2 = 36$, this equation can be rewritten as follows:

374

$$\left(x - \underbrace{3}_{\substack{h \\ \downarrow}}\right)^2 + \left(y - \underbrace{(-4)}_{\substack{k \\ \downarrow}}\right)^2 = \underbrace{6^2}_{\substack{r \\ \downarrow}}$$

$$(x-h)^2 + (y-k)^2 = r^2$$

Since $h = 3$, $k = -4$, and $r = 6$, the circle whose equation is $(x - 3)^2 + (y + 4)^2 = 36$ has its center at $(3, -4)$ and its radius is 6 units.

Example

The coordinates of the endpoint of a diameter of a circle are $A(1, 2)$ and $B(-7, -4)$.
(a) Find an equation of this circle.
(b) Determine whether the circle passes through point $P(-6, -5)$.

Solution: (a) $\mathbf{(x + 3)^2 + (y + 1)^2 = 25}$

The center $O(h, k)$ of the circle is found by finding the midpoint of diameter $\overline{AB}$:

$$h = \frac{1+(-7)}{2} = \frac{-6}{2} = -3, \quad k = \frac{2+(-4)}{2} = \frac{-2}{2} = -1$$

The coordinates of the center of the circle are $(-3, -1)$. to find the length of the radius, find the distance between the center and *any* point on the circle, say point A.

Let $(x_1, y_1) = A(1, 2)$ and $(x_2, y_2) = (-3, -1)$. Then:

$$\begin{aligned}
OA &= \sqrt{(x_2 - x_1)^2 + (y_2 - y_1)^2} \\
&= \sqrt{(-3-1)^2 + (-1-2)^2} \\
&= \sqrt{(-4)^2 + (-3)^2} \\
&= \sqrt{16+9} \\
&= \sqrt{25} \\
&= 5
\end{aligned}$$

An equation of a circle whose center is at $(-3, -1)$ and whose radius is 5 is

$$(x-[-3])^2 + (y-[-1])^2 = 5^2$$
$$(x+3)^2 + (y+1)^2 = 25$$

(b) **The circle passes through $P\,(-6, -5)$.**

Point $P(-6, -5)$ lies on the circle if its coordinates satisfy the equation of the circle:

$$(x+3)^2 + (y+1)^2 = 25$$
$$(-6+3)^2 + (-5+1)^2 \overset{?}{=} 25$$
$$(-3)^2 + (-4)^2 \overset{?}{=} 25$$
$$9 + 16 \overset{?}{=} 25$$
$$25 \overset{\checkmark}{=} 25$$

Dilations of Circles

If two circles have the same center, then either one of the circles may be considered to be the **dilation** of the other circle. The **center of dilation** is the common center of the two circles, and the **constant of dilation** is the ratio of their radii.

For example, the circle $x^2 + y^2 = 64$ may be considered a dilation of the circle $x^2 + y^2 = 16$ since each circle has the origin as its center. In this case, since the radius of the larger circle is 8 and the radius of the smaller circle is 4, the constant of dilation is $8 \div 4$ or 2. To illustrate further, the dilation of the circle $x^2 + y^2 = 9$, using a constant of dilation of 4, is the circle $x^2 + y^2 = 144$.

Translations of Circles

The circles $(x - h)^2 + (y - k)^2 = r^2$ and $x^2 + y^2 = r^2$ have the same size and shape, but differ in their locations in the coordinate plane. Compared to the center of the circle $x^2 + y^2 = r^2$, the center of the circle $(x - h)^2 + (y - k)^2 = r^2$ is shifted h units in the horizontal direction and k units in the vertical direction.

For example, since the center of the circle $(x - 1)^2 + (y + 3)^2 = 100$ is $(1, -3)$, the circle $(x - 1)^2 + (y + 3)^2 = 100$ is a **translation** of the circle $x^2 + y^2 = 100$, shifted 1 unit horizontally to the right and 3 units vertically down.

Check Your Understanding of Section 14.1

A. Multiple Choice

1. Which is an equation of the circle with center at $(4, -2)$ and radius length of 3?
   (1) $(x + 4)^2 + (y - 2)^2 = 9$
   (2) $(x - 4)^2 + (y + 2)^2 = 9$
   (3) $(x + 4)^2 + (y - 2)^2 = 3$
   (4) $(x - 4)^2 + (y + 2)^2 = 3$

2. When a circle whose equation is $x^2 + y^2 = 25$ is shifted 3 units horizontally to the right and 2 units vertically down, it coincides with the circle whose equation is
   (1) $(x + 3)^2 + (y - 2)^2 = 25$
   (2) $(x - 3)^2 + (y + 2)^2 = 25$
   (3) $(x - 3)^2 + (y - 2)^2 = 25$
   (4) $(x + 3)^2 + (y + 2)^2 = 25$

3. Which could be an equation of a circle that touches the x-axis at one point and whose center is at $(-6, 8)$?
   (1) $(x - 6)^2 + (y - 8)^2 = 36$
   (2) $(x - 6)^2 + (y + 8)^2 = 64$
   (3) $(x + 6)^2 + (y - 8)^2 = 36$
   (4) $(x + 6)^2 + (y - 8)^2 = 64$

4. What is the area, in terms of π, between the circles whose equations are $x^2 + y^2 = 25$ and $x^2 + y^2 = 39$?
   (1) 8π
   (2) 14π
   (3) 64π
   (4) $\left(\sqrt{39} - 5\right)\pi$

5. What is an equation of the locus of points 7 units from point $(-2, 3)$?
   (1) $(x + 2)^2 + (y - 3)^2 = 49$
   (2) $(x + 2)^2 + (y - 3)^2 = 7$
   (3) $(x - 2)^2 + (y + 3)^2 = 49$
   (4) $(x - 2)^2 + (y + 3)^2 = 7$

6. Which pair of graphs intersect in exactly one point?
   (1) $x^2 + y^2 = 4$ and $x = -3$
   (2) $x^2 + y^2 = 4$ and $y = -2$
   (3) $x^2 + y^2 = 4$ and $y = x$
   (4) $x^2 + y^2 = 4$ and $x = 4$

7. A circle whose equation is $(x + 1)^2 + (y - 2)^2 = 36$ is translated 1 unit horizontally to the left and 3 units vertically down. What is an equation of the image of this circle?
   (1) $(x + 2)^2 + (y + 1)^2 = 36$
   (2) $(x + 2)^2 + (y - 5)^2 = 36$
   (3) $x^2 + (y + 1)^2 = 36$
   (4) $x^2 + (y - 5)^2 = 36$

8. Circle O has its center at the origin and passes through point $(-3, 4)$. What is an equation of the image of circle O after a dilation with a scale factor of 2?

(1) $x^2 + y^2 = 100$ (3) $x^2 + y^2 = 25$

(2) $(x + 3)^2 + (y - 4)^2 = 100$ (4) $x^2 + y^2 = 50$

9. What is an equation of the circle with a diameter that has endpoints at $(2, 5)$ and $(-6, 5)$?

(1) $(x - 2)^2 + (y + 5)^2 = 64$ (3) $(x + 2)^2 + (y - 5)^2 = 64$

(2) $(x - 2)^2 + (y + 5)^2 = 16$ (4) $(x + 2)^2 + (y - 5)^2 = 16$

10. The center of a circle that has a radius of 6 is $(-1, -2)$. What is an equation of the image of this circle after it is shifted 2 units horizontally to the right and 1 unit vertically down?

(1) $(x - 3)^2 + (y - 1)^2 = 36$ (3) $(x - 1)^2 + (y + 3)^2 = 36$

(2) $(x + 1)^2 + (y + 3)^2 = 36$ (4) $(x - 3)^2 + (y - 3)^2 = 36$

B. *In each case, show how you arrived at your answer by clearly indicating all of the necessary steps, formula substitutions, diagrams, graphs, charts, etc.*

11. What is an equation of the circle that has its center at O $(2, -3)$ and passes through P $(-2, 0)$?

12. After the circle $(x - 1)^2 + (y + 1)^2 = 4$ is dilated, an equation of the image is $(x - 1)^2 + (y + 1)^2 = 36$. Under the same dilation, circle O' is the image of circle O, whose equation is $x^2 + y^2 = 9$.
(a) What is an equation of circle O'?
(b) What is an equation of the locus of all points equidistant from circles O and O'?
(c) What percent of the area of circle O', correct to the *nearest tenth*, is *not* enclosed by circle O?

13. (a) What is an equation of the circle that touches the line $x = 4$ at exactly one point and whose center is at $(-6, 7)$?
(b) What is an equation of the line that intersects the circle at $(0, -1)$ and is perpendicular to the radius drawn to that point?

14. The vertices of $\triangle ABC$ are $A(-4, -2)$, $B(2, 6)$, and $C(2, -2)$.
(a) Write an equation of the locus of points equidistant from vertex B and vertex C.
(b) Write an equation of the locus of points 4 units from vertex C.
(c) What is the total number of points that satisfy the loci described in parts (a) and (b)?

15. (a) On graph paper, draw the locus of a point 5 units from point $(-1, -5)$.
(b) On the same set of axes, draw the image of the graph drawn in part (a) after a reflection in the x-axis. Label this graph (b).
(c) On the same set of axes, draw the image of the graph drawn in part (b) after a translation that moves point (x, y) to point $(x + 1, y - 5)$. Label this graph (c).

14.2 GRAPHING VERTICAL PARABOLAS

 KEY IDEAS

The graph of a quadratic equation that has the form

$$y = ax^2 + bx + c,$$

where a, b, and c stand for numbers, provided that $a \neq 0$, is called a **parabola**. Since the parabola shown in the accompanying diagram has a *vertical* **axis** (line) **of symmetry**, it is called a "vertical" parabola. The axis of symmetry intersects a parabola at a point called the **turning point** or **vertex** of the parabola.

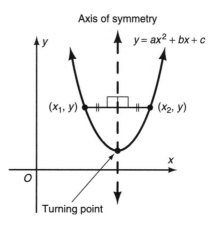

Axis of Symmetry and Turning Point

In the equation $y = ax^2 + bx + c$, the constants a and b determine the location of the axis of symmetry and the coordinates of the turning point of the parabola.

- As shown in Figure 14.1, an equation of the axis of symmetry is $x = -\dfrac{b}{2a}$.

- The x-coordinate of the turning point is $-\dfrac{b}{2a}$.

- The y-coordinate of the turning point is found by substituting $-\dfrac{b}{2a}$ for x in the equation

$$y = ax^2 + bx + c.$$

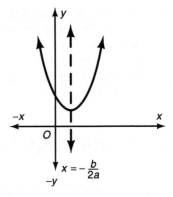

Figure 14.1 Axis of Symmetry of a Parabola

Example 1

For the parabola $y = 3x^2 - 6x + 2$, find:
(a) an equation of the axis of symmetry
(b) the coordinates of the turning point of the parabola

Solution: (a) $x = 1$

Let $a = 3$ and $b = -6$. Then

$$x = -\frac{b}{2a} = -\frac{-6}{2 \cdot 3} = \frac{6}{6} = 1$$

(b) **(1, −1)**
Since the axis of symmetry contains the turning point of the parabola, the x-coordinate of the turning point is 1. To find the y-coordinate of the turning point, replace x by 1 in the quadratic equation $y = 3x^2 - 6x + 2$:

$$
\begin{aligned}
y &= 3x^2 - 6x + 2 \\
&= 3 \cdot 1^2 - 6 \cdot 1 + 2 \\
&= 3 \quad - \quad 6 + 2 \\
&= -3 \qquad + 2 \\
&= -1
\end{aligned}
$$

As shown in Figure 14.2, the sign of the x^2-term determines whether the turning point of a parabola is the lowest (minimum) point or the highest (maximum) point on the curve.

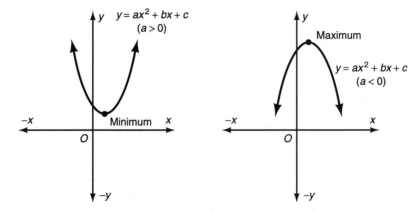

Figure 14.2 Effect of Coefficient *a* on the Turning Point of a Parabola

For a parabola whose equation is written in the form $y = ax^2 + bx + c$:
- If the x^2-term is positive ($a > 0$), the turning point is the lowest point on the parabola.
- If the x^2-term is negative ($a < 0$), the turning point is the highest point on the parabola.

Graphing $y = ax^2$ ($a \neq 0$)

The turning point of the parabola $y = ax^2$ is the origin, and its axis of symmetry is the *y*-axis, as illustrated in Figure 14.3.

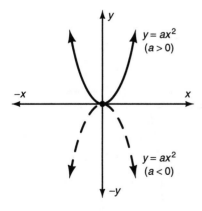

Figure 14.3 Parabola $y = ax^2$

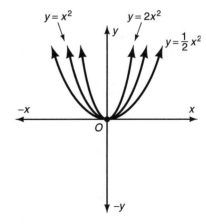

Figure 14.4 Comparing the Width of $y = ax^2$ for Different Values of *a*

Figure 14.4 shows how the width of the parabola is affected by the coefficient of the x^2-term. For example, the parabola $y = ax^2$ $(a > 0)$ becomes narrower as the coefficient a gets larger.

Graphing $y = ax^2 + c$ $(a > 0)$

The parabola $y = ax^2 + c$ has the same shape as the parabola $y = ax^2$ except that it is shifted vertically up or down, depending on whether c is positive or negative. Figure 14.5 shows the parabolas obtained by shifting $y = ax^2$ vertically up 2 units $(y = ax^2 + 2)$, and vertically down 2 units $(y = ax^2 - 2)$.

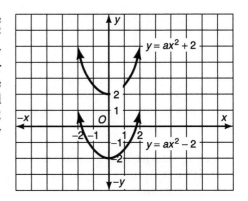

Figure 14.5 Shifting $y = ax^2$

In general:

- If $c > 0$, you can obtain the graph of $y = ax^2 + c$ by shifting the parabola $y = ax^2$ vertically up c units *above* the origin.
- If $c < 0$, you can obtain the graph of $y = ax^2 + c$ by shifting the parabola $y = ax^2$ vertically down $|c|$ units *below* the origin.

Example 2

What is an equation of the horizontal line that intersects the graph of

$$y = -2x^2 + 1$$

Solution: $y = 1$

Since, as shown in the accompanying diagram, the turning point of the parabola $y = -2x^2 + 1$ is $(0, 1)$, and equation of the horizontal line is $y = 1$.

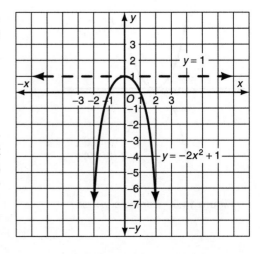

382

Graphing a Parabola on a Given Interval

To graph the parabola $y = x^2 - 4x + 1$ from $x = -1$ to $x = 5$:

- For each integer value of x from -1 to 5, find the corresponding value of y that lies on the curve. Organize the work in a table, as shown below.

x	$x^2 - 4x$ $+1$	$= y$
-1	$(-1)^2 - 4(-1)+1$	$= 6$
0	$0^2 - 4 \cdot 0 + 1$	$= 1$
1	$1^2 - 4 \cdot 1 + 1$	$= -2$
2	$2^2 - 4 \cdot 2 + 1$	$= -3$
3	$3^2 - 4 \cdot 3 + 1$	$= -2$
4	$4^2 - 4 \cdot 4 + 1$	$= 1$
5	$5^2 - 4 \cdot 5 + 1$	$= 6$

- Notice in the table that corresponding points above and below $x = 2$ have the same y-coordinates. Therefore, $(2, -3)$ is the turning point of the parabola.
- Plot the points from the table. Then connect the points with a smooth, U-shaped curve, as shown in Figure 14.6.

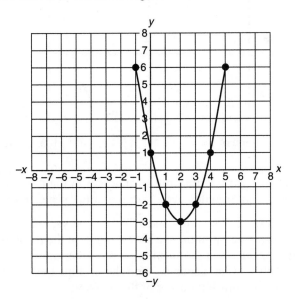

Figure 14.6   Graph of $y = x^2 - 4x + 1$

- Label the curve with its equation.

$$\boxed{\text{TIP}}$$

If the interval for x is not given, find the x-coordinate of the axis of symmetry of $y = ax^2 + bx + c$ using the formula $x = -\dfrac{b}{2a}$. Then make a table of values that includes three x-values above and below the x-coordinate of the turning point.

Graphing a Parabola Using a Graphing Calculator

If you have the Texas Instruments TI-83 graphing calculator, the parabola whose equation is $y = x^2 - 4x + 1$ can be graphed by proceeding as follows:

- Enter the right side of the equation $y = x^2 - 4x + 1$ in the $\boxed{Y=}$ editor by pressing these keys:

$$\boxed{x, T, \theta, n} \quad \boxed{\wedge} \quad \boxed{2} \quad \boxed{-} \quad \boxed{4} \quad \boxed{x, T, \theta, n} \quad \boxed{+} \quad \boxed{1}$$

- Press the $\boxed{\text{ZOOM}}$ key, and then press $\boxed{4}$. The graph shown in Figure 14.7 should appear in the display window.

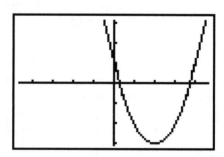

Figure 14.7   Graph of $y = x^2 - 4x + 1$ in [–4.7, 4.7] by [–3.1, 3.1] Window

- Use the **minimum** (or **maximum**) function in the CALCULATE menu to determine the coordinates of the turning point, if needed.

  1. After pressing the $\boxed{\text{2nd}}$ and $\boxed{\text{TRACE}}$ keys, select 3: **minimum** when the turning point is the lowest point on the parabola, and select 4: **maximum** when the turning point is the highest point. The current graph will be displayed with the prompt **Left Bound?** appearing in the bottom-left corner of the window.
  2. Press the left or right arrow key to pick a value to the left of the turning point, and then press $\boxed{\text{ENTER}}$. The prompt **Right Bound?** will appear in the bottom-left corner of the window.

3. Press the left or right arrow key to pick a value to the right of the turning point, and then press ENTER. The prompt **Guess?** will appear in the bottom-left corner of the window.
4. Press the left or right arrow key to pick a guess value that lies between the left bound and the right bound. Then press ENTER. The coordinates of the turning point will appear at the bottom of the window.

Making a Table Using a Calculator

After you enter the equation $y = x^2 - 4x + 1$ in the Y= editor of your calculator, you can have the calculator build a table of values from $x = -1$ to $x = 5$.

- Enter the table setup [TBLSET] mode by pressing the 2nd key followed by the WINDOW key. A window opens that allows you to set some key values of your table.
- Set the initial x-value to -1 by letting **TblStart = -1**. If necessary, set **ΔTbl = 1** so that the value of x in the table increases in steps of 1.

- Look at the table by pressing the 2nd key followed by the GRAPH key. The table in Figure 14.8 should appear in the display window. You can bring other values of x into view by using the up or down cursor keys. The corresponding values of y will be automatically calculated.

X	Y1	
-1	6	
0	1	
1	-2	
2	-3	
3	-2	
4	1	
5	6	
X=-1		

Figure 14.8   Table of Values for $y = x^2 - 4x + 1$

Check Your Understanding of Section 14.2

A. Multiple Choice

1. Which represents an equation for the axis of symmetry of the graph of $y = 2x^2 - 7x - 5$?

   (1) $x = -\dfrac{5}{4}$ (3) $x = \dfrac{7}{4}$

   (2) $x = \dfrac{5}{4}$ (4) $x = -\dfrac{7}{4}$

2. What are the coordinates of the highest point on the parabola whose equation is $y = -x^2 + 2x - 7$?
   (1) $(-1, -10)$ (2) $(-1, 6)$ (3) $(1, -6)$ (4) $(1, 10)$

3. What is the minimum value of y in the equation $y = x^2 - 6x + 5$?
   (1) -14 (2) -4 (3) 3 (4) 4

4. Which is an equation of the parabola that has the y-axis as its axis of symmetry and its minimum point at $(0, 3)$?
   (1) $y = x^2 - 3$ (3) $y = x^2 + 3$
   (2) $y = -x^2 + 3$ (4) $y = -x^2 - 3$

5. Which is an equation of the parabola that has the y-axis as its axis of symmetry and its maximum point at $(0, -5)$?
   (1) $y = x^2 - 5$ (3) $y = x^2 + 5$
   (2) $y = -x^2 + 5$ (4) $y = -x^2 - 5$

6. Which is an equation of the parabola graphed in the accompanying diagram?
   (1) $y = x^2 - 4$
   (2) $y = x^2 + 4$
   (3) $y = -x^2 + 4$
   (4) $y = -x^2 - 4$

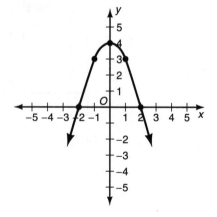

7. A parabola intersects the x-axis at $x = -1$ and $x = 3$, and intersects the y-axis at $y = 6$. Which is an equation of the parabola?
   (1) $y = -2x^2 + 4x + 6$ (3) $y = 6x^2 - 2x + 4$
   (2) $y = 2x^2 - 4x + 6$ (4) $y = 6x^2 + 2x - 4$

8. How many points are 2 units from the turning point of the parabola $y = x^2 + 5$ and 3 units from the origin?
   (1) 1   (2) 2   (3) 3   (4) 4

9. When drawn on the same set of axes, the graphs of $x^2 + y^2 = 16$ and $y = x^2 - 4$ intersect at
   (1) 1 point   (2) 2 points   (3) 3 points   (4) 4 points

10. When graphed on the same set of axes, which pair of graphs intersect at exactly two points?
    (1) $y = 4$ and $y = x^2 + 5$ (3) $y = 5$ and $y = -x^2 + 5$
    (2) $y = x^2 - 1$ and $y = -x^2 + 1$ (4) $x^2 + y^2 = 25$ and $y = -x^2 + 5$

B. *In each case, show how you arrived at your answer by clearly indicating all of the necessary steps, formula substitutions, diagrams, graphs, charts, etc. If you use a graphing calculator, show each of the following: (1) a sketch of the viewing window; (2) scales indicated on the x- and y-axes; (3) clearly labeled x- and y-intercepts and point(s) of intersection.*

11. What is an equation of a line that contains the turning point of the parabola $y = 2x^2 + 8x + 3$ and is parallel to the x-axis?

12. (a) Describe fully the locus of all points that are 3 units from axis of symmetry of the parabola $y = \dfrac{1}{2}x^2 - x + 6$.
    (b) What is an equation of the line that is parallel to the x-axis and passes through the turning point?

13. (a) What is an equation of the locus of all points that are 3 units from the turning point of the parabola $y = -x^2 + 6x - 7$?
    (b) What is an equation of the line that is perpendicular to the axis of symmetry at the turning point?

14. What is an equation of the line that contains the y-intercept and the turning point of the parabola $y = x^2 - 6x + 9$?

15. What is an equation of the line that contains the turning point of the parabola $y = -x^2 + 2x + 5$ and is parallel to the line whose equation is $y - 3x = 4$?

16. In the accompanying figure, parabola $y = -2x^2 + 12x$, and $\overline{AB}$ and $\overline{BC}$ are drawn so that $\overline{AB}$ touches the parabola at its turning point with $\overline{AB} \perp \overline{BC}$. What is the perimeter of rectangle $OABC$?

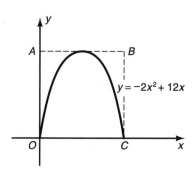

17. (a) On graph paper, draw the graph of the equation $y = x^2 + 2x - 3$ for all values of x in the interval $-4 \leq x \leq 2$.
    (b) On the same set of axes, draw the graph of the equation $(x + 1)^2 + (y + 4)^2 = 16$.
    (c) How many different ordered pairs satisfy both the equation graphed in part (a) and the equation graphed in part (b)?

18. (a) Draw the graph of the equation $y = \frac{1}{2}x^2 - 3x + 4$, including all values of x in the interval $0 \leq x \leq 6$.
    (b) What are the coordinates of the turning point of the image of the graph drawn in part (a) after a reflection in line $y = 4$?

14.3 LEARNING MORE ABOUT PARABOLAS

=== **KEY IDEAS** ===

Quadratic equations can be solved graphically as well as algebraically. Geometric transformations can be applied to parabolas.

Solving Quadratic Equations by Graphing

At each x-intercept of a graph, $y = 0$. The x-intercepts of the graph of $y = ax^2 + bx + c$ must, therefore, satisfy the equation $ax^2 + bx + c = 0$. If a parabola has no x-intercepts, then the related quadratic equation has no real roots.

Example 1

Find the roots of $(x + 5)(x - 3) = 20$ by graphing.

Solution: **−7, 5**

Rewrite the given equation in an equivalent form so that it can be graphed. Then find the *x*-intercepts of the graph.

Method 1: Put the equation in the form $ax^2 + bx + c = 0$.

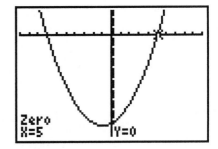

- Since $(x + 5)(x - 3) = 20$, $x^2 + 2x - 15 = 20$, so $x^2 + 2x - 35 = 0$.
- Graph $Y_1 = x^2 + 2x - 35$ in an appropriate viewing window such as $[-10, 10]$ by $[-40, 10]$.
- Find the *x*-intercepts of the graph. Select **2:zero** from the CALCULATE menu, and then follow a procedure similar to the last bulleted item on page 384.

Method 2: Rewrite the equation as $(x + 5)(x - 3) - 20 = 0$.

Rather than multiplying the binomials of the original equation, set $Y_1 = (x + 5)(x - 3) - 20$. Then find the *x*-intercepts of the graph of Y_1.

Method 3: Use two graphs.

Change your point of view, and think of solving $(x + 5)(x - 3) = 20$ as finding the *x*-coordinates of the points at which two graphs intersect:

$$\overbrace{(x+5)(x-3)}^{\text{Graph } Y_1} = \overbrace{20}^{\text{Graph } Y_2}$$

By selecting **5:intersect** from the CALC menu and using an appropriate viewing window, you can determine that −7 and 5 are the *x*-coordinates of the points at which the graphs of $Y_1 = (x + 5)(x - 3)$ and $Y_2 = 20$ intersect.

Transformations of Parabolas

A parabola may be reflected in either coordinate axis or translated. For example, if you create a table of values to draw a parabola, you can sketch a translation of that parabola by applying the translation rule to the set of ordered pairs in the table and then plotting the image points.

Example 2

(a) On graph paper, draw the graph of the equation $y = x^2 - 6x + 8$ for all values of *x* in the interval $0 \le x \le 6$.

(b) On the same set of axes, draw the image of $y = x^2 - 6x + 8$ after the translation $(x, y) \rightarrow (x - 3, y + 1)$. Label this graph (b), and state its equation.

Solution: (a) See the accompanying figure.

Use your calculator to generate a table of values for $y = x^2 - 6x + 8$ from $x = 0$ to $x = 6$. Plot points $(0, 8)$, $(1, 3)$, $(2, 0)$, $(3, -1)$, $(4, 0)$, $(5, 3)$, and $(6, 8)$ on graph paper. Then connect these points with a smooth, U-shaped curve, as shown in the accompanying graph.

(b) See the accompanying figure; $y = x^2$. Under the translation $(x, y) \rightarrow (x - 3, y + 1)$:

$$
\begin{aligned}
(0, 8) &\rightarrow (0 - 3, 8 + 1) &&= (-3, 9) \\
(1, 3) &\rightarrow (1 - 3, 3 + 1) &&= (-2, 4) \\
(2, 0) &\rightarrow (2 - 3, 0 + 1) &&= (-1, 1) \\
(3, -1) &\rightarrow (3 - 3, -1 + 1) &&= (0, 0) \\
(4, 0) &\rightarrow (4 - 3, 0 + 1) &&= (1, 1) \\
(5, 3) &\rightarrow (5 - 3, 3 + 1) &&= (2, 4) \\
(6, 8) &\rightarrow (6 - 3, 8 + 1) &&= (3, 9)
\end{aligned}
$$

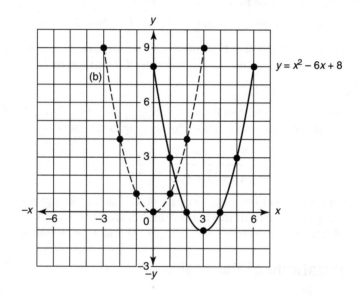

Plot the image points, connect them, and label this graph (b), as shown in the accompanying figure. Since the translated parabola has a minimum turning point at $(0, 0)$, its equation is $y = x^2$.

Check Your Understanding of Section 14.3

In each case, show how you arrived at your answer by clearly indicating all of the necessary steps, formula substitutions, diagrams, graphs, charts, etc. If you use a graphing calculator, show each of the following: (1) a sketch of the viewing window; (2) scales indicated on the x- and y-axes; (3) clearly labeled x- and y-intercepts and point(s) of intersection.

1. (a) Describe at least two different ways in which $(2x - 3)(x - 1) = 10$ can be solved using a graphing calculator.
   (b) Solve the quadratic equation in part (a) using one of the methods described in this section.
   (c) Check your solution by solving the quadratic equation algebraically.

2. (a) Using graph paper, graph $y = x^2 + 2x$ for $-4 \leq x \leq 2$. Label this graph (a).
   (b) On the same set of axes, graph the image of the parabola drawn in part (a) after a translation that shifts each point (x, y) to $(x + 1, y - 2)$. Label this graph (b), and state its equation.

3. (a) On graph paper, draw the graph of the equation $y = x^2 - 4x + 3$, including all values of x in the interval $-1 \leq x \leq 5$. Label this graph (a).
   (b) On the same set of axes, draw the image of the graph drawn in part (a) after the translation that moves (x, y) to $(x + 3, y + 2)$. Label this graph (b).
   (c) On the same set of axes, draw the image of the graph drawn in part (b) after a reflection in the x-axis. Label this graph (c).

4. After t seconds, a ball tossed into the air from ground level reaches a height of h feet, where $h = 144t - 16t^2$ and both h and t are greater than or equal to 0.
   (a) Graph $h = 144t - 16t^2$ in the first quadrant using your graphing calculator. From the graph determine the maximum height of the ball and the number of seconds it takes to reach the maximum height.
   (b) After how many seconds will the ball hit the ground before rebounding? Confirm your answer algebraically.

5. Amy tossed a ball into the air in such a way that the path of the ball was modeled by the equation $y = -x^2 + 6x$. In the equation, y $(y \geq 0)$ represents the height, in feet of the ball, and x is the time, in seconds.
   (a) Using graph paper, graph $y = -x^2 + 6x$ for $0 \leq x \leq 6$.
   (b) From the graph determine the maximum height of the ball and the number of seconds it took to reach the maximum height.

6. (a) On graph paper, draw the graph of the equation $y = x^2 - 4x + 4$, including all values of x in the interval $-1 \leq x \leq 5$. Label this graph (a).
(b) On the same set of axes, draw the image of the graph drawn in part (a) after a translation that maps (x, y) onto $(x - 2, y + 3)$. Label this graph (b).
(c) On the same set of axes, draw the image of the graph drawn in part (b) after a reflection in the x-axis. Label this graph (c).
(d) Write an equation that could represent the graph drawn in part (c).

7. A parabola that passes through the origin has its turning point at $(-2, 12)$. What is an equation of the parabola?

8. (a) On graph paper, draw the graph of the equation $y = x^2 - 6x + 5$, including all values of x in the interval $0 \leq x \leq 6$.
(b) On the same set of axes, sketch the image of the graph drawn in part (a) after a reflection in the y-axis. Label this graph (b).

9. An arch is built so that it has the shape of a parabola with the equation $y = -3x^2 + 21x$, where y represents the number of feet in the height of the arch.
(a) What is the number of feet in the width of the arch at its base?
(b) What is the number of feet in the maximum height of the arch?

14.4 SOLVING LINEAR-QUADRATIC SYSTEMS

KEY IDEAS

The solution set of a linear-quadratic system of equations, as illustrated in the accompanying figure, consists of: (a) two ordered pairs of numbers; (b) one ordered pair of numbers; or (c) no ordered pair.

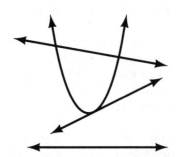

To solve a linear-quadratic system of equations *algebraically*, solve for one of the variables in the linear equation. Eliminate that variable in the quadratic equation, which can then be solved in the usual way.

Solving a Linear-Quadratic System by Graphing

When a parabola and a line are drawn on the same set of axes, the points of intersection, if any, represent the solution set to the system of equations used to graph the parabola and the line.

Example 1

(a) Draw the graph of $y = -x^2 + 4x - 3$ for all values of x such that $-1 \le x \le 5$.

(b) On the same set of axes, draw the graph of $x + y = 1$.

(c) Determine the solution set of the system

$$y = -x^2 + 4x - 3$$
$$x + y = 1$$

(d) Check the answer obtained in part (c) algebraically.

Solution: (a) Make a table of values, using all integral values of x such that $-1 \le x \le 5$ as shown below, and then draw the parabola, as shown below the table.

x	$-x^2 + 4x - 3$	$=$	y
−1	$-(-1)^2 + 4(-1) - 3$	$=$	−8
0	$-(0)^2 + 4 \cdot 0 - 3$	$=$	−3
1	$-(1)^2 + 4 \cdot 1 - 3$	$=$	0
2	$-(2)^2 + 4 \cdot 2 - 3$	$=$	1
3	$-(3)^2 + 4 \cdot 3 - 3$	$=$	0
4	$-(4)^2 + 4 \cdot 4 - 3$	$=$	−3
5	$-(5)^2 + 4 \cdot 5 - 3$	$=$	−8

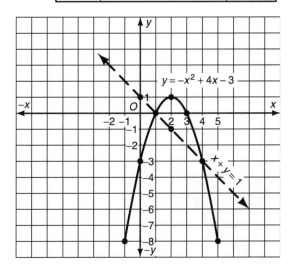

393

Note: $(2, 1)$ is the turning point of the parabola.

(b) Solve $x + y = 1$ for y: $y = -x + 1$. Make a table of values, using any three convenient values for x, as shown on the right, and then draw the line as shown in part (a).

x	$-x+1$	$=$	y
0	$0 + 1$	$=$	1
1	$-1 + 1$	$=$	0
2	$-2 + 1$	$=$	-1

(c) **$(1, 0), (4, -3)$**
Since the graphs drawn in parts (a) and (b) intersect at $(1, 0)$ and $(4, -3)$, the solution set of the system of equations $y = -x^2 + 4x - 3$ and $x + y = 1$ is $\{(1, 0), (4, -3)\}$.

(d) To check algebraically, show that each ordered pair of the solution set satisfies each of the *original* equations.

To check $(1, 0)$, let $x = 1$ and $y = 0$:

$$y \overset{?}{=} -x^2 + 4x - 3 \qquad x + y \overset{?}{=} 1$$
$$0 \overset{?}{=} -(1)^2 + 4 \cdot 1 - 3 \qquad 1 + 0 \overset{?}{=} 1$$
$$0 \overset{?}{=} -1 + 4 - 3 \qquad 1 \overset{\checkmark}{=} 1$$
$$0 \overset{?}{=} \quad 3 \quad - 3$$
$$0 \overset{\checkmark}{=} 0$$

To check $(4, -3)$, let $x = 4$ and $y = -3$:

$$y \overset{?}{=} -x^2 + 4x - 3 \qquad x + y \overset{?}{=} 1$$
$$-3 \overset{?}{=} -(4)^2 + 4 \cdot 4 - 3 \qquad 4 + (-3) \overset{?}{=} 1$$
$$-3 \overset{?}{=} -16 + 16 - 3 \qquad 1 \overset{\checkmark}{=} 1$$
$$-3 \overset{?}{=} \quad 0 \quad - 3$$
$$-3 \overset{\checkmark}{=} -3$$

Using a Graphing Calculator

To solve the system $y = -x^2 + 4x - 3$ and $x + y = 1$ using the TI-83 graphing calculator, proceed as follows:

- Rewrite $x + y = 1$ as $y = -x + 1$. Then enter the right sides of the equations $y = -x^2 + 4x - 3$ and $y = -x + 1$ in the $\boxed{\text{Y=}}$ editor.
- In the $\boxed{\text{ZOOM}}$ menu, press $\boxed{4}$.
- Since one of the points of intersection is at the bottom edge of the screen, resize the viewing rectangle. Enter the $\boxed{\text{WINDOW}}$ menu, and multiply Ymin and Ymax each by 2. Then press the $\boxed{\text{GRAPH}}$ key.

- Select **5:intersect** from the CALCULATE menu. You can then determine that the coordinates of the points of intersection of the two graphs are $(1, 0)$ and $(4, -3)$, as shown in Figure 14.9.

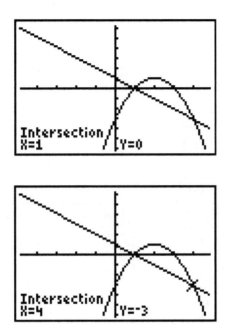

Figure 14.9   Using the **intersect** Function in the CALCULATE Menu to Find the Points of Intersection of $y = -x^2 + 4x - 3$ and $x + y = 1$

Solving a Linear-Quadratic System Algebraically

A linear-quadratic system of equations such as $x + y = 1$ and $y = -x^2 + 4x - 3$ may be solved algebraically as follows:

Step	Example
1. If necessary, rewrite the linear equation so that one of the variables is expressed in terms of the other. When the quadratic equation has the form $y = ax^2 + bx + c$, it is usually easier to solve the system by solving the linear equation for y in terms of x.	1. $x + y = 1$ $\quad y = -x + 1$
2. Substitute into the quadratic equation the expression obtained by solving for the variable in the linear equation.	2. Since $y = -x + 1$, replace y by $-x + 1$: $\quad y = -x^2 + 4x - 3$ $-x + 1 = -x^2 + 4x - 3$
3. Express the quadratic equation in standard form. Solve the resulting quadratic equation, usually by factoring.	3. $0 = -x^2 + 4x + x - 3 - 1$ $\quad 0 = -x^2 + 5x - 4$ Multiply each term by -1: $\quad 0 = x^2 - 5x + 4$ $\quad 0 = (x - 1)(x - 4)$ $x - 1 = 0$ or $x - 4 = 0$ $\quad x = 1 \qquad\quad x = 4$
4. Find the value of the remaining variable of each ordered pair by substituting each root of the quadratic equation into the linear equation.	4. Let $\;x = 1.\qquad\qquad$ Let $\;x = 4.$ $\;$ Then $y = -x + 1 \qquad$ Then $y = -x + 1$ $\qquad\quad = -1 + 1 \qquad\qquad\qquad = -4 + 1$ $\qquad\quad = 0 \qquad\qquad\qquad\quad\; = -3$ The solution set is $\{(1, 0), (4, -3)\}$
5. Check algebraically by verifying that each ordered pair satisfies each of the *original* equations.	5. The check is left for you. (See solution by graphing in Example 1.)

Check Your Understanding of Section 14.4

A. *Multiple Choice*

1. The graphs of the equations $y = x^2$ and $x = 2$ intersect in
   (1) 1 point   (2) 2 points   (3) 3 points   (4) 0 points

2. How many points do the graphs of $x^2 + y^2 = 9$ and $y = 4$ have in common?
   (1) 1   (2) 2   (3) 0   (4) 4

3. Which is a point of intersection of the graphs of the line $y = x$ and the parabola $y = x^2 - 2$?
   (1) (1, 1)   (2) (2, 2)   (3) (0, 0)   (4) (2, −1)

4. What is a solution for the following system of equations?

   $$y = x^2$$
   $$y = -4x + 12$$

   (1) (−2, 4)   (2) (6, 36)   (3) (2, 4)   (4) (−6, 24)

5. Which is a solution for the following system of equations?

   $$y = 2x - 15$$
   $$y = x^2 - 6x$$

   (1) (3, −9)   (2) (0, 0)   (3) (5, 5)   (4) (6, 0)

6. When the graphs of the equations $y = x^2 - 5x + 6$ and $x + y = 6$ are drawn on the same set of axes, at which point do the graphs intersect?
   (1) (4, 2)   (2) (5, 1)   (3) (3, 3)   (4) (2, 4)

B. *In each case, show how you arrived at your answer by clearly indicating all of the necessary steps, formula substitutions, diagrams, graphs, charts, etc. If you use a graphing calculator, show each of the following: (1) a sketch of the viewing window; (2) scales indicated on the x- and y-axes; (3) clearly labeled x- and y-intercepts and point(s) of intersection.*

7–9. Solve each system of equations algebraically for x and y.

7. $y = x^2 - 4x + 9$
      $y - x = 5$

397

8. $y = x^2 - 6x + 6$

$y - x = 4$

9. $x^2 + y^2 = 97$

$y = x + 5$

10. There are two pairs of integers that satisfy both of these conditions:

The smaller integer is 10 less than the larger integer.
The sum of the squares of the two integers is 250.

What are the two pairs of integers?

11–13. Solve each system of equations either algebraically or graphically for x and y.

11. $y = -x^2 - 2x + 8$

$y = x + 4$

12. $y = x^2 - 6x + 5$

$y + 7 = 2x$

13. $(x + 2)^2 + (y - 1)^2 = 16$

$x - y = 1$

14. A rocket is launched from the ground and follows a parabolic path represented by the equation $y = -x^2 + 8x$. At the same time, a flare is launched from a height of 14 feet and follows a straight path represented by the equation $y = -x + 14$. Find the coordinates of the point at which the rocket and flare first collide.

15. Over a 12-month period the price of stock A increased according to the equation $y = 0.5x^2 + 2x + 18$, where y is the average price of the stock during month x. The price of stock B fell according to the equation $y = -3.75x + 96$ over the same 12-month period.
(a) During which month, by number, were the average prices of the two stocks the same? What was the average price of each stock during this month?
(b) During which month, by number, did the average price of stock A first increase to more than twice its value at the beginning of the 12-month period?

PROBABILITY AND STATISTICS

CHAPTER 15

SIMPLE PROBABILITY AND STATISTICS

15.1 FINDING SIMPLE PROBABILITIES

KEY IDEAS

The **probability** of an event is a number from 0 to 1 that reflects the likelihood that the event will happen. If an event E can happen in r out of n equally likely ways, then the *probability* that event E will occur is the ratio of r to n, which can be written as $P(E) = \dfrac{r}{n}$. The closer the fractional value $\dfrac{r}{n}$ is to 1, the greater the likelihood that the event will happen.

The Language of Probability

The spinner shown in Figure 15.1 consists of five equal regions, numbered from 1 through 5.

Figure 15.1 Spinner with Five Equally Likely Outcomes

- An activity whose outcome is uncertain, such as predicting where the spinner will land, is called a **probability experiment**.
- Although we do not know the region in which the spinner will stop, we know the set of all possible outcomes, which is called the **sample space**. For this experiment, the sample space is the set of region numbers 1, 2, 3, 4, and 5.
- The outcomes in the sample space are **equally likely** to occur since each of the five regions is the same size.

- A possible *event* of the probability experiment is spinning a 2. Another possible event is spinning an odd number. An **event** is a particular outcome or set of outcomes that are contained in the sample space.

Probability Facts

The outcomes that will make an event happen are the **favorable outcomes** or **successes** for that event. In Figure 15.1, for the event of spinning an odd number, the favorable outcomes are 1, 3, and 5. Since three of the five regions have odd numbers, the probability of spinning an odd number is $\frac{3}{5}$, where 3 is the number of favorable outcomes and 5 is the total number of possible outcomes. Thus:

$$P(E) = \frac{\text{number of favorable outcomes for event } E}{\text{total number of possible outcomes}}$$

Since the numerator of this probability fraction cannot be greater than the denominator or less than 0, $P(E)$ must range in value from 0 to 1.

- $P(E) = 0$ if event E is an impossibility. In Figure 15.1, the probability of spinning a 6 is 0.
- $P(E) = 1$ if event E is a certainty. In Figure 15.1, the probability of spinning a whole number that is less than 6 is 1.
- $P(\text{not } E) = 1 - P(E)$. In Figure 15.1, the probability of spinning an odd number is $\frac{3}{5}$. Hence, the probability of *not* spinning an odd number (or of spinning an even number) is $1 - \frac{3}{5} = \frac{2}{5}$. If there is a 30% chance of rain tomorrow, then the probability that it will *not* rain tomorrow is $1 - .30 = .70$ or 70%.

Probability and Area

To find the probability of an event, it may be necessary to compare the areas of two regions.

Example 1

In the accompanying diagram, a fair spinner is placed at the center of circle O.

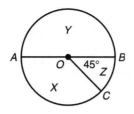

Diameter $\overline{AOB}$ and radius $\overline{OC}$ divide the circle into three regions labeled X, Y, and Z. If m$\angle BOC = 45$, what is the probability that the spinner will land in region X?

Solution: $\frac{3}{8}$

The fractional part of the area of circle O that is represented by region X represents the probability that the spinner will land in region X.

- Since $\overline{AOB}$ is a line, the degree measures of adjacent angles AOC and BOC must add up to 180. Hence, m$\angle AOC = 180 - 45 = 135$.

- A circle contains 360°. Thus, region X represents $\dfrac{135°}{360°} = \dfrac{3}{8}$ of the area of circle O, so the probability that the spinner will land in region X is $\frac{3}{8}$.

Example 2

In the accompanying diagram, circle O is inscribed in square $ABCD$. A dart is thrown and lands inside the square. What is the probability, correct to the *nearest tenth*, that the dart does *not* land inside the circle?

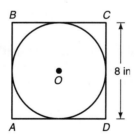

Solution: **.2**

To find the probability that the dart does *not* land inside the circle, compare the area of the region of the square that is not inside the circle with the area of the square.

The area of the square is $8 \times 8 = 64\,\text{in}^2$.

- Since the radius of the circle is 4, the area of the circle is approximately $\pi \times 4^2 = 3.14 \times 16 = 50.24\,in^2$. Hence, the area of the region of the square that is *not* inside the circle is $64 - 50.24 = 13.76\,in^2$.

- $P(\text{dart does } not \text{ land inside circle}) = \dfrac{13.76}{64} = .215 \approx .2$

Check Your Understanding of Sectioning 15.1

A. Multiple Choice

1. During a half hour of television programming, 8 minutes is used for commercials. If a television set is turned on at a random time during the half hour, what is the probability that a commercial is *not* being shown?

   (1) 1 $\qquad$ (2) $\dfrac{11}{15}$ $\qquad$ (3) $\dfrac{4}{15}$ $\qquad$ (4) 0

2. A bag has five green marbles and four blue marbles. If one marble is drawn at random, what is the probability that it is *not* green?

   (1) $\dfrac{1}{9}$ $\qquad$ (2) $\dfrac{4}{9}$ $\qquad$ (3) $\dfrac{5}{9}$ $\qquad$ (4) $\dfrac{5}{20}$

3. An urn contains four yellow marbles and three blue marbles. How many blue marbles must be added to the urn so that the probability of picking a yellow marble will be $\dfrac{1}{3}$?

   (1) 5 $\qquad$ (2) 2 $\qquad$ (3) 3 $\qquad$ (4) 4

4. A jar contains only red, white, and blue marbles. If the ratio of red to white to blue marbles is 1 to 3 to 8, what is the probability that a marble drawn at random will *not* be white?

   (1) $\dfrac{1}{4}$ $\qquad$ (2) $\dfrac{1}{3}$ $\qquad$ (3) $\dfrac{2}{3}$ $\qquad$ (4) $\dfrac{3}{4}$

5. The probability of drawing a red marble from a sack of marbles is $\dfrac{2}{5}$.

   Which set of marbles could the sack contain?
   (1) 2 red marbles and 5 green marbles
   (2) 4 red marbles and 6 green marbles
   (3) 6 red marbles and 15 green marbles
   (4) 2 red marbles, 1 blue marble, and 4 white marbles

6. If x represents a number picked at random from the set

$$\{-2, -1, 0, 1, 2, 3\},$$

what is the probability that x will satisfy the inequality $4 - 3x < 6$?

(1) $\dfrac{1}{3}$ (2) $\dfrac{1}{2}$ (3) $\dfrac{2}{3}$ (4) $\dfrac{5}{6}$

7. If the probability that an event A will occur is $\dfrac{x}{4}$ and $0 \le x \le 4$, what is the probability that event A will *not* occur?

(1) $\dfrac{1-x}{4}$ (2) $\dfrac{4-x}{4}$ (3) $\dfrac{4-x}{x}$ (4) $\dfrac{4}{x}$

8. When a number is chosen at random from the set $\{1, 2, 3, 4, 5, 6\}$, which event has the greatest probability of occurring?
(1) choosing an even number
(2) choosing a prime number
(3) choosing a number greater than 3
(4) *not* choosing either 1 or 6

9. The probability of guessing the correct answer to a certain test question is $\dfrac{x}{12}$. If the probability of *not* guessing the correct answer to the same question is $\dfrac{2}{3}$, what is the value of x?
(1) 9   (2) 3   (3) 8   (4) 4

10. A dart is thrown and lands in square $ABCD$. If E is the midpoint of $\overline{AD}$ and F is the midpoint of $\overline{CD}$, as shown in the accompanying diagram, what is the probability that the dart will land in the shaded region?

(1) $\dfrac{1}{4}$ (2) $\dfrac{1}{3}$ (3) $\dfrac{3}{8}$ (4) $\dfrac{1}{2}$

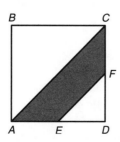

B. *In each case, show how you arrived at your answer by clearly indicating all of the necessary steps, formula substitutions, diagrams, graphs, charts, etc.*

11. The numbers from 1 to 20, inclusive, are written on individual slips of paper and placed in a hat. If a slip of paper is picked out of the hat without looking, what is the probability of picking each of the following?
    (a) a number that is divisible by 5   (c) a number that is at most 13
    (b) a prime number   (d) a number that is at least 7

12. From a jar of red jellybeans and white jellybeans, the probability of picking a white jellybean is $\frac{2}{3}$. If the jar contains 24 jellybeans, how many red jellybeans are in the jar?

13. One marble is drawn at random from a jar that contains x red marbles, $2x - 1$ blue marbles, and $2x + 1$ white marbles.
    (a) Express, in terms of x, the probability of drawing a blue marble.
    (b) If the probability of drawing a blue marble is $\frac{1}{3}$, find the value of x.
    (c) What is the probability of *not* drawing a red marble?

14. In the accompanying diagram, both circles have the same center, O. The radii of the circles are 3 and 5.
    (a) Find, in terms of π, the number of square units in the area of the shaded region.
    (b) A dart is thrown and lands on the diagram. Find the probability that the dart will land:

    (1) on the shaded area   (2) on the unshaded area

Exercise 14

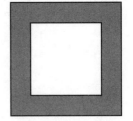

Exercise 15

15. In the accompanying diagram of a square dartboard, the length of a side of the outer square is 1.5 times the length of a side of the inner square. If a dart is thrown and lands on the square, what is the probability that the dart will land in the shaded region?

16. A square dartboard is placed in Quadrant I from $x = 0$ to 6 and $y = 0$ to 6, as shown in the accompanying figure. A triangular region on the dartboard is enclosed by the graphs of the equations $y = 2$, $x = 6$, and $y = x$. Find the probability that a dart that randomly hits the dartboard will land in the triangular region formed by the three lines.

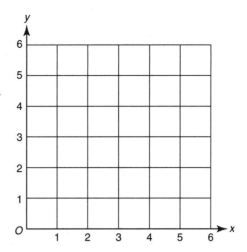

15.2 STATISTICAL MEASURES

KEY IDEAS

Data are facts and figures. Numerical data may be called *values, scores*, or *measures*. **Statistics** involves collecting, organizing, displaying, and interpreting data. The *mean* (average), *mode*, and *median* are single numbers that help describe how the individual scores in a set are distributed in value.

The Mean

The **mean** or average of a set of scores is the sum of the scores divided by the number of scores. For example, to find the mean of 76, 82, and 85, add the three scores and then divide their sum by 3:

$$\text{Mean} = \frac{76 + 82 + 85}{3} = \frac{243}{3} = 81$$

In the set 76, 82, and 85, the scores are fairly close in value so the mean of 81 represents a central value about which the individual scores are clustered. The mean is an example of a *measure of central tendency*.

If a set of scores includes a value that differs greatly from the other numbers in the set, the mean will not be a good measure of central tendency. For example, the mean of the set of numbers 1, 2, 5, and 200 is 52 since

$$\text{Mean} = \frac{1 + 2 + 5 + 200}{4} = \frac{208}{4} = 52$$

405

In this case, the mean does not provide useful information about how the individual values in the set are distributed, so it is not a good measure of central tendency.

In some problems, the mean (average) of a set of scores is given and you must determine one of the numbers in the set.

Example 1

Raymond's first four test grades are 85, 89, 87, and 96. What is the lowest grade Raymond can get on his next text so that the average of the five test grades will be at least 90?

Solution: **93**

For the average to be at least 90, the average must be equal to or greater than 90. Let x represent Raymond's grade on the next test. Then

$$\text{Average} \geq 90$$
$$\frac{x+85+89+87+96}{5} \geq 90$$
$$\frac{x+357}{5} \geq 90$$
$$x+357 \geq 90 \times 5$$
$$x+357 \geq 450$$
$$x \geq 450-357$$
$$\geq 93$$

Since the least possible value of x is 93, that is the lowest grade Raymond can receive on the next test for an average of at least 90.

The Mode

The **mode** of a set of scores is the number in the set that appears the most number of times. For example, in the unordered set of numbers

$$19, 23, 19, 18, 27, 19, 15, 23, 16$$

the mode is 19 since it occurs three times, which is more often than any other number in the list.

A set of scores may have more than one mode. The unordered set of numbers

$$15, 9, 8, 9, 11, 15$$

has two modes, 15 and 9.

The Median

When the scores in a set are listed in size order, the center value in the list is called the **median**.

- If an ordered list of scores contains an *odd* number of values, the median is the middle number. For example:

$$\overbrace{11,\ 12,\ 18,\ \boxed{23}\ ,\ 37,\ 46,\ 50}^{\text{7 data values}}$$

$$\uparrow$$

Median is the middle score.

Notice that there are three scores below the median and three scores above the median. The median always divides a set of numbers into two groups that have the same number of scores.

- If an ordered list of scores contains an *even* number of values, the median is the average of the two middle numbers. For example, the median of the set of eight numbers 7, 15, 18, 23, 29, 37, 46, and 50 is the average of the two middle numbers:

$$\overbrace{7,\ 15,\ 18,\ 23,}^{\text{4 data values}}\ \boxed{?}\ ,\ \overbrace{29,\ 37,\ 46,\ 50}^{\text{4 data values}}$$

$$\uparrow$$

$$\text{Median} = \frac{23 + 29}{2}$$

$$= \frac{52}{2}$$

$$= 26$$

Example 2

Multiple Choice:
If a set of data consists of the numbers 3, 3, 5, 6, and 8, which of the following statements is true?

(1) Median > mean
(2) Mean = mode

(3) Mode > median
(4) Median = mean

Solution: **(4)**

The mode is 3, the median or middle value is 5, and the

$$\text{Mean} = \frac{3 + 3 + 5 + 6 + 8}{5} = \frac{25}{5} = 5$$

Since both the median and the mean are equal to 5, the correct choice is (4).

Example 3

A number is randomly selected from the set 8, 5, 13, 5, 7, and 10. Find the probability that the number will be:

(a) greater than the mean   (b) equal to the median

Solution: (a) $\frac{2}{6}$

$$\text{Mean} = \frac{8+5+13+5+7+10}{6} = \frac{48}{6} = 8$$

The set of data values consists of six numbers, two of which (10 and 13) are greater than 8. Hence, the probability that the number selected will be greater than the mean is $\frac{2}{6}$.

(b) **0**

To find the median, first arrange the numbers in size order:

$$5, 5, 7, 8, 10, 13$$

Since there is an even number of scores, the median is the average of the two middle scores, 7 and 8. Thus:

$$\text{Median} = \frac{7+8}{2} = 7.5$$

Since no number in the set is equal to 7.5, selecting a number from the set that is equal to 7.5 is an impossible event. Hence, the probability that the number selected will be equal to the median is 0.

Check Your Understanding of Section 15.2

A. Multiple Choice

1. The mean (average) of a set of seven numbers is exactly 81. If one of the numbers is discarded, the average of the remaining numbers is exactly 78. Which number was discarded?
   (1) 89   (2) 92   (3) 97   (4) 99

2. The accompanying table shows how the cost of a specific notebook varied over a 5-week period. According to the table, which statement is true about the cost of this notebook over this period?
   (1) The mode was $3.00.
   (2) The mean was $4.30.
   (3) The median was $4.50.
   (4) The median was $3.00.

Week	Cost
1	$5.00
2	$5.25
3	$3.00
4	$3.50
5	$4.75

3. A man drove a car at an average rate of speed of 45 miles per hour for the first 3 hours of a 7-hour trip. If the average rate of speed for the entire trip was 53 miles per hour, what was the average rate of speed, in miles per hour, for the remaining part of the trip?
   (1) 55   (2) 57   (3) 59   (4) 62

4. For the set of numbers 8, 13, 8, 7, 4, 8, which statement is true?
   (1) Median > mean   (3) Mode < median
   (2) Mean = mode   (4) Median = mode

5. Five members of a team reported the numbers of boxes of cookies that they sold to raise money as follows: 20, 20, 40, 50, and 70. Which statement is true?
   (1) The median is 20.   (3) Median = mean.
   (2) The mean is 20.   (4) Median = mode.

6. For which set of numbers do the mean, median, and mode all have the same value?
   (1) 1, 3, 3, 3, 5   (3) 1, 1, 1, 2, 5
   (2) 1, 1, 2, 5, 6   (4) 1, 1, 3, 5, 10

B. *In each case, show how you arrived at your answer by clearly indicating all of the necessary steps, formula substitutions, diagrams, graphs, charts, etc.*

7. The average of five test grades is 84. If four of the test grades are 71, 81, 94, and 77, what is the other test grade?

8. Vanessa's average for five mathematics tests was 86. The median score was 87 and the mode was 80. What was the highest score she could have received on any one test?

9. The mean (average) weight of three dogs is 38 pounds. One of the dogs, Sparky, weighs 46 pounds. The other two dogs, Spot and Eddie, have the same weight. Find Eddie's weight.

10. Susan received 78, 89, and 82 on her first three exams. What is the lowest grade she can receive on her next exam and have an average grade for the four exams of at least 85?

11. What is the number of square units in the area, in terms of π, of the circle whose radius is the average of the radii of two circles with areas of 16π and 100π?

12. A number is selected at random from the set 2, 2, 2, 2, 3, 3, 3, and 7. Find the probability that the number selected will be:
(a) greater than the mean   (b) equal to the median

13. The average of a set of five different positive integers is 360. The two smallest integers in the set are 99 and 102. What is the largest possible integer in this set?

14. Find all possible values of x such that, when the five numbers x, 14, 11, 6, and 17 are arranged in order, the mean is equal to the median.

15. The median of seven test scores is 82, the mode is 87, the lowest score is 70, and the mean (average) is 80. If each of the scores is a whole number, what is the greatest possible test score?

15.3 PERCENTILES AND QUARTILES

KEY IDEAS

If on a math test 75% of the test scores are 81 or less, then 81 is the 75th *percentile* for that test. A **percentile** tells the score at or below which a specified percent of the scores in a set of data values fall.

A **box-and-whisker plot** shows at a glance the spread of data above and below the median.

Percentiles

After the data values in a set have been put in size order, half or 50% of the values fall at or below the median. The median is, therefore, the 50th percentile.

In general, the **pth percentile** is the value at or below which $p\%$ of the data values fall. Suppose, for example, that 20 of 25 students have test scores less than or equal to 92. Since $\dfrac{20}{25} = 80\%$, 80% of the students have test scores of 92 or less. Thus, for this set of data 92 represents the 80th percentile.

Quartiles

Quartiles are the numbers that separate a set of data values arranged in size order into four equal groups, as shown in Figure 15.2.

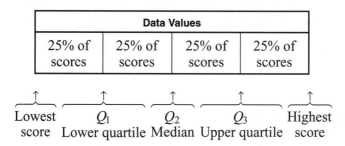

Data Values			
25% of scores	25% of scores	25% of scores	25% of scores

Lowest score Q_1 Lower quartile Q_2 Median Q_3 Upper quartile   Highest score

Figure 15.2   Quartile Measures

- Q_2 is the median or **middle quartile**. The median separates a set of ordered data values into two groups that have the same number of scores.
- Q_1 is the first or **lower quartile**. The lower quartile is the median of the data *below* Q_2.
- Q_3 is the third or **upper quartile**. The upper quartile is the median of the data *above* Q_2.

For example, the quartiles for the scores 10, 21, 27, 33, 45, 48, 55, 57, 65, 65, 75, 78, and 80 are found as follows:

Lower half of data Q_2 Upper half of data

10, 21, 27,   33, 45, 48, $\boxed{55}$,   57, 65, 65,   75, 78, 80

↑   Median   ↑

$$\frac{27+33}{2} = Q_1 \qquad\qquad Q_3 = \frac{65+75}{2}$$

$$\frac{60}{2} = Q_1 \qquad\qquad Q_3 = \frac{140}{2}$$

$$30 = Q_1 \qquad\qquad Q_3 = 70$$

Notice that:

- Since the number of data values is an odd number, the median is the middle or seventh value. Thus, the median is 55. When figuring out quartiles for this set of scores, 55 is not included in the group of data that is below the median or above it.
- The *lower half* of the data consists of the six values *below* 55: 10, 21, 27, 33, 45, and 48. Since the number of data values is an even number,

411

the median of these six values is the average of the two middle values. Hence, the lower quartile is $\dfrac{27+33}{2}$ or 30.

- The *upper half* of the data consists of the six values *above* 55: 57, 65, 65, 75, 78, 80. Since the number of data values is an even number, the median of these six values is the average of the two middle values. Hence, the upper quartile is $\dfrac{65+75}{2}$ or 70.

Box-and-Whisker Plots

A **box-and-whisker plot** shows pictorially how spread out the data are about the median. To draw a box-and-whisker plot for the scores 10, 21, 27, 33, 45, 48, 55, 57, 65, 65, 75, 78, and 80;

- Arrange the data scores from lowest to highest, as shown below.
- Find these five key values: lowest score, lower quartile, median, upper quartile, highest score.
- Draw a number line, using a convenient scale that begins with the lowest score and ends with the highest score.
- Below the number line, graph the five key values, as shown in Figure 15.3.

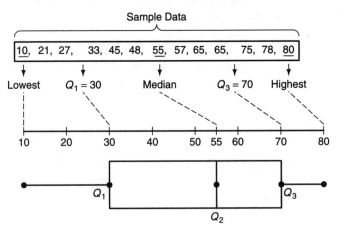

Figure 15.3   Drawing a Box-and-Whisker Plot

- Draw a **box** from Q_1 to Q_3. Then draw a vertical line through the median.
- Draw horizontal lines, called **whiskers**, from the sides of the box to the lowest and highest scores.

Check Your Understanding of Section 15.3

A. Multiple Choice

1. On a mathematics test, Bob scored at the 80th percentile. Which statement is true?
   (1) Bob scored 80% on the test.
   (2) Twenty percent of the students who took the same test had scores equal to or less than Bob's score.
   (3) Eighty percent of the students who took the same test had the same score as Bob.
   (4) Eighty percent of the students who took the same test had scores equal to or less than Bob's score.

2. On a test taken by 24 students, the 75th percentile was 84. How many students scored higher than 84?
   (1) 6   (2) 12   (3) 18   (4) 24

3. On a quiz taken by the students in a mathematics class, a score of 78 is at the 25th percentile. If 8 students scored 78 or less on the test, how many students scored higher than 78?
   (1) 12   (2) 16   (3) 24   (4) 32

4. Nine students scored 75 or less on a mathematics test. If 75 is the 25th percentile, what is the number of students who took this test?
   (1) 12   (2) 27   (3) 36   (4) 45

B. In each case, show how you arrived at your answer by clearly indicating all of the necessary steps, formula substitutions, diagrams, graphs, charts, etc.

5. In a set of test scores, nine scores are above the 40th percentile. How many test scores are in this set?

6. The accompanying box-and-whisker plot summarizes the results on a Math A test taken by 24 students.

   (a) What percent of the students scored above 84?
   (b) How many students scored above 91?

413

(c) How many students scored between 62 and 73?

(d) A test score of 80 could correspond to which of the following percentiles?

(1) 47   (2) 20   (3) 53   (4) 80

7. The midterm exam scores for a class were 60, 65, 65, 67, 71, 70, 73, 75, 76, 76, 79, 81, 83, 84, 85, 85, 88, 89, 90, 92, 95, 96, 99, 100, and 100.

(a) What are the lower, middle (median), and upper quartiles?

(b) Which score corresponds to the 68th percentile?

(c) Summarize the data by drawing a box-and-whisker plot.

8. The accompanying box-and-whisker plots summarize Regents test scores for two different Math A classes.

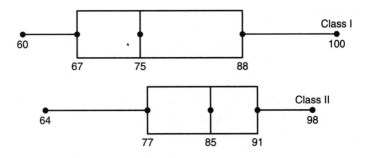

(a) By how much does the median Regents test score for class II exceed the median Regents test score for class I?

(b) Which class had the greater range in Regents test scores? Justify your answer.

(c) If a student is picked at random from class II, what is the probability the student's Regents test score is at least 92?

15.4 HISTOGRAMS AND STEM-AND-LEAF PLOTS

KEY IDEAS

The **frequency** of a score is the number of times that score appears in a list of data values. A **histogram** is a type of bar graph that shows how the data are grouped. The height of each bar represents the frequency or number of data items contained in the interval at the base of the bar. A **stem-and-leaf plot** shows how the individual data items are ordered and grouped.

Making a Frequency Table Using Intervals

For example, suppose that a class of 25 students received the following scores on a Math A test: 58, 70, 60, 65, 68, 70, 90, 70, 72, 74, 70, 70, 75, 78, 80, 96, 75, 80, 83, 80, 88, 83, 90, 65, 75

Table 15.1 shows how the data can be tabulated and organized by using a three-column *frequency table*.

TABLE 15.1   FREQUENCY TABLE WITH INTERVALS

Interval	Tally	Frequency
50–59	/	1
60–69	////	4
70–79	LHT LHT /	11
80–89	LHT /	6
90–99	///	3
	Sum = 25	

The intervals were selected so that they have a convenient uniform width of 10 while accommodating both the lowest and highest scores. Since the lowest score is 58, the first interval is 50–59. The highest score is 96, so the last interval is 90–99. A quick glance at Table 15.1 gives a good idea of how the test scores are distributed, with most of them falling between 70 and 79.

Drawing a Frequency Histogram

To draw a *frequency histogram* based on Table 15.1, follow these steps:

- Using graph paper, draw the coordinate axes in the first quadrant. Label the vertical axis "Frequency," and the horizontal axis "Test scores."
- Mark off the vertical axis in units of 1.
- Mark off the horizontal axis so that each interval has the same width. The width may be of any convenient size.
- For each interval, draw vertical bars next to one another. The frequency of each interval determines the bar height.

Figure 15.4 shows the completed frequency histogram.

When drawing a histogram based on a frequency table in which the data are organized into intervals, remember that:

- The intervals are located on the horizontal axis and, for the same histogram, always have the same width.
- The *frequency* of an interval is the number of data values that fall into that interval. Frequencies are always located on the vertical axis.
- The height of each rectangle represents the number of data values contained in the interval noted at the base of that rectangle. For example, in

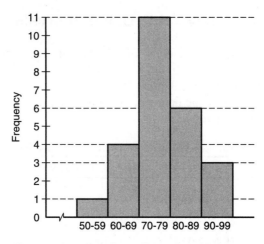

Figure 15.4 Drawing a Frequency Histogram

Figure 15.4 the interval 70–79 has the greatest height, indicating that this interval contains more data values than any other interval. Since the height of this rectangle corresponds to a frequency of 11, the interval 70–79 contains 11 of the original data values.

Example 1

For the accompanying frequency table, answer the following questions:
(a) Which interval contains the median?
(b) Which interval contains the lower quartile?
(c) Which interval contains the upper quartile?

Interval	Frequency
1–15	1
16–30	2
31–45	6
46–60	4
61–75	2
76–90	5

Solution: (a) **46–60**

There is a total of 20 scores. The median is the middle value, so it lies between the 10th and 11th scores. By accumulating frequencies, you know that the fourth interval, 46–60, contains the 10th, 11th, 12th, and 13th scores.

(b) **31–45**

The lower quartile is the number at or below which 25% of the scores fall. Since there is a total of 20 scores:

$$25\% \text{ of } 20 = \frac{1}{4} \times 20 = 5$$

Adding frequencies shows that the 5th score is found in the third interval, 31–45.

(c) **61–75**
The upper quartile is the number at or below which 75% of the scores fall. Since there is a total of 20 scores:

$$75\% \text{ of } 20 = \frac{3}{4} \times 20 = 15$$

Adding frequencies reveals that the 15th score is found in the fifth interval, 61–75.

Example 2

The table at the right represents the distribution of ages of principals in a school district.
(a) Using this table, draw a frequency histogram.
(b) Find the interval that contains the lower quartile.
(c) What percent of the principals are older than 43?

Interval	Frequency
68–75	2
60–67	6
52–59	5
44–51	8
36–43	5
28–35	2

Solution: (a) See the accompanying figure.

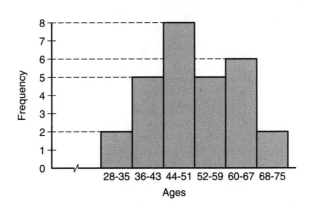

(b) **36–43**
The frequency table contains 28 data values. In an ordered list of data, the lower quartile is the value at or below which $\frac{1}{4}$ of the data values fall. Since

$\frac{1}{4} \times 28 = 7$, the lower quartile is the 7th lowest data value in the frequency

table. Adding the frequencies in the lowest two intervals of the table gives 7. Thus, the lower quartile is contained in the second interval, 36–43.

(c) **75**
Summing the frequencies in the intervals above 36–43 gives $8 + 5 + 6 + 2 = 21$. Thus, the ages of 21 of the 28 principals fall in the intervals above 36–43.

Since $\frac{21}{28} = \frac{3}{4} = 75\%$, 75% of the principals are older than 43.

Constructing a Stem-and-Leaf Plot

In a **stem-and-leaf plot** each data value is split into a "stem" and a "leaf." The "leaf" is usually the last digit of the number, and the digits to the left of the leaf form the "stem." Thus, the data value 145 would be split in this way:

Data value

$\underbrace{14}\ \underbrace{5}$

Stem Leaf

Suppose, for example, that the top 30 scores of a bowling team are as follows:

140, 140, 144, 145, 148, 150, 154, 154, 156, 156, 158, 159, 160, 160, 162, 163, 163, 164, 165, 168, 169, 170, 175, 178, 184, 188, 188, 190, 192, 205

These 30 bowling scores can be better organized by creating a stem-and-leaf plot in which, for each data value, the stem consists of the hundreds and tens digits and the leaf is the units digit, as shown in Figure 15.5.

Legend: 14|8 means 148.

```
14 | 0  0  4  5  8
15 | 0  4  4  6  6  8  9
16 | 0  0  2  3  3  4  5  8  9
17 | 0  5  8
18 | 4  8  8
19 | 0  2
20 | 5
```

Figure 15.5   Stem-and-Leaf Plot of Bowling Scores

Each line of the stem-and-leaf plot consists only of scores that have the same stem. The leaves, which consist of the units digits, are arranged in increasing order to the right of each stem. For example, the first line of the stem-and-leaf plot lists the five bowling scores that have 14 as their stem: **140, 140, 144, 145,** and **148.**

Stem-and-Leaf Plots Versus Histograms

A stem-and-leaf plot resembles a histogram turned on its side so that the bars are horizontal rather than vertical. The stem value represents an interval, and the number of leaves for that stem is the frequency for that interval. For example, in Figure 15.5 the stem 14 represents the interval 140–149. Since there are five leaves on the first line, the interval 140–149 has a frequency of 5. The interval 160–169 has the greatest frequency: 9.

A histogram gives a more visually appealing picture of how data are grouped than a stem-and-leaf plot. When data are grouped into intervals to form a histogram, however, the identity of each of the individual values is lost. A stem-and-leaf plot includes the individual data values and also displays how the data are grouped.

Check Your Understanding of Section 15.4

A. Multiple Choice

1 and 2. The accompanying histogram shows the distribution of scores of 30 students on a mathematics test.

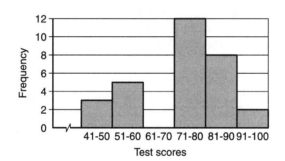

1. Which interval contains the median score?
   (1) 51–60   (2) 61–70   (3) 71–80   (4) 81–90

2. Which interval contains the lower quartile?
   (1) 41–50   (2) 51–60   (3) 71–80   (4) 81–90

3. The accompanying frequency table shows data collected by the police in an automobile speed check. Which interval contains the median speed?
   (1) 36–45   (3) 56–65
   (2) 46–55   (4) 66–75

Speed Interval	Frequency
66–75	45
56–65	110
46–55	120
36–45	25

4. The accompanying histogram shows the distribution of SAT I scores at Oceanview High School. What percent of students in this school scored at least 610?
   (1) 25   (3) 62.5
   (2) 37.5   (4) 75

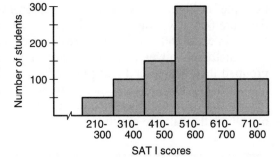

B. *In each case, show how you arrived at your answer by clearly indicating all of the necessary steps, formula substitutions, diagrams, graphs, charts, etc.*

5. For the stem-and-leaf plot in Figure 15.5, construct:
   (a) a frequency table   (b) a histogram

6. The accompanying table summarizes the results of a test.
   (a) Which interval contains the upper quartile?
   (b) If a score is picked at random, what is the probability that the score will be at most 80?

Interval	Frequency
91–100	3
81–90	5
71–80	4
61–70	5
51–60	3

7. For the data in the accompanying table:
   (a) What is the median score?
   (b) What percentile corresponds to a score of 88?

Score	Frequency
68	3
75	9
82	4
88	2
93	6

8. The accompanying histogram shows the distribution of test scores for a class.
   (a) What percent of students had test scores less than 76?
   (b) Which interval contains the 80th percentile?
   (c) What is the probability that a score picked at random will be between 65 and 70?

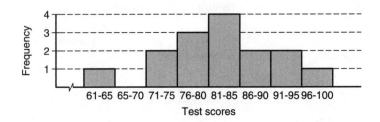

9. The accompanying table shows the distribution of misspellings in 25 student essays.

Misspelled Words	Frequency (Number of Essays)
0	1
1	0
2	3
3	5
4	4
5	9
6	3

   (a) On graph paper, construct a frequency histogram based on the data.
   (b) Find the mode number of misspelled words.
   (c) Find the mean number of misspelled words.
   (d) Find the median number of misspelled words.

10. The following data are test scores for a class of 16 students:

    96, 83, 91, 77, 58, 88, 80, 62, 89, 100, 87, 93, 64, 98, 88, 86.

    (a) Copy and complete the accompanying table.

Interval	Frequency
91–100	
81–90	
71–80	
61–70	
51–60	

    (b) On graph paper, construct a frequency histogram based on the data.
    (c) Which interval contains the median?
    (d) Which interval contains the lower quartile?

11. The accompanying table represents the distribution of grades in a college mathematics class. A, B, C, and D are passing grades, and x represents a positive integer.

Grade	Frequency
A	x
B	$2x - 3$
C	$x + 1$
D	5
F	$x - 4$

    (a) If 29 students are in this class, find x.
    (b) What is the median grade?
    (c) If a student is selected at random from the class, find the probability that the student's grade is:
    (1) *not* a passing grade
    (2) equal to the mode
    (3) at least a C

12. The accompanying histogram shows the daily temperature intervals at 8:00 A.M. for the month of June at a local weather station.

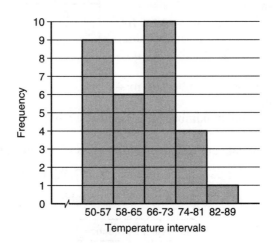

(a) What is the maximum temperature possible at the 30th percentile?
(b) What is the probability that on a day chosen at random the temperatures at 8:00 A.M. was less than 74°?
(c) Explain why you could not construct a stem-and-leaf plot from the histogram.

15.5 CUMULATIVE FREQUENCY TABLES AND HISTOGRAMS

KEY IDEAS

Cumulative frequency tables and histograms display, for each given interval, the sum of the number of scores in all preceding intervals up to and including those contained in the interval being studied.

Creating a Cumulative Frequency Table

The first two columns of Table 15.2 represent a frequency table that shows the distribution of 25 student test scores. The last two columns form the cumulative frequency table for this set of scores.

TABLE 15.2 COMPLETING A CUMULATIVE FREQUENCY TABLE

Interval	Frequency	Cumulative Frequency		Interval
50–59	1	1		50–59
60–69	4	5	(4 + 1 = 5)	50–69
70–79	11	16	(11 + 5 = 16)	50–79
80–89	6	22	(6 + 16 = 22)	50–89
90–99	3	25	(3 + 22 = 25)	50–99

Notice that, for each interval after the first, the entries in the cumulative frequency column are obtained by adding the frequency entry that appears on the same line to the cumulative frequency of the preceding line. Since the first frequency has no entry before it, nothing is added to 1 to obtain the cumulative frequency for the first interval.

Drawing a Cumulative Frequency Histogram

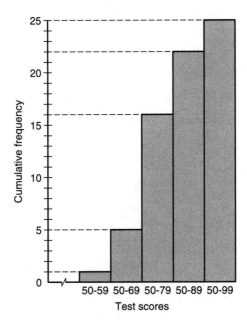

Figure 15.6 Cumulative Frequency Histogram for Table 15.2

Figure 15.6 is the cumulative frequency histogram for the data contained in the cumulative frequency table (Table 15.2). The difference in the heights of consecutive rectangles indicates the number of scores in the corresponding frequency interval with the same right endpoint. For example, since the difference in the heights of the rectangles for the cumulative frequency intervals 50–79 and 50–89 is $22 - 16 = 6$, there are six scores in the frequency interval 80–89.

Example 1

Table 1 below represents the distribution of SAT math scores for 60 students at State High School. Table 2 is the cumulative frequency table for the same set of scores.

TABLE 1

Scores	Frequency
710–800	4
610–700	10
510–600	15
410–500	18
310–400	11
210–300	2

TABLE 2

Scores	Cumulative Frequency
210–800	
210–700	
210–600	
210–500	
210–400	
210–300	2

(a) Copy and then complete the cumulative frequency table (Table 2).
(b) Using the table completed in part (a), draw a cumulative frequency histogram.
(c) What percent of students scored above 700 or below 310?

Solution: (a) On each line of the cumulative frequency table after the first, add the frequency from the corresponding line of the frequency table to the cumulative frequency on the preceding line.

TABLE 2

Scores	Cumulative Frequency
210–800	$56 + 4 = 60$
210–700	$46 + 10 = 56$
210–600	$31 + 15 = 46$
210–500	$13 + 18 = 31$
210–400	$2 + 11 = 13$
210–300	2

(b) See the accompanying figure.

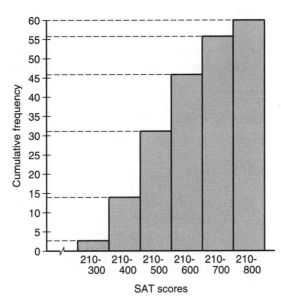

SAT scores

(c) **10**

Table 1 shows that 4 students scored above 700 and 2 students scored below

310. Thus, 6 of 60 students scored above 700 or below 310. Since $\dfrac{6}{60} = \dfrac{1}{10}$

= 10%, 10% of the students scored above 700 or below 310.

Example 2

Table 1 below shows the cumulative frequency of the ages of 35 people standing on a movie theater line.

TABLE 1

Ages	Cumulative Frequency
10–19	2
10–29	17
10–39	27
10–49	32
10–59	32
10–69	35

TABLE 2

Ages	Frequency
10–19	2
20–29	
30–39	
40–49	
50–59	
60–69	

(a) Using the data given in the cumulative frequency table (Table 1), copy and then complete the frequency table (Table 2).

(b) Using the frequency table obtained in part (a), determine the interval in which the median occurs.

Solution: (a) The frequency on each line after the first is obtained by taking the difference of the cumulative frequency that appears on the corresponding line of the cumulative frequency table and the entry that appears below it.

TABLE 2

Ages	Frequency
10–19	2
20–29	$17 - 2 = 15$
30–39	$21 - 17 = 10$
40–49	$32 - 27 = 5$
50–59	$32 - 32 = 0$
60–69	$35 - 32 = 3$

(b) **30–39**

Since there are 35 people, the median age is the age of the 18th person when the people are arranged in order of their ages. Counting down from the top line of the frequency table shows that the first two intervals contain $2 + 15 = 17$ of the lowest ages. Hence, the 18th age is contained in the next interval, 30–39.

Check Your Understanding of Section 15.5

A. Multiple Choice

1. The accompanying cumulative frequency histogram shows the distribution of scores that students received on an English test. How many students had scores between 71 and 80?
   (1) 16   (2) 8   (3) 20   (4) 4

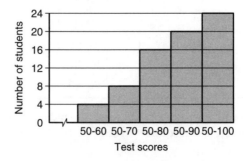

2. The accompanying cumulative frequency histogram shows the distribution of scores of 20 students on a Math A test.

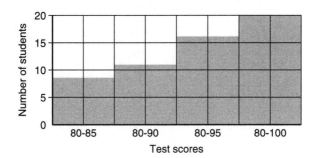

What percent of the students had scores above 90?
(1) 20   (2) 25   (3) 45   (4) 55

B. *In each case, show how you arrived at your answer by clearly indicating all of the necessary steps, formula substitutions, diagrams, graphs, charts, etc.*

3. The accompanying cumulative frequency histogram shows the distribution of mistakes 28 students in a French language class made on a text.

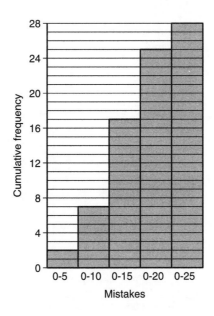

(a) What percent of the French class made fewer than 11 mistakes?
(b) What is the probability that a student selected at random made at least 16 mistakes?

4. The table on the left below shows the distribution of scores that 100 students received on a standardized test.

(a) Copy and complete the cumulative frequency table on the right below.

Scores	Frequency
91–100	15
81–90	26
71–80	23
61–70	15
51–60	11
41–50	5
31–40	3
21–30	2

Scores	Cumulative Frequency
21–100	
21–90	
21–80	
21–70	
21–60	
21–50	
21–40	
21–30	

(b) Using the table completed in part (a), draw a cumulative frequency histogram.

(c) According to the frequency table, which interval contains the upper quartile?

(1) 31–40   (2) 61–70   (3) 71–80   (4) 81–90

(d) What percent of the students scored less than 51?

5. The accompanying cumulative frequency histogram shows the distribution of the heights, in inches, of 24 high school students.

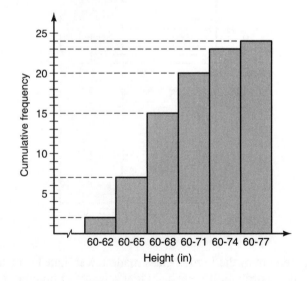

428

(a) Using the cumulative frequency histogram, copy and then complete the frequency table below.

Heights	Number of Students
60–62	
63–65	
66–68	
69–71	
72–74	
75–77	

(b) Using the table completed in part (a):
(1) Draw a frequency histogram.
(2) Determine the interval that contains the lower quartile,
(3) Determine the interval that contains the upper quartile.
(c) If a student is selected at random, what is the probability that the student is at most 68 inches tall?

6. (a) Using the frequency histogram given in Exercise 12 on page 422, copy and complete the accompanying cumulative frequency table.

Temperature Intervals	Cumulative Frequency
50–57	
50–65	
50–73	
50–81	
50–89	

(b) Using the information from part (a), construct a cumulative frequency histogram.

CHAPTER 16

PROBABILITY AND COUNTING METHODS

16.1 PROBABILITY OF A COMPOUND EVENT

KEY IDEAS

Sometimes we want to know the probability that a sequence of two or more events will occur. One way of finding the probability of such a **compound event** is to systematically write down all of the possible outcomes, and then compare the number of favorable outcomes to the total number of possible outcomes.

Listing Outcomes

Suppose a probability experiment consists of tossing a fair coin and then rolling a six-sided cube on which each side is marked with a different number from 1 to 6. The set of possible outcomes for this experiment can be described as a set of 12 ordered pairs:

$$\text{Sample space} = \begin{Bmatrix} (H,1),(H,2),(H,3),(H,4),(H,5),(H,6) \\ (T,1),\ (T,2),\ (T,3),\ (T,4),\ (T,5),\ (T,6) \end{Bmatrix}$$

The first member of each ordered pair represents a possible outcome of tossing the coin; H is a head and T is a tail. The second member of each pair is a possible outcome of rolling the cube.

- What is the probability of getting a head *and* rolling a 4? Of the 12 possible outcomes, only one, (H, 4), satisfies the given condition. Thus, $P(\text{H and 4}) = \dfrac{1}{12}$.

- What is the probability of getting a tail *and* rolling an odd number? Three of the 12 possible outcomes satisfy the given condition: (T, 1), (T, 3), and (T, 5). Hence, $P(\text{T and odd}) = \dfrac{3}{12}$.

- What is the probability of getting a head *or* rolling a 6? Of the 12 possible outcomes, seven are favorable: (H, 1), (H, 2), (H, 3), (H, 4), (H, 5), (H, 6), and (T, 6). Hence, $P(\text{H or 6}) = \dfrac{7}{12}$.

Tree Diagrams

The sample space for a compound event can also be described by using a **tree diagram** in which the path along the branches gives the set of all possible outcomes. For example, suppose Mr. and Mrs. Anderson plan on having three children born in different years. The accompanying tree diagram shows all of the possible arrangements of boy and girl children, where B represents a boy and G represents a girl.

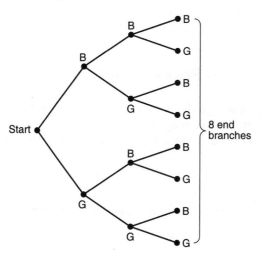

Tree Diagram of Possible Arrangements of Three Children

Notice in the tree diagram that:

- There are two branches at the start since there are two possibilities for the first child. If the first child is a boy (B), there are two possibilities for the second child; similarly, if the first child is a girl (G), there are two possibilities for the second child. For each of these four possibilities, branches are attached to represent the possible boy and girl outcomes for the third child.
- The tree ends in eight branches that correspond to the eight possible arrangements of three children.
- The eight arrangements can also be described as a set of ordered triples:

$$(B, B, B), (B, B, G), (B, G, B), (B, G, G),$$
$$(G, B, B), (G, B, G), (G, G, B), \text{ and } (G, G, G)$$

Example 1

According to the tree diagram given above, what is the probability that Mr. and Mrs. Anderson will have:

(a) all girls?   (b) *at least* two boys?

Solution: (a) $\dfrac{1}{8}$

Of the eight possible arrangements of boy and girl children, only one has all girls (G, G, G).

(b) $\dfrac{4}{8}$

Of the eight possible arrangements of boy and girl children, four have two or three boys: (B, B, B), (B, B, G), (B, G, B), and (G, B, B).

The Multiplication Principle of Counting

You may have noticed that the number of total possible arrangements of boy and girl children in the tree diagram could have been predicted by multiplying $2 \times 2 \times 2$ to obtain 8 since each of the three events has two possible outcomes.

MATH FACTS

MULTIPLICATION RULE OF COUNTING

If event *A* has *a* possible outcomes and event *B* has *b* possible outcomes, then there are *a* × *b* possible outcomes for event *A* followed by event *B*. This counting principle works for a sequence of two or more events.

Example 2

If a man has five different shirts, four different neckties, and three different sport jackets, how many different possible outfits, each consisting of a shirt, tie, and sports jacket, are possible?

Solution: **60**

There are three different events: (1) picking a shirt, with 5 possible outcomes, (2) picking a necktie, with 4 possible outcomes, and (3) picking a sport jacket, with 3 possible outcomes. Hence, $5 \times 4 \times 3 = 60$ different outfits, each consisting of one sport jacket, one shirt, and one tie, are possible.

Check Your Understanding of Section 16.1

Show how you arrived at your answer by clearly indicating all of the necessary steps, formula substitutions, diagrams, graphs, charts, etc.

1. An ice cream parlor makes a sundae using one of six different flavors of ice cream, one of three different flavors of syrup, and one of four different toppings. What is the total number of different sundaes that this ice cream parlor sells?

2. Marcy has five skirts, six blouses, and three scarves. How many outfits, each consisting of one skirt, one blouse, and one scarf, can Marcy create?

3. Meghan will be applying to three different colleges. She is choosing from among six 4-year colleges in New York, three out-of-state colleges, and a certain number of 2-year colleges. If Meghan finds that there are 90 different ways in which she can apply to one 4-year college in New York, one 2-year college, and one out-of-state college, how many choices of 2-year colleges does she have?

4. A penny, a nickel, and a dime are in a box. Bob randomly selects a coin, notes its value, and returns it to the box. He then randomly selects another coin from the box.
   (a) Show all possible outcomes by drawing a tree diagram or by representing the possible outcomes as a set of ordered pairs.
   (b) What is the probability that the same coin will be drawn both times?
   (c) What is the probability that a nickel will be drawn *at least* once?
   (d) What is the probability that the total value of both coins that are selected will exceed 11¢?

5. The assembly committee of the River High School student council consists of four students whose ages are 14, 15, 16, and 17, respectively. One student will be chosen at random to be chairperson, and then, from the remaining three, one will be chosen at random to be the recording secretary.
   (a) Draw a tree diagram or list the sample space showing all possible outcomes after both drawings.
   (a) Find the probability that:
       (1) the chairperson is older than the recording secretary
       (2) both students chosen are under the age of 16
       (3) both students chosen are the same age

6. A book bag contains one novel, one biography, and one poetry book, all of the same size. Henry selects one book at random from the bag, looks at the book, and then places the book back in the bag. Henry then selects another book without looking.

(a) What is the probability that Henry will select a novel both times?

(b) What is the probability that Henry will selected a poetry book *at least* once?

(c) What is the probability that a biography will *not* be selected?

(d) What is the probability that Henry will select the same book both times?

16.2 PROBABILITY FORMULAS

Listing and counting all possible outcomes for two or more events can be time consuming, if not impractical. To find the probability that event *A* **or** event *B* will happen, we *add* probabilities. To find the probability that event *A* **and** event *B* will happen, we *multiply* probabilities. The notation P(*A*, *B*) indicates the probability that event *B* will occur after event *A*.

Adding Probabilities

When calculating the probability that event A or event B will occur, you need to know whether the two events can happen at the same time. Suppose each side of a cube has on it a different whole number from 1 to 6. When the cube is tossed, each of the numbers from 1 to 6 has an equal chance of showing on the top face.

• If event A is tossing a 2 and event B is tossing an odd number, then the two events cannot happen at the same time. The probability that event A or event B will occur is the sum of their individual probabilities:

$$P(2 \text{ or odd}) = P(2) + P(\text{odd})$$
$$= \frac{1}{6} + \frac{3}{6}$$
$$= \frac{4}{6}$$

- Let event A be tossing a number greater than 4 and event B be tossing an odd number. Since 5 is greater than 4 and is also odd, the two events have one favorable outcome in common. Thus:

$$P(\text{greater than 4 odd}) = P(\text{greater than 4}) + P(\text{odd}) - P(\text{greater than 4 and odd})$$

$$= \frac{2}{6} \qquad + \frac{3}{6} \qquad - \frac{1}{6}$$

$$= \frac{4}{6}$$

The probability value $\frac{1}{6}$ is subtracted since one outcome of the six possible outcomes, namely 5, is counted when calculating P (greater than 4) and counted again when calculating P (odd), which is one extra time.

=== **MATH FACTS** ===

- If events A and B cannot happen at the same time, then

$$P(A \text{ or } B) = P(A) + P(B)$$

- If events A and B can happen at the same time, then

$$P(A \text{ or } B) = P(A) + P(B) - P(A \text{ and } B)$$

Multiplying Probabilities

To find the probability that two events will both occur, multiply their probabilities together. For example, if a fair coin is tossed two times, the probability it will land head up on each toss is $\frac{1}{2} \times \frac{1}{2} = \frac{1}{4}$. If a fair coin is tossed three times, the probability it will land head up on each toss is $\frac{1}{2} \times \frac{1}{2} \times \frac{1}{2} = \frac{1}{8}$.

Suppose that, when a fair coin was tossed three times, it landed head up on each toss. The probability that it will land head up on the fourth toss remains $\frac{1}{2}$. Past results have no effect on a future outcome provided that the coin is fair.

Probabilities *With* and *Without* Replacement

When finding the probability that a sequence of two events will occur, you need to know whether the outcome of the first event affects the outcome of the second event.

Example 1

From a jar that contains three white marbles and five green marbles, a marble is picked without looking and its color is noted (event A). Then a second marble is picked without looking and its color is noted (event B). Find the probability of picking two green marbles if:

(a) the first marble is replaced in the jar before the second marble is picked
(b) the first marble is *not* replaced

> *Solution*: (a) $\dfrac{25}{64}$

Since the first marble is replaced, the outcome of the first pick does *not* affect the outcome of the next selection, so events A and B are *independent*.

- Five of the eight marbles are green, so the probability of picking a green marble on the first draw is $\dfrac{5}{8}$. Hence, $P(A) = \dfrac{5}{8}$.
- Because the first marble is replaced in the jar, the probability of picking a green marble on the second draw is also $\dfrac{5}{8}$. Hence, $P(B) = \dfrac{5}{8}$.
- Thus, if the two marbles are picked *with* replacement:

$$P(A \text{ and } B) = P(A) \times P(B)$$
$$= \frac{5}{8} \times \frac{5}{8}$$
$$= \frac{25}{64}$$

> (b) $\dfrac{20}{56}$

If the first marble is *not* replaced, then the outcome of the first draw affects the outcome of the next pick, so events A and B are *dependent*.

- $P(A) = \dfrac{5}{8}$.

- Since the first marble is not replaced, four of the remaining seven marbles are green. Hence, $P(B) = \dfrac{4}{7}$.

- Thus, if the two marbles are picked *without* replacement:

$$P(A \text{ and } B) = P(A) \times P(B/\text{given } A \text{ occurs first})$$

$$= \frac{5}{8} \times \frac{4}{7}$$

$$= \frac{20}{56}$$

MATH FACTS

- If event *A* does not affect the outcome of event *B*, as is the case when finding probability with replacement, then

$$P(A \text{ and } B) = P(A) \times P(B)$$

- If event *A* affects the outcome of event *B*, as is the case when finding probability without replacement, then

$$P(A \text{ and } B) = P(A) \times P(B/\text{given } A \text{ occurs first})$$

Example 2

A sock drawer contains 12 navy socks, eight black socks, and no other socks. Find the probability that, if you reach in the drawer and take two socks, one at a time without replacement, the socks will be the same color.

Solution: $\dfrac{188}{380}$

To find the probability of selecting two socks of the same color, find the probability of picking two navy or two black socks.

- Of the $12 + 8 = 20$ socks, 12 are navy. On the first pick, $P(\text{first navy}) = \dfrac{12}{20}$.

  On the second pick there is 1 less navy sock, so $P(\text{second navy}) = \dfrac{11}{19}$.

  Thus:

$$P(\text{navy, navy}) = \frac{12}{20} \times \frac{11}{19} = \frac{132}{380}$$

- Of the 20 socks, 8 are black. On the first pick, $P(\text{first black}) = \frac{8}{20}$. On the second pick there is 1 less black sock, so $P(\text{second black}) = \frac{7}{19}$. Thus:

$$P(\text{black, black}) = \frac{8}{20} \times \frac{7}{19} = \frac{56}{380}$$

- Hence:

$$P(\text{same color}) = P(\text{navy, navy}) + P(\text{black, black})$$
$$= \frac{132}{380} + \frac{56}{380}$$
$$= \frac{188}{380}$$

Check Your Understanding of Section 16.2

A. *Multiple Choice*

1. A bag of marbles contains two green, one blue, and three red marbles. If two marbles are chosen at random without replacement, what is the probability that both will be red?

   (1) $\frac{1}{5}$ (2) $\frac{1}{6}$ (3) $\frac{1}{10}$ (4) $\frac{1}{12}$

2. A pencil holder contains six blue pencils and three red pencils. If two pencils are picked at random and without replacement, what is the probability that both are blue?

   (1) $\frac{2}{9}$ (2) $\frac{6}{9}$ (3) $\frac{30}{72}$ (4) $\frac{30}{81}$

3. A whole number from 1 to 12, inclusive, is picked at random. What is the probability that the number is less than 7 or is prime?

   (1) $\frac{1}{2}$ (2) $\frac{7}{12}$ (3) $\frac{2}{3}$ (4) $\frac{11}{12}$

B. *In each case, show how you arrived at your answer by clearly indicating all of the necessary steps, formula substitutions, diagrams, graphs, charts, etc.*

4. There are 11 girls and 19 boys in a mathematics class. If the teacher randomly selects two students to come to the board to show their work, what is the probability both students will be girls?

5. A letter is selected at random from the word "TRAPEZOID." Find the probability that the letter selected will have vertical or horizontal line symmetry.

6. John's sock drawer contains 10 identical navy blue socks, 14 identical black socks, and four identical white socks with no other socks. If John selects two socks at random from the drawer without replacement, what is the probability that the socks will *not* be the same color?

7. A softball team plays two games each weekend, one on Saturday and the other on Sunday. The probability of winning the game scheduled for next Saturday is $\frac{3}{5}$, and the probability of winning the following game, scheduled for Sunday, is $\frac{4}{7}$. What is the probability that the team will win *at least* one of the two games?

8. A bookshelf contains six mysteries and three biographies. Two books are selected at random without replacement.
   (a) What is the probability that both books are biographies?
   (b) What is the probability that one book is a mystery and the other is a biography?

9. A candy jar has four red gumdrops, five green gumdrops, and one black gumdrop. Without looking, Kim reaches into the jar and chooses one gumdrop. Without replacing this gumdrop, Kim chooses a second gumdrop. The accompanying tree diagram represents all possible outcomes with the probability value on each branch.
   (a) Find the values of x, y, and z.
   (b) Find the probability that
       (1) both gumdrops chosen are green
       (2) neither of the gumdrops chosen is red

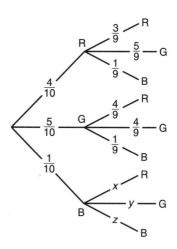

10. There are only three flavors of gumdrops in a jar containing 40 gumdrops. There are three times as many cherry gumdrops as lemon gumdrops. There are 4 more than twice as many orange gumdrops as lemon gumdrops.
(a) How many gumdrops of each flavor are in the jar?
(b) Two gumdrops are drawn at random and without replacement. Find the probability that both are the same flavor.

11. Tyrone randomly picks two bills from his wallet, which contains three $1 bills, four $5 bills, and one $10 bill. If the two bills are removed without replacement, determine:
(a) the probability that the two bills will have the same face values.
(b) whether the probability that the two bills will total exactly $6 is greater than the probability that the two bills will total *at least* $10.

12. In a certain class, there are four students in the first row: three girls (Ann, Barbara, and Cathy), and one boy (David). The teacher will call on one of these students to solve a problem at the board. When the problem is completed, the teacher will call on one of the remaining students in the first row to do a second problem at the board. What is the probability that:
(a) the teacher will call Ann first, and Barbara second?
(b) the teacher will call two girls to the board?
(c) David will be one of the two students called?

13. Elton bought a pack of 16 baseball cards. He sorted them by position and noted that he had pictures of four pitchers, five outfielders, and seven infielders. Two cards are randomly chosen from the pack of 16 cards without replacement. Find the probability that:
(a) *at least* one of the cards shows a pitcher
(b) neither card shows a pitcher

14. Adam bought a package of marbles and sorted them by color as shown by the accompanying graph.

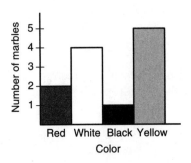

(a) If one marble is selected at random, what is the probability it will be red, black, or yellow?

(b) If two marbles are selected at random, without replacement, what is the probability that a black marble will be selected?

(c) If two marbles are selected at random, without replacement, what is the probability that neither marble selected will be red or black?

16.3 PERMUTATIONS

KEY IDEAS

The product of consecutive positive integers from n down to 1, inclusive, is called n **factorial** and is written as $n!$, where $0!$ is defined to be equal to 1. For example, $4! = 4 \times 3 \times 2 \times 1 = 24$. Because of the counting principle, $4!$ represents the number of different ways in which four people or objects can be lined up. Each different arrangement is called a *permutation*. The number of different ways in which n objects can be arranged in a line is $n!$.

Arranging n Objects in n Slots

The lock combination 5-19-34 is different from the lock combination 34-19-5. Each of these lock combinations represents a *permutation* of the same three numbers: 5, 19, and 34. A **permutation** is an arrangement of objects in which the order of the objects matters. To figure out the total number of different lock combinations that consist of the three numbers 5, 19, and 34, use the counting principle and reason as follows:

- Any one of the three numbers can fill the first position of the lock combination:

$$\boxed{3} \times \boxed{?} \times \boxed{?}$$

total number of choices for each
position of the lock combination

- After one of the three numbers is used, any one of the remaining two numbers can fill the second position of the lock combination:

$$\boxed{3} \times \boxed{2} \times \boxed{?}$$

- Since the remaining number must be used to fill the third position, there is only one possible choice for the third position of the lock combination:

$$\boxed{3} \times \boxed{2} \times \boxed{1} = 6$$

Thus, the *permutation of three objects* (the numbers 5, 19, and 34) *taken three at a time* (all three numbers are used in every lock combination) is 6. The shorthand notation $_nP_n$ represents the permutation of n objects taken n at a time. Thus,

$$_3P_3 = 3! = 3 \times 2 \times 1 = 6$$

Arranging *n* Objects in Fewer Than *n* Slots

The number of objects or people to be arranged may be greater than the available number of slots. For example, if seven students run in a race in which there are no ties, then the number of different arrangements of first, second, and third place that are possible is

$$\boxed{7} \times \boxed{6} \times \boxed{5} = 210$$

Thus, the total number of possible arrangements of seven students taken three at a time is the product of the three greatest factors of 7. This product can be represented by the shorthand notation $_7P_3$; that is, $_7P_3 = 7 \times 6 \times 5 = 210$. In general, $_nP_r$ means the product of the r greatest factors of n.

Example 1

Shari remembers the last four digits of a seven-digit telephone number but knows only that each of the first three digits is an odd number. Find the maximum number of telephone calls she must make before she dials the correct number when:
  (a) the first three digits are different
  (b) digits can be repeated

Solution: (a) **60**

There are five odd digits: 1, 3, 5, 7, and 9. Hence, you must find the number of arrangements of the five digits in three available slots. Any one of the five digits can be used as the first digit of the telephone number. Since no digit can be repeated, any of the other four odd digits can serve as the second digit, and any of the remaining three odd digits can fill the third slot. According to the counting principle, the first three positions of the telephone number can be selected from the five odd digits, without repetition of digits, in $_5P_3 = 5 \times 4 \times 3 = 60$ ways.

  (b) **125**

If a digit can be repeated, then there are five ways in which each of the first three positions of the telephone number can be filled. According to the counting principle, the first three positions of the telephone number can be selected from the five odd digits, with repetition of digits allowed, in $5 \times 5 \times 5 = 125$ ways.

================================ **MATH FACT** ================================

- To find the number of different arrangements when n objects are used to fill n available slots, calculate $n!$ or, using permutation notation, $_nP_n$.
- To find the number of different arrangements when n objects are used to fill r available slots, where r is less than or equal to n, multiply together the r greatest factors of n. This product is denoted by $_nP_r$.

Counting Arrangements of a Set of Objects Some of Which Are Identical

If in a set of n objects some are exactly alike, then the n objects can be arranged in fewer ways than n objects that are all different.

- If a out of n objects are identical, the number of different arrangements of the n objects is $\dfrac{n!}{a!}$. For example, the word "BETWEEN" has seven letters, and three of these letters (E) are identical. Thus, the number of different arrangements of these seven letters is as follows:

$$\frac{7!}{3!} = \frac{7 \times 6 \times 5 \times 4 \times \cancel{3} \times \cancel{2} \times \cancel{1}}{\cancel{3} \times \cancel{2} \times \cancel{1}}$$

$$= 7 \times 6 \times 5 \times 4$$

$$= 840$$

- If a set of n objects includes a identical objects, b identical objects, c identical objects, and so forth, the number of different arrangements of the n objects is $\dfrac{n!}{a! \times b! \times c! \times \cdots}$. For example, to find the number of ways in which four red flags, three blue flags, and one green flag can be arranged on a vertical flag pole, use the formula $\dfrac{n!}{a! \times b! \times c!}$ when $n = 4 + 3 + 1 = 8$, $a = 4$, $b = 3$, and $c = 1$:

$$\frac{8!}{4! \times 3! \times 1!} = \frac{8 \times 7 \times \cancel{6} \times 5 \times \cancel{4} \times \cancel{(3 \times 2 \times 1)}}{\cancel{(4!)} \times \cancel{(3 \times 2 \times 1)} \times (1)}$$

$$= 8 \times 7 \times 5$$

$$= 280$$

443

Counting Arrangements of a Set of Objects Subject to Conditions

When counting the arrangements of a set of objects, there may be conditions that require certain objects in the set to fill particular positions in each arrangement.

Example 2

How many three-digit numbers greater than 500 can be formed using the digits 1, 2, 3, 4, 5 and 6 if no digit is repeated?

Solution: **40**

Since the three-digit number must be greater than 500, there are two digits that can fill the first position, 5 or 6. After the first position is filled, there is no restriction on how to fill each of the remaining positions, provided that the same digit is not used more than once. Since five digits are available for the second position, four digits remain to fill the last position. Thus, the number of three-digit numbers greater than 500 that can be formed is:

$$\boxed{2} \times \boxed{5} \times \boxed{4} = 40$$

Evaluating n! and $_nP_r$ Using a Calculator

With a scientific calculator, you can use the "second" function keys to evaluate factorials and permutations, as outlined below.

- To evaluate 7!, follow these steps:
  1. Enter 7.
  2. Locate the key that has n! (or x!) above it. To access this "second" function, first press the $\boxed{\text{INV}}$ or $\boxed{\text{2nd}}$ or $\boxed{\text{SHIFT}}$ key, depending on your particular calculator.
  3. Then press the key that has n! (or x!) above it.
  4. Copy the number in the display window, 5040, which is the value of 7!.
- To evaluate $_7P_3$, follow these steps:
  1. Enter 7.
  2. Locate the key that has $_nP_r$ above it. To access this "second" function, first press the $\boxed{\text{INV}}$ or $\boxed{\text{2nd}}$ or $\boxed{\text{SHIFT}}$ key, depending on your particular calculator.
  3. Then press the key that has $_nP_r$ above it.
  4. Enter 3.
  5. Press the $\boxed{=}$ key.

6. Copy the number in the display window, 210, which is the value of $_7P_3$.

Not all calculators work in the same way. If this procedure does not work with your calculator, you will need to read the instruction manual that came with the calculator.

⎯⎯⎯⎯⎯⎯⎯⎯⎯⎯⎯(**TIP**)⎯⎯⎯⎯⎯⎯⎯⎯⎯⎯⎯

To evaluate 7! or $_7P_3$ using a TI-graphing calculator:

- Press 2nd [QUIT] to return to the home screen. Then enter 7.
- Press MATH ▷ ▷ ▷ to access the MATH PRoBability menu.
- Press 4 ENTER to calculate 7!, or calculate $_7P_3$ by pressing 2 to select $_nP_r$. Then enter 3 and press ENTER.

Probability and Arrangements

In order to write the numerator or denominator of a probability fraction, it may be necessary to first figure out the number of arrangements of objects that are possible according to the conditions of the problem.

Example 3

Roberta has a lock for which the combination consists of a sequence of four single digits. Roberta forgets the combination, but remembers that it contains the digits 2, 4, 6, and 9. What is the probability that in the correct combination the digit 9 comes before the other digits?

Solution: $\dfrac{6}{24}$

Since the combination consists of four digits and must contain the digits 2, 4, 6, and 9, each of these digits appears exactly once in the combination.

- The total number of different combinations that begin with 9 is

$$\boxed{1} \times \boxed{3} \times \boxed{2} \times \boxed{1} = 6$$
↑
Must be 9.

- The total number of different four-digit combinations is

$$\boxed{4} \times \boxed{3} \times \boxed{2} \times \boxed{1} = 24$$

- Thus, six of the 24 possible lock combinations begin with 9. If it is assumed that each of the 24 possible combinations may be the correct one, the probability that the digit 9 comes before the other three digits in the correct combination is $\frac{6}{24}$.

Check Your Understanding of Section 16.3

A. Multiple Choice

1. How many different six-letter arrangements can the made from the letters of the name "JENNIE"?
   (1) 15   (2) 30   (3) 180   (4) 720

2. The greatest number of different five-letter arrangements can be made from the letters in the word
   (1) "ANGLE"   (2) "DADDY"   (3) "ORDER"   (4) "ADAPT"

3. A locker combination system uses four digits from 0 to 9. How many different four-digit locker combinations are possible if the first and the last digit of the combination cannot be 0 and no digit can be repeated?
   (1) 3024   (2) 4032   (3) 5040   (4) 7200

B. In each case, show how you arrived at your answer by clearly indicating all of the necessary steps, formula substitutions, diagrams, graphs, charts, etc.

4. In how many different ways can a chemistry book, a calculus book, a history book, a poetry book, and a dictionary be arranged on a shelf so that the chemistry book or the history book appears first?

5. How many four-digit odd numbers can be formed using the digits 0, 1, 2, 3, 4, and 5 if no digit is used in any number more than once?

6. In how many different ways can three black flags, four red flags, and two green flags be arranged on a vertical flagpole if a green flag is always on top?

7. (a) In how many different ways can the letters of the word "COMMITTEE" be arranged?
   (b) If the letters of the word "COMMITTEE" are randomly arranged, what is the probability that the letter C, O, or I appears first?

8. Three digits selected at random and without repetition from the set {2, 4, 5, 6, 8} are used to form a three-digit whole number. What is the probability that the number formed will be less than 600?

9. If the letters of the word "PARABOLA" are randomly arranged to form a new word, what is the probability that the three A's will appear together at the beginning or at the end of the new word?

10. Five students, all of different heights, are to be randomly arranged in a line. What is the probability that the tallest student will be first and the shortest student will be last in line?

11. What is the probability that when Allan, Barbara, Jose, Steve, Charles, and Maria line up:
    (a) all four boys are before the two girls?
    (b) a girl is first and a boy is last?

12. The telephone company has run out of seven-digit telephone numbers for an area code. To fix this problem, the telephone company will introduce a new area code. Find the maximum number of new seven-digit telephone numbers that can be generated for the new area code if both of the following conditions must be met:
    1. The first digit cannot be 0 or 1.
    2. The first three digits cannot be the emergency number 911 or the number 411, which is used for information.

16.4 COMBINATIONS

KEY IDEAS

If from a group of five people a committee consisting of Joe, Susan, and Elizabeth is selected, this *combination* of three people is the *same* as a committee consisting of Elizabeth, Joe, and Susan. A **combination** is a selection of people or objects in which the *identity*, rather than the order, of the people or objects is important.

In a *permutation*, unlike a combination, order is considered. For example, the arrangement in a line of Joe followed by Susan followed by Elizabeth is *different* from the arrangement in a line of Elizabeth followed by Joe followed by Susan.

Combination Versus Permutation

The letters A, B, and C may be arranged in six different ways:

(1) ABC, (2) ACB, (3) BAC, (4) BCA, (5) CAB, (6) CBA

Although there are six permutations of the letters A, B, and C, there is only one distinct set, or combination, of the three letters: $\{A, B, C\}$. A *combination* is an unordered set of objects, while a *permutation* is any ordered arrangement of the objects in that set.

Combination Notation

If a set consists of n objects, then the *number* of different subsets that contain r of those n objects, without regard to their order, is denoted by $_nC_r$. The notation $_nC_r$ is read as "the number of combinations of n objects selected r at a time." The different combinations of the three letters A, B, and C ($n = 3$) selected two at a time ($r = 2$) are the subsets $\{A, B\}$, $\{A, C\}$, and $\{B, C\}$. Since there are three subsets for $n = 3$ and $r = 2$, $_3C_2 = 3$.

Combination Formula

It is usually not practical to evaluate $_nC_r$ by listing and then counting the number of different subsets with r objects that can be formed from a set of n objects. Instead, the following formula can be used to evaluate $_nC_r$:

MATH FACTS

If $0 \leq r \leq n$, then

$$_nC_r = \frac{n!}{r!(n-r)!} = \frac{_nP_r}{r!}$$

For example, to find the total number of committees that can be formed from five people, evaluate $_5C_3$ by letting $n = 5$ and $r = 3$ in the combination formula:

$$_nC_r = \frac{_nP_r}{r!}$$

$$_5C_3 = \frac{_5P_3}{3!}$$

$$= \frac{5 \cdot 4 \cdot 3}{3 \cdot 2 \cdot 1}$$

$$= 5 \cdot 2$$

$$= 10$$

Combinational Relationships

The following formulas can often save you time in evaluating combinations that have the indicated forms:

MATH FACTS

- $_nC_n = 1$ *Example:* $_9C_9$ $= 1$
- $_nC_1 = n$ *Example:* $_{13}C_1$ $= 13$
- $_nC_0 = 1$ *Example:* $_8C_0$ $= 0$
- $_nC_k = {}_nC_{n-k}$ $(n \geq k)$ *Example:* $_{15}C_{13}$ $= {}_{15}C_2$

Example 1

A jar contains two red marbles, three white marbles, and five blue marbles. In how many ways can a set of seven marbles be selected?

Solution: **120**

A group of seven marbles ($r = 7$) can be selected from a group of 10 marbles ($n = 2 + 3 + 5 = 10$) in

$$_{10}C_7 = {}_{10}C_{10-7} = \frac{_{10}P_3}{3!} = \frac{10 \cdot \overset{3}{\cancel{9}} \cdot \overset{4}{\cancel{8}}}{\cancel{3} \cdot \cancel{2} \cdot 1} = 120 \text{ ways}$$

Example 2

The coach of a team plans to select five players at random from a group of 11 students trying out for the team. If Lois is one of the 11 students trying out, how many five-player combinations will:
(a) include Lois?   (b) not include Lois?

Solution: (a) **210**

Since Lois must be on the team, the other four team players can be selected from the remaining 10 students in $_{10}C_4$ ways. Therefore, the number of five-player combinations that include Lois is:

$$_{10}C_4 = \frac{_{10}P_4}{4!} = \frac{10 \cdot \overset{3}{\cancel{9}} \cdot \overset{2}{\cancel{8}} \cdot 7}{\underset{1}{\cancel{4} \cdot \cancel{3} \cdot \cancel{2} \cdot 1}} = 210$$

(b) **252**

Since Lois cannot be on the team, the five team members must be selected from the remaining 10 students. Therefore, the number of five-player combinations that do *not* include Lois is:

$$_{10}C_5 = \frac{_{10}P_5}{5!} = \frac{\cancel{10} \cdot 9 \cdot \overset{2}{\cancel{8}} \cdot 7 \cdot \overset{2}{\cancel{6}}}{\cancel{5} \cdot \cancel{4} \cdot \cancel{3} \cdot \cancel{2} \cdot 1} = 252$$

Evaluating $_nC_r$ Using a Calculator

With a scientific calculator, you can use the "second" function keys to evaluate combinations. For example, to evaluate $_7C_3$, follow these steps:

1. Enter 7.
2. Locate the key that has $_nC_r$ above it. To access this "second" function, first press the INV or 2nd or SHIFT key, depending on your particular calculator.
3. Then press the key that has $_nC_r$ above it.
4. Enter 3.
5. Press the = key.
6. Copy the number in the display window, 35, which is the value of $_7C_3$.

Not all calculators work in the same way. If this procedure does not work with your calculator, you will need to read the instruction manual that came with the calculator.

(**Tip**)

To evaluate $_7C_3$ using a TI-83 graphing calculator:

- Press $\boxed{\text{2nd}}$ [QUIT] to return to the home screen. Then enter 7.
- Press $\boxed{\text{MATH}}$ $\boxed{\triangleright}$ $\boxed{\triangleright}$ $\boxed{\triangleright}$ $\boxed{3}$ to select $_nC_r$ from the MATH PRoBability menu.
- Enter 3. Then press $\boxed{\text{ENTER}}$.

Multiplying Combinations

To figure out the total number of selections that are possible when selecting from two or more groups, multiply together the numbers of possible selections for all of the groups.

Example 3

There are six pens and seven books on a desk. In how many different ways can four pens and three books be selected from the desk?

Solution: **525**

Figure out the number of possible selections of four from six pens and the number of possible selections of three from seven books. Then multiply the results together.

- Four pens can be selected from six pens in $_6C_4$ ways. Use your calculator to find out that $_6C_4 = 15$.
- Three books can be selected from seven books in $_7C_3$ ways. Use your calculator to find out that $_7C_3 = 35$.
- The number of different ways four pens and three books can be selected is $15 \times 35 = 525$.

Combination Problems with *At Least* and *At Most* Conditions

At least means "is equal to or is greater than," and *at most* means "is equal to or is less than." Example 4 illustrates how to solve probability problems with combinations involving these two types of conditions.

Example 4

A four-member subcommittee is selected at random from a U.S. Senate committee consisting of five Republicans and three Democrats.
(a) What is the probability that the four-member subcommittee will include *at least* two Democrats?
(b) What is the probability that the four-member subcommittee will include *at most* three Republicans?

Solution:

The number of four-member subcommittees that can be formed from eight people (5 Republicans + 3 Democrats) is $_8C_4 = 70$.

(a) $\dfrac{35}{70}$

Since the committee includes three Democrats, any subcommittee cannot include more than three Democrats. The four-member subcommittee will include *at least* two Democrats if it consists of two Democrats and two Republicans *or* three Democrats and one Republican.

Committee Makeup	Number of Committees
• 2 Democrats and 2 Republicans	$_3C_2 \times _5C_2 = 3 \times 10 = 30$
• 3 Democrats and 1 Republican	$_3C_3 \times _5C_1 = 1 \times 5 = 5$

Thus:

$$P(\textit{at least } 2 \text{ Democrats}) = \frac{30}{70} + \frac{5}{70} = \frac{35}{70}$$

(b) $\dfrac{65}{70}$

Method 1. The four-member subcommittee will include at most three Republicans if it consists of three Republicans and one Democrat, two Republicans and two Democrats, or one Republican and three Democrats. The subcommittee cannot consist of no Republican and four Democrats, because there are only three Democrats.

Committee Makeup	Number of Committees
• 3 Republicans and 1 Democrat	$_5C_3 \times _3C_1 = 10 \times 3 = 30$
• 2 Republicans and 2 Democrats	$_5C_2 \times _3C_2 = 10 \times 3 = 30$
• 1 Republicans and 3 Democrats	$_5C_1 \times _3C_3 = 5 \times 1 = 5$

Thus:

$$P(at\ most\ 3\ Republicans) = \frac{30}{70} + \frac{30}{70} + \frac{5}{70} = \frac{65}{70}$$

<u>Method 2.</u> You could also adopt a different point of view by reasoning as follows:

- $P(at\ most\ 3\ Republicans) + P(4\ Republicans) = 1$.
- $P(4\ Republicans) = \frac{{}_5C_4}{70} = \frac{5}{70}$.
- $P(at\ most\ 3\ Republicans) = 1 - P(4\ Republicans) = 1 - \frac{5}{70} = \frac{70}{70} - \frac{5}{70} = \frac{65}{70}$.

Check Your Understanding of Section 16.4

A. *Multiple Choice*

1. What is the value of $\frac{{}_4P_2}{{}_4C_2}$?

(1) $\frac{1}{2}$ (2) 2 (3) 6 (4) 72

2. How many different three-member teams can be formed from six students?
(1) 20 (2) 120 (3) 216 (4) 720

3. If ${}_xC_{10} = {}_xC_2$, what is the value of x?
(1) 5 (2) 8 (3) 12 (4) 20

4. How many committees of four can be selected from a group of seven if the same person must be included in each committee?
(1) ${}_7C_3$ (2) ${}_6C_3$ (3) ${}_6C_4$ (4) ${}_7C_4 - 1$

5. A classroom has 12 girls and 15 boys. A committee of two students is selected at random. Which expression represents the probability that both students are girls?

(1) $\frac{{}_{12}C_2}{{}_{27}C_2}$ (2) $\frac{{}_{12}C_2}{{}_{15}C_2 + {}_{12}C_2}$ (3) $\frac{{}_{15}C_2}{{}_{27}C_2}$ (4) $\frac{{}_{12}P_2}{{}_{15}P_2 \cdot {}_{12}P_2}$

B. *In each case, show how you arrived at your answer by clearly indicating all of the necessary steps, formula substitutions, diagrams, graphs, charts, etc.*

6. If $_xC_2 = 45$, what is the value of x?

7. A committee of five is to be chosen from six freshmen and eight sophomores. What is the probability that the committee will include two freshmen and three sophomores?

8. A dinner menu at a restaurant offers a choice of four appetizers, three soups, seven entrees, and five desserts. If Roberto orders an appetizer *or* a soup, one entree, and two different desserts, how many different choices can he make?

9. Jennifer has five sweaters, six vests, four jackets, and seven pieces of jewelry. Find the number of different outfits consisting of a sweater *or* a vest, one jacket, and three pieces of jewelry that she can wear.

10. Emily receives a box of chocolates containing 10 candies: four nut clusters, one peppermint, two jellies, and three caramels. What is the probability that a five-candy selection will contain:
    (a) four nut clusters and one peppermint?
    (b) two nut clusters, two jellies, and one caramel?
    (c) all caramels?

11. A committee of five is to be selected from five sophomores, four juniors, and three seniors. What is the probability that the majority of the committee will be seniors?

12. The student government at Central High School consists of four seniors, three juniors, three sophomores, and two freshmen.
    (a) What is the probability that a committee of four students will have *at least* three seniors?
    (b) If nine students will be chosen from the student government to attend a convention, what is the probability that a senior will be chosen to go?

13. Four books are selected at random from a shelf that contains five mysteries, three biographies, and two history books.
    (a) What is the probability that in the four-book selection a mystery is *not* included?
    (b) What is the probability that in the four-book selection *at most* two mysteries are included?

14. On a math test a student is to select any four of nine problems of equal difficulty. The test contains one geometry, three algebra, one locus, and four probability problems.
(a) What is the probability that a four-problem selection will include the same number of algebra problems as probability problems?
(b) What is the probability that a four-problem selection will include *at least* two probability problems?

15. Kayla, Alyssa, Juanita, Dominic, and Troy want to go to a concert, but they have only three tickets. They decide to select randomly who will go.
(a) What is the probability that Dominic will be selected?
(b) What is the probability that Alyssa will *not* be selected?
(c) How many different three-person groups can be selected if Kayla must be a member?

16. Theresa has five mysteries, four adventure stories, and three historical novels. She can choose eight books for her summer reading.
(a) What is the probability that an eight-book selection will contain three mysteries, two adventure stories, and three historical novels?
(b) What is the probability that an eight-book selection will contain no adventure stories?
(c) What is the probability that an eight-book selection will contain no mysteries?

Answers and Solution Hints

CHAPTER 1

Section 1.1
1. (4) **2.** (2) **3.** (3) **4.** (3) **5.** (3) **6.** −7
7. 4 **8.** −2.5 **9.** $\dfrac{1}{6}$ **10.** −4 **11.** 4 **12.** −2
13. 13 **14.** 12 **15.** 60 **16.** −4.5 **17.** 2 **18.** −3
19. 3 **20.** −4 **21.** 4 **22.** −24 **23.** −8 **24.** 30

Section 1.2
1. (3) **2.** (3) **3.** (1) **4.** (1) **5.** (4) **6.** (1)
7. (2) **8.** (3) **9.** (4) **10.** (2) **11.** (2) **12.** (2)
13. (3) **14.** (3) **15.** (1) **16.** (1) **17.** (2)

Section 1.3
1. (1) **2.** (1) **3.** (2) **4.** (2) **5.** (3) **6.** (3)
7. (2) **8.** (1) **9.** (3) **10.** $-32y^{15}$ **11.** $\dfrac{11}{4}$
12. 1.25×10^{-10} **13.** (a) $2^4 \times 4^2 \neq 8^6$ (b) 8
14. $x^2 > y^2$ when $x > y > 0$

Section 1.4
1. (3) **2.** (1) **3.** (1) **4.** (2) **5.** (3) **6.** (2)
7. 100 **8.** 21 **9.** −7 **10.** −32 **11.** 32 **12.** 32
13. 3 **14.** 0

Section 1.5
1. (4) **2.** (4) **3.** $2x - 5$ **4.** $2x - 4 < 11$
5. $3x + 6 \leq 21$ **6.** $2(x + 3) \geq 12$ **7.** $x - 2 = \dfrac{1}{2}x + 1$
8. $2(x - 5) \leq 24$ **9.** $x + 7 = 2x - 3$ **10.** $3(13 - x) + 2x = 33$
11. $5x - x = 8 + 2x$ **12.** $x(x + 7) = 5[x + (x + 7)] + 19$
13. $c + 3m$ **14.** $\$12.95 + \$0.25[60(x - 1) + y]$

CHAPTER 2

Section 2.1
1. −14 **2.** −4 **3.** −6 **4.** 70 **5.** −6 **6.** −15
7. −9.2 **8.** −2.3 **9.** $-\dfrac{2}{3}$ **10.** 24 **11.** −24 **12.** −8.5

13. $-\dfrac{5}{3}$ **14.** -25.2 **15.** 6   **16.** $\dfrac{3}{8}$ **17.** $\dfrac{28}{3}$ **18.** 7

19. -4.8 **20.** $\dfrac{11}{6}$ **21.** -5.7 **22.** 4   **23.** -20 **24.** -6

25. 18

Section 2.2
1. (1)   **2.** (1)   **3.** -10 **4.** -5 **5.** 10   **6.** 9

7. 10   **8.** 39   **9.** $\dfrac{5}{3}$ **10.** $-\dfrac{27}{2}$ **11.** 0   **12.** 9

13. $\dfrac{5}{7}$ **14.** 8   **15.** 8   **16.** 50   **17.** $-\dfrac{1}{3}$ **18.** -2

19. 5   **20.** -5 **21.** 28   **22.** 22   **23.** 95   **24.** 15

25. 7   **26.** 43   **27.** 16   **28.** 20   **29.** 42

Section 2.3
1. (3)   **2.** (2)   **3.** (2)   **4.** -4 **5.** 9   **6.** -4

7. -7 **8.** 2   **9.** -13 **10.** -6 **11.** 10   **12.** 3

13. -5 **14.** 18   **15.** 8   **16.** 31   **17.** 225   **18.** 48

19. 20   **20.** 80   **21.** 24   **22.** 3.6

23. 5 nickels,   2 quarters,   1 dime   **24.** 315   **25.** 7

26. 17 apple pies,   34 cheesecakes

Section 2.4
1. (2)   **2.** (4)   **3.** (2)   **4.** (3)   **5.** (3)   **6.** (1)

7. (3)   **8.** (4)   **9.** (2)   **10.** 300   **11.** 560   **12.** 57.78

13. 18   **14.** $\dfrac{5}{9}$ **15.** \$195   **16.** \$45   **17.** 13   **18.** 50

19. \$167.50   **20.** Yes, \$107   **21.** No, 40%

Section 2.5
1. (2)   **2.** (3)   **3.** (2)   **4.** (3)   **5.** (4)   **6.** (2)

7. (2)   **8.** (2)   **9.** 144   **10.** (a) 50   (b) 92.5   **11.** 16

12. 11, 13   **13.** 30   **14.** (a) 80   (b) 64   **15.** \$42   **16.** 54

17. 12   **18.** 4.8   **19.** \$144   **20.** (a) 50 in   (b) 108 in^2

21. 22   **22.** 84

Section 2.6
1. (4)   **2.** (2)   **3.** (3)   **4.** (2)   **5.** (3)   **6.** 24

7. $-\dfrac{1}{2}$ **8.** -6 **9.** $\dfrac{47}{8}$ **10.** -22 **11.** 5

12. 49,200;   114,800;   32,800   **13.** 1280   **14.** \$47.25   **15.** 14

16. 21 quarters,   28 dimes,   14 nickels   **17.** 216   **18.** 8

19. $\dfrac{11}{18}$ **20.** 15

Section 2.7
1. (2) **2.** (1) **3.** (1) **4.** (2) **5.** (4) **6.** (4)
7. (1) **8.** (4) **9.** (4) **10.** (2) **11.** (1) **12.** (4)
13. $x < 4$ **14.** $x > -3$ **15.** $x < 3$ **16.** $x \geq 3$ **17.** $x \leq -6$ **18.** $x \leq 3$
19. 1 **20.** $\{-2, -1, 0\}$ **21.** 9 **22.** 16
23. 10, 18 **24.** 25 **25.** 274
26. George: 30 Edward: 15, Robert: 11 **27.** 9 in, 19 in

CHAPTER 3

Section 3.1
1. (4) **2.** (4) **3.** (3) **4.** (2) **5.** (1) **6.** (2)
7. (3) **8.** (1) **9.** \$37.20 **10.** 72 **11.** 2.5
12. 18, 14, 10, 6, 2 **13.** 250,000 **14.** \$120
15. (a) 0.5 (b) −1 (c) 0 (d) −0.5 **16.** 32 **17.** 6
18. Sandy: 5, Ariela: 6, Vanessa: 9, Rick: 10, Mark: 14 **19.** 18
20. 14 **21.** 266 **22.** 15 **23.** 6 **24.** 60
25. 105 **26.** 311 **27.** 7 **28.** 46 **29.** \$32,000
30. Mindy: Drama, Arturo: Chorus, Renee: Tennis, Joan: Chess
31. Allan, Vickie; Barry, Debbie; Craig, Stephanie
32. Dennis: country, Joni: jazz, Elvis: classical, Taisha: rock

Section 3.2
1. (4) **2.** (1) **3.** (2) **4.** (3) **5.** (3)
6. 50 mph **7.** Freight: 60 mph Passenger: 80 mph

8. 2:30 P.M. **9.** 48 **10.** 9.6 **11.** $2\frac{1}{2}$ hr

12. 12:50 P.M. **13.** 10

Section 3.3
1. (1) **2.** (4) **3.** 5 **4.** 2.75 **5.** 4 **6.** 20
7. Disagree, $F = C = -40°$ **8.** (a) 3 (b) \$24 **9.** 85
10. (a) 4.8 (b) 31.2 **11.** (a) $x = 6$ **12.** (a) August; \$240

Section 3.4
1. (3) **2.** (1) **3.** 16% **4.** 48 **5.** (a) 12 (b) 84
6. 26 **7.** 85 **8.** 33

Section 3.5
1. (3) **2.** (3) **3.** (4) **4.** (4) **5.** (3) **6.** (2)
7. (2) **8.** (2) **9.** (2) **10.** (1) **11.** (3) **12.** False
13. False **14.** False

Section 3.6

1. (1)	**2.** (1)	**3.** (2)	**4.** (2)	**5.** (1)	**6.** (1)
7. (2)	**8.** (3)	**9.** (1)	**10.** (4)	**11.** (2)	**12.** (3)
13. $5.4 < x < 7.8$		**14.** (a) 400 (b) (2)		**15.** 6	

CHAPTER 4

Section 4.1

1. (3)	**2.** (4)	**3.** (a) Z (b) 3	**4.** 10	**5.** 134
6. 20	**7.** 18	**8.** 38		

Section 4.2

1. (3)	**2.** (2)	**3.** (3)	**4.** (1)	**5.** (2)	**6.** (2)
7. 9	**8.** 40	**9.** 19	**10.** 72	**11.** 73.2	**12.** 45
13. 38	**14.** 101	**15.** 143			

Section 4.3

1. (3)	**2.** (4)	**3.** (1)	**4.** (3)	**5.** (4)

6. Two lines perpendicular to the same line are always parallel to each other since interior angles on the same side of the transversal (the given line) are supplementary, which makes the lines parallel.

7. 50	**8.** 50	**9.** 22	**10.** (a) 31 (b) $\ell \parallel m$

Section 4.4

1. (3)	**2.** (2)	**3.** (3)	**4.** (1)	**5.** (2)	**6.** (3)
7. $18x + 12$		**8.** $3x^2 - 5x$		**9.** 60	

Section 4.5

1. (2)	**2.** (2)	**3.** (1)	**4.** (2)	**5.** (3)	**6.** (3)
7. (3)	**8.** (4)	**9.** (1)	**10.** (3)	**11.** (3)	**12.** 80
13. 110	**14.** 97	**15.** 270°	**16.** 75		

Section 4.6

1. (4)	**2.** (1)	**3.** (2)	**4.** (4)	**5.** (1)	**6.** (3)
7. (3)	**8.** 1980	**9.** (8)	**10.** 15		

CHAPTER 5

Section 5.1

1. (2)	**2.** (2)	**3.** (3)	**4.** (1)	**5.** (2)	**6.** (4)
7. (3)					

Section 5.2
1. (3) **2.** (3) **3.** (2) **4.** (1) **5.** (3) **6.** (3)
7. (3) **8.** (3) **9.** (2) **10.** 100 **11.** 34

Section 5.3
1. (1) **2.** (4) **3.** (3) **4.** (4) **5.** (1) **6.** (2)
7. (2) **8.** (4) **9.** (1) **10.** (4) **11.** (1) **12.** (3)
13. (1) **14.** $\overline{BC}$ **15.** (a) $\angle R$ (b) $\angle S$ **16.** 15

Section 5.4
1. (1) **2.** (4) **3.** (1) **4.** (4) **5.** (3) **6.** (2)
7. (3) **8.** (3) **9.** (1) **10.** (3) **11.** (2) **12.** (4)
13. (3) **14.** (4)

Section 5.5
1. (4) **2.** (4) **3.** (2) **4.** (2) **5.** (2) **6.** (3)
7. (2) **8.** (4) **9.** 6 **10.** 76
11. (a) Yes, by SAS $\cong$ SAS (b) 116 **12.** 55

CHAPTER 6

Section 6.1
1. (1) **2.** (3) **3.** (1) **4.** $8y - 10$ **5.** $-3n^2 - 5$
6. $2x^3 + 8x^2 - 6x - 9$ **7.** $5x - 7y + 7z$ **8.** $-2x^2 - 2x - 6$
9. $4x^3 + 6x^2 - 14x + 6$ **10.** $-4x^3 + 5x^2 - 13$ **11.** $a = b = \dfrac{5}{2}$

Section 6.2
1. (3) **2.** (3) **3.** (2) **4.** (1) **5.** (2) **6.** (1)
7. $4a^3b^4$ **8.** $-24x^6$ **9.** $-0.06xy^5$ **10.** $\dfrac{y}{x^2}$ **11.** $-\dfrac{x}{9}$
12. $\dfrac{2}{3b^2}$ **13.** $\dfrac{3x^3}{y}$ **14.** $3x^2y$ **15.** $5y^4 - 40y^2 - 20y$
16. $6x^2 - 13x - 63$ **17.** $25w^2 - 64$ **18.** $4b^2 + 5b - 6$
19. $0.09y^4 - 1$ **20.** $-3x^2 - 5x + 2$ **21.** $4x^2 - 12x + 9$
22. $\dfrac{2r^3}{s} - 3r^2s$ **23.** $-7c^2 + 4c - 1$ **24.** $0.2a^2 - 1.5ab$
25. Greater by 36 **26.** Agree
27. (a) $-x^2 + 9x + 15$ (b) 2 (c) 29 (d) 17.1

Section 6.3
1. (2) **2.** (3) **3.** $3x(5x - 2)$ **4.** $7(p^2 + q^2)$
5. $x(x^2 + x - 1)$ **6.** $3y^3(y^4 - 2y^2 + 4)$ **7.** $-4(a + b)$

8. $8u^2w^2(u^3 - 5w^3)$ **9.** $(y - 12)(y + 12)$ **10.** $0.12x(2x - 3y)$

11. $\dfrac{ab}{4}(a - 3b^2)$ **12.** $(9 - x)(9 + x)$ **13.** $\left(p - \dfrac{1}{3}\right)\left(p + \dfrac{1}{3}\right)$

14. $(b - 0.6)(b + 0.6)$ **15.** $\left(\dfrac{2}{3}c - 1\right)\left(\dfrac{2}{3}c + 1\right)$

16. $(10a - 7b)(10a + 7b)$ **17.** $(x - 1)(4x^2 - 7)$

Section 6.4
1. (2)   **2.** (3)   **3.** (3)   **4.** (3)   **5.** (1)   **6.** (4)
7. (4)   **8.** (4)   **9.** (2)   **10.** $(x + 5)(x + 3)$
11. $(x - 7)(x - 3)$ **12.** $(y + 3)(y + 3)$ **13.** $(a - 9)(a + 5)$
14. $(b + 8)(b - 5)$ **15.** $(w - 6)(w - 7)$ **16.** $(3x - 7)(x + 3)$
17. $(4n - 1)(n + 3)$ **18.** $(5s + 1)(s - 3)$ **19.** $2y(y - 5)(y + 5)$
20. $-5(t - 1)(t + 1)$ **21.** $4(m - n)(m + n)$ **22.** $8xy(y - 3)(y + 3)$
23. $-2(y + 2)(y + 5)$ **24.** $2x(x + 8)(x - 7)$ **25.** $10y^2(y + 10)(y - 5)$

26. $2x(3x - 5y)(3x + 5y)$ **27.** $\dfrac{1}{2}(x - 6)(x + 6)$

Section 6.5
1. (3)   **2.** (1)   **3.** (1)   **4.** (1)   **5.** $0, \dfrac{1}{2}$

6. $-1, -2$ **7.** $1, 4$ **8.** $-3, 4$ **9.** $0, 13$ **10.** $-\dfrac{1}{2}, 3$

11. $-2, \dfrac{9}{2}$ **12.** $\dfrac{2}{3}$ **13.** $0, 6$ **14.** $-\dfrac{1}{3}, \dfrac{3}{2}$ **15.** $-5, \dfrac{1}{3}$

16. $-5, \dfrac{2}{5}$ **17.** $-3, 7$ **18.** $1, 6$ **19.** 5 **20.** 4 in by 8 in

Section 6.6
1. 54   **2.** 13, 8   **3.** 10 ft by 12 ft   **4.** 81, 117, 162
5. 7, 8   **6.** 7, 8, 9   **7.** 7   **8.** 6 ft   **9.** 10 in by 20 in
10. 6 ft   **11.** 3 cm   **12.** 24 cm, 18 cm   **13.** 2 in
14. 26   **15.** 2.5

CHAPTER 7

Section 7.1
1. (2)   **2.** (2)   **3.** (4)   **4.** $\dfrac{2}{x + 8}$ **5.** $\dfrac{a - 4}{3}$ **6.** $\dfrac{b - a}{2}$

7. $0.6x - 0.2$ **8.** $\dfrac{r(3 - r)}{2}$ **9.** $\dfrac{10}{y - 3x}$ **10.** $\dfrac{-5}{x + 1}$ **11.** $\dfrac{x - 5}{x + 2}$

12. $\dfrac{x + y}{x - y}$ **13.** $\dfrac{x - 7}{x + 1}$ **14.** $\dfrac{2x - 3y}{5xy}$ **15.** $\dfrac{3y(2x - 1)}{x + 3}$

Section 7.2

1. $\dfrac{2a-1}{3b}$ **2.** $\dfrac{y}{2x}$ **3.** $\dfrac{x}{(x-y)y}$ **4.** $\dfrac{2}{3}$ **5.** $\dfrac{x-3}{x}$ **6.** $2x$

7. $\dfrac{-25y^2}{y+2}$ **8.** $\dfrac{3}{y-2}$ **9.** 1 **10.** $x+2$

Section 7.3

1. (1)   **2.** (3)   **3.** (1)   **4.** (3)   **5.** (3)   **6.** 27

7. 178.2   **8.** 900   **9.** 4.03   **10.** 12 hr

Section 7.4

1. (2)   **2.** (3)   **3.** (4)   **4.** (2)   **5.** (1)   **6.** (1)

7. (2)   **8.** (4)   **9.** $\dfrac{b}{2x}$ **10.** $\dfrac{y-2}{2xy}$ **11.** $\dfrac{b+1}{6b}$ **12.** $\dfrac{x+1}{2x}$

13. $\dfrac{y+10}{10y}$ **14.** $\dfrac{7(4x-1)}{12x}$ **15.** $\dfrac{1}{a-1}$ **16.** $\dfrac{2}{x+3}$

17. $\dfrac{3x-2}{4}$ **18.** $\dfrac{y-4x}{2xy^2}$ **19.** $\dfrac{17y+40}{4(y-4)(y+4)}$ **20.** $\dfrac{xy}{2(x-2y)(x+2y)}$

Section 7.5

1. 10   **2.** 2   **3.** 3   **4.** 3   **5.** 6, −5   **6.** −2, 4

7. $\dfrac{1}{3}, -2$ **8.** $-\dfrac{3}{2}, 3$ **9.** −5, 2   **10.** $\dfrac{2}{3}$ **11.** 6   **12.** 4

13. $\dfrac{bx+1}{x}$ **14.** 147   **15.** 28

CHAPTER 8

Section 8.1

1. (3)   **2.** (3)   **3.** (1)   **4.** (3)   **5.** (2)   **6.** (1)

7. (3)   **8.** (4)   **9.** (3)   **10.** 0.8   **11.** $4\sqrt{3}$ **12.** $-3\sqrt{7}$

13. $12\sqrt{3}$ **14.** $\dfrac{3}{5}$ **15.** $4\sqrt{3}$ **16.** $2\sqrt{6}$ **17.** -1 **18.** $\dfrac{41}{99}$

19. $\dfrac{7}{18}$ **20.** $\dfrac{19}{11}$ **21.** $4\sqrt{2}rs^3$ **22.** $10\sqrt{2}xy^2$

23. $5\sqrt{3}a^3$ **24.** $2s^2\sqrt{3t}$ **25.** $2\sqrt{2}a^2$ **26.** $\dfrac{3\sqrt{2}x}{y^2}$

Section 8.2

1. (3)   **2.** (1)   **3.** (3)   **4.** (1)   **5.** (2)   **6.** (1)

7. $-7\sqrt{11}$ **8.** $12\sqrt{2}$ **9.** $2\sqrt{5}$ **10.** 45   **11.** $5\sqrt{2}$ **12.** $\dfrac{3}{2}$

13. $30\sqrt{35}$ **14.** $9\sqrt{2}$ **15.** 20 **16.** 3 **17.** 2 **18.** $\dfrac{7}{2}$

19. -1 **20.** Greater by $2\sqrt{ab}$

Section 8.3

1. (4)	**2.** (2)	**3.** (4)	**4.** (2)	**5.** (1)	**6.** (2)
7. (2)	**8.** (3)	**9.** 26	**10.** 17.7	**11.** 87.7	**12.** $2\sqrt{10}$
13. 24 ft	**14.** 25	**15.** 20	**16.** 18	**17.** 3	**18.** 26

Section 8.4

1. (1) **2.** (3) **3.** (3) **4.** (1) **5.** (1) **6.** 12.2

7. $34+\dfrac{32}{\sqrt{3}}$ **8.** $34+\dfrac{32}{\sqrt{2}}$ **9.** (a) 10 (b) $10\sqrt{3}$ **10.** 4.4

CHAPTER 9

Section 9.1

1. (3)	**2.** (4)	**3.** (4)	**4.** (3)	**5.** (4)	**6.** (3)
7. (2)	**8.** (2)	**9.** (2)	**10.** (3)	**11.** (2)	**12.** (2)
13. (1)	**14.** (2)	**15.** 210	**16.** 18	**17.** 90	**18.** 4.5

19. (a) $9\sqrt{3}$ in^2 (b) $27\sqrt{3}$ in^2 (c) $13.5\sqrt{3}$ in^2 **20.** 27,342

Section 9.2

1. (2)	**2.** (3)	**3.** (1)	**4.** (1)	**5.** (3)	**6.** (3)
7. (3)	**8.** (2)	**9.** (a) 157.1 (b) 61.1			**10.** 8.6

11. 15 cm **12.** 2.7

Section 9.3

1. $144-36\pi$ **2.** $12.5\pi-24$ **3.** $392-98\pi$

4. $256-64\pi$ **5.** 14.5 **6.** 5.5 **7.** (a) 314 (b) 21

8. (a) 66.0 (b) 67 **9.** 6586 **10.** (a) 103 (b) 40

Section 9.4

1. (1) **2.** (4) **3.** (3) **4.** 392π **5.** $\dfrac{64}{3}\sqrt{3}\pi$ **6.** 130

7. 12 **8.** 343 **9.** 87.5π cm^3 **10.** 3 **11.** 18

12. 1250π **13.** 640 **14.** (a) 6 (b) 96 **15.** 10 **16.** 47

CHAPTER 10

Section 10.1
1. (2)   **2.** (3)   **3.** (2)   **4.** 10
5. (a) Right angle $B \cong$ right angle E, and $\angle ACB \cong \angle DCE$ (b) 20
6. (a) $\angle SEF \cong \angle SRT$, and $\angle SFE \cong \angle STR$ (b) 20   **7.** 10
8. 10   **9.** 6   **10.** 28   **11.** 9

Section 10.2
1. (1)   **2.** (1)   **3.** (4)   **4.** (4)   **5.** (1)   **6.** $\dfrac{100}{48}$
7. 19.5   **8.** 185   **9.** 140
10. (a) m$\angle A = 103$, m$\angle C = 77$ (b) 13.6   **11.** 2058   **12.** 76°
13. 79.4   **14.** (a) 11   (b) 21

Section 10.3
1. (1)   **2.** (2)   **3.** 34   **4.** 72.0   **5.** 754   **6.** 53
7. 42   **8.** 669   **9.** 106   **10.** 37   **11.** 92   **12.** 63.0
13. (a) 16.2   (b) 81   **14.** (a) 119   (b) No, since $\overline{ME}$ is not a diagonal.
15. 136   **16.** 46

CHAPTER 11

Section 11.1
1. (2)   **2.** (2)   **3.** (1)   **4.** (2)   **5.** (3)   **6.** 15
7. 7.5   **8.** 20   **9.** 5   **10.** 4   **11.** 16
12. (a) $h = 8$, $k = 5$ (b) 40   **13.** 90   **14.** 76
15. (b)   (1) 16   (2) 24   (3) 40

Section 11.2
1. (2)   **2.** (2)   **3.** (1)   **4.** (3)   **5.** (4)   **6.** (2)
7. (2)   **8.** 169π square units   **9.** (a) $4\sqrt{5}, 6\sqrt{5}$ (b) 56°
10. $x = 9$, $y = 5$
11. Since midpoint $\overline{AS}$ = midpoint = $\overline{RT}$ = 2, 1, $STAR$ is a parallelogram.
12. 16   **13.** $(-2, 13)$ or $(-2, -5)$
14. (b) No, $\overline{AC}$ and $\overline{BD}$ have different midpoints.

Section 11.3
1. (2)   **2.** (3)   **3.** (1)   **4.** (1)   **5.** (2)   **6.** (3)
7. (1)   **8.** -4
9. (a) $m_{\overline{AB}} = -1$ and $m_{\overline{CD}} = 2$
   (b) No, since -1 and 2 are not negative reciprocals.
10. 1

11. (a) −3

(b) $\overline{BH} \perp \overline{AC}$ since their slopes, 4 and $-\dfrac{1}{4}$, are negative reciprocals.

(c) 64°

12. (a) *TEAM* is a trapezoid since $\overline{TM} \parallel \overline{AE}$ and $\overline{TE} \nparallel \overline{MA}$ (b) $h = 3, k = 2$

Section 11.4

1. (1)	**2.** (3)	**3.** (4)	**4.** (3)	**5.** (1)	**6.** (4)
7. (2)	**8.** (3)	**9.** (3)	**10.** (4)	**11.** (4)	**12.** (4)

13. 6   **14.** (a) $y = 2x + 2$ (b) $y = -x + 2$ **15.** $y = 2x - 4$

16. (a) $D = 9h + 12$ (b) $9

17. (a) $y = 2x - 2$ (b) $y = 2x - 1$ (c) $y - 2 = -\dfrac{1}{2}(x - 2)$

18. (a) $F = 1.8C + 32$ (b) 95   **19.** 10; $y + 3 = -\dfrac{3}{4}(x - 1)$

20. (a) $y = 2x + 6$ (b) $y = -\dfrac{1}{2}x + 3$ (c) $2y + 12 = 4x$

Section 11.5

7. 32   **9.** 9
8. 8   **10.** 24

Section 11.6

1. (4)	**2.** (3)	**3.** $\dfrac{3}{2}$	**4.** 1440	**5.** 9	**6.** 93.12

Section 11.7

1. (a) $A = 6h + 7$ (c) $34   **2.** (a) Martin (b) 40
3. (a) 8 (b) 6 (c) 4 mph
4. (a) $25 (b) May (c) $15 (d) $y = 10x + 75$
5. (a) Electric (b) Electric by $7000 (c) $C = Y + 4$

CHAPTER 12

Section 12.1

1. (2)	**2.** (4)	**3.** (3)	**4.** (1)	**5.** (3)	**6.** (1)
7. (2)	**8.** (4)	**9.** (3)	**10.** (1)		

Section 12.2

1. (2)	**2.** (4)	**3.** (1)	**4.** (2)	**5.** (4)	**6.** (3)
7. (2)	**8.** (2)	**9.** (4)	**10.** (4)		

Section 12.3

1. (4)	**2.** (4)	**3.** (2)	**4.** (1)	**5.** (3)	**6.** (3)
7. (3)	**8.** (−4, 7)	**9.** (−5, −3)	**10.** (1, −10)		

11. Graph $R'(-2, -1)$, $S'(2, -6)$, $T'(8, 4)$

12. (a) $h = 5, k = 2$ (b) 22.5 (c) 90
13. (a) Graph $B'(-4, -2)$, $I'(2, 8)$, $G'(10, -2)$ (b) 4 to 1
14. (a) Graph $A'(-1, 3)$, $B'(6, -2)$, and $C'(8, 5)$
  (b) Graph $A''(1, -3)$, $B''(-6, 2)$, and $C''(-8, -5)$
15. Yes; $A'(1, -4)$ and $B'(3, 1)$

Section 12.4
1. Concentric circles with radii of 7 and 9
2. Two parallel lines each at a distance of 3 cm from the given line
3. Intersection of the perpendicular bisectors of $\overline{AB}$ and $\overline{AC}$
4. Concentric circle with radius of 9
5. Line parallel to the table and $\dfrac{1}{2}$ in above it
6. $x = -2$ and $x = 8$
7. $y = -3$ and $y = 1$
8. $y = 3x + 4$

Section 12.5
1. (2) **2.** (4) **3.** (2) **4.** (1) **5.** (1) **6.** (3)
7. (2) **8.** (1) **9.** (1) **10.** (3) **11.** (b) 2 **12.** 4
13. (a) (1) Parallel lines 3 units from $\overleftrightarrow{AB}$
  (2) Circle with radius = 5, center at P
  (b) 4
14. (a) (1) Line 3 units from both $\overleftrightarrow{AB}$ and $\overleftrightarrow{CD}$
  (2) Circle with radius = 2, center at P
  (b) 1
15. (a) $x = 1, x = 5$ (b) Circle with center at P and radius of n units (c) 2

CHAPTER 13

Section 13.1
1. (3) **2.** (3) **3.** (4) **4.** (1, -6) **5.** (-1, 2)
6. (b) 45 **7.** (b) (3) **8.** 3
9. Graph $x \le 14$, $y \le 8$, and $x + y \le 18$. **10.** (b) 16
11. (b) 24 **12.** (a) $C = 15 + 2n$ (b) $C = 3.50n$ (d) 10

Section 13.2
1. (4) **2.** (2) **3.** (1) **4.** -39 **5.** (2, -10)
6. $\left(\dfrac{7}{5}, \dfrac{28}{5}\right)$ **7.** $a = -1, b = -1$ **8.** (-2, -1) **9.** (3, 4) **10.** (3, 9)
11. \$0.70 **12.** 70° **13.** $\dfrac{5}{12}$ **14.** 7 **15.** (a) $10t + u$ (b) 23
16. 12 **17.** (a) \$1.25 (b) \$0.45 **18.** 60°

19. (a) $x = y + 100$, $4x + 12y = 3056$ where x = balcony tickets
and y = main-floor tickets
  (b) 266
20. 15 hats, 45 T-shirts   **21.** $x = 10$, $y = 20$

Section 13.3
  1. (1)   **2.** (2)   **3.** (2)   **4.** 0   **5.** 5
  6. (3, –4)   **7.** (9, 2)   **8.** (6, 2)   **9.** (2.5, 4) **10.** $a = 5$, $b = -2$
  11. (9, 7)   **12.** $26   **13.** 15   **14.** 115   **15.** (a) $4 (b) $7
  16. 75°   **17.** $x = 10$, $y = 20$ **18.** (a) 3 (b) 2.5
  19. (a) $6x + 6y = 90$, $4x + 8y = 120$ (b) $10   **20.** (a) $3 (b) $5
  21. $6.15   **22.** $10   **23.** $6.85

CHAPTER 14

Section 14.1
  1. (2)   **2.** (2)   **3.** (4)   **4.** (2)   **5.** (1)   **6.** (2)
  7. (1)   **8.** (1)   **9.** (4)   **10.** (3)
  11. $(x - 2)^2 + (y + 3)^2 = 25$
  12. (a) $x^2 + y^2 = 81$ (b) $x^2 + y^2 = 36$ (c) 88.9
  13. (a) $(x + 6)^2 + (y - 7)^2 = 100$ (b) $y = \dfrac{3}{4}x - 1$
  14. (a) $y = 2$ (b) $(x - 2)^2 + (y + 2)^2 = 16$ (c) 1
  15. (a) Using a compass, draw a circle with center at $(-1, -5)$ and
      radius 5.
    (b) Using a compass, draw a circle with center at $(-1, 5)$ and
      radius 5.
    (c) Using a compass, draw a circle with center at
      $(-1 + 1, 5 - 5) = (0, 0)$ and radius 5.

Section 14.2
  1. (3)   **2.** (3)   **3.** (2)   **4.** (3)   **5.** (4)   **6.** (3)
  7. (1)   **8.** (1)   **9.** (3)   **10.** (2)   **11.** $y = -5$
  12. (a) Lines $x = -2$ and $x = 4$ (b) $y = \dfrac{11}{2}$
  13. (a) $(x - 3)^2 + (y - 2)^2 = 9$ (b) $y = 2$ **14.** $y = -3x + 9$
  15. $y = 3x + 3$ **16.** 48   **17.** (c) 2   **18.** (b) $\left(3, \dfrac{17}{2}\right)$

Section 14.3
1. (b) −1, 3.5
2. (a) See graph.   (b) $y = x^2 - 3$

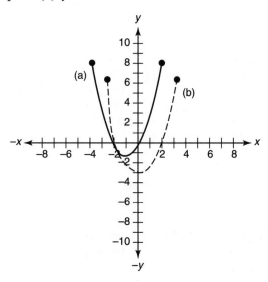

3. (a)–(c)

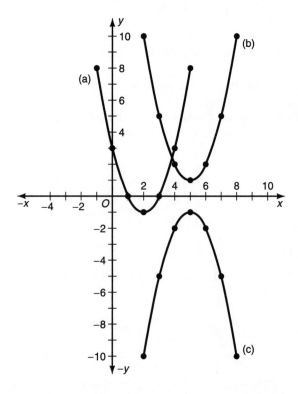

4. (a) 324 ft, 4.5
   (b) 9
5. (b) 9 ft, 3
6. (a)–(c)

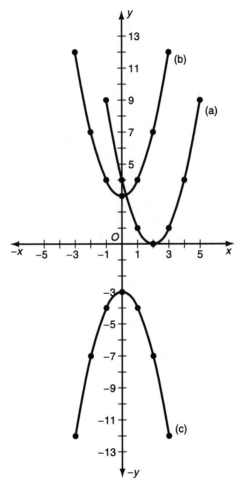

   (d) $y = -x^2 - 3$

7. $y = -3x^2 - 12x$
8. (a) and (b)

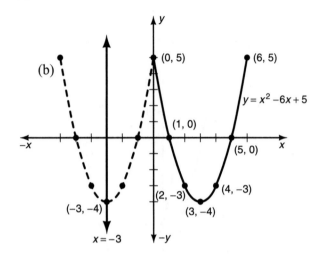

9. (a) 7   (b) 36.75

Section 14.4
1. (1)   **2.** (3)   **3.** (2)   **4.** (3)   **5.** (1)   **6.** (1)
7. (1, 6), (4, 9)   **8.** (2, −2), (5, 1)   **9.** (−9, −4), (4, 9)
10. (15, 5), (−5, −15)   **11.** (−4, 0), (1, 5)   **12.** (2, −3), (6, 5)
13. (2, 1), (−2, −3)   **14.** (2, 12)   **15.** (a) 8; $66   (b) 6

CHAPTER 15

Section 15.1
1. (2)   **2.** (2)   **3.** (1)   **4.** (4)   **5.** (2)   **6.** (3)
7. (2)   **8.** (4)   **9.** (4)   **10.** (3)

11. (a) $\dfrac{4}{20}$ (b) $\dfrac{8}{20}$ (c) $\dfrac{13}{20}$ (d) $\dfrac{14}{20}$ **12.** 8

13. (a) $\dfrac{2x-1}{5x}$ (b) 3 (c) $\dfrac{12}{15}$ **14.** (a) 16π (b) (1) $\dfrac{16}{25}$ (2) $\dfrac{9}{25}$

15. $\dfrac{5}{9}$ **16.** $\dfrac{8}{36}$

Section 15.2
1. (4)   **2.** (2)   **3.** (3)   **4.** (2)   **5.** (3)   **6.** (1)
7. 97   **8.** 95   **9.** 34 lb   **10.** 91   **11.** 49π

12. (a) $\dfrac{1}{8}$ (b) 0   **13.** 1392   **14.** 7, 12, 22   **15.** 91

Section 15.3
1. (4)   2. (1)   3. (3)   4. (3)   5. 15
6. (a) 50   (b) 6   (c) 6   (d) (1)
7. (a) $Q_1 = 71.5$, $Q_2 = 83$, $Q_3 = 91$ (b) 88

8. (a) 10   (b) I   (c) $\dfrac{1}{4}$

Section 15.4
1. (3)   2. (2)   3. (3)   4. (1)
5. (a)

Score Interval	Frequency
140–149	5
150–159	7
160–169	9
170–179	3
180–189	3
190–199	2
200–209	1

6. (a) 81–90   (b) $\dfrac{12}{20}$
7. (a) 78.5   (b) 75
8. (a) 20   (b) 86–90   (c) 0

9. (a)

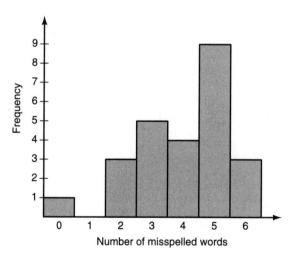

(b) 5   (c) 4   (d) 4

10. (a)

Interval	Frequency
91–100	5
81–90	6
71–80	2
61–70	2
51–60	1

(b)

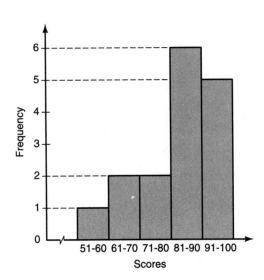

(c) 81–90   (d) 71–80

11. (a) 6   (b) B   (c) (1) $\dfrac{2}{29}$ (2) $\dfrac{9}{29}$ (3) $\dfrac{22}{29}$

12. (a) 57   (b) $\dfrac{25}{30}$

Section 15.5

1. (2)   **2.** (3)   **3.** (a) 25   (b) $\dfrac{11}{28}$

4. (a)

Scores	Cumulative Frequency
21–100	100
21–90	85
21–80	59
21–70	36
21–60	21
21–50	10
21–40	5
21–30	2

(b)

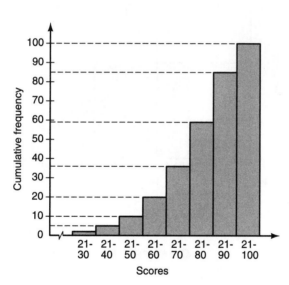

(c) (4)   (d) 10

5. (a)

Height (in)	Number of Students
60–62	2
63–65	5
66–68	8
69–71	5
72–74	3
75–77	1

(b) (1)

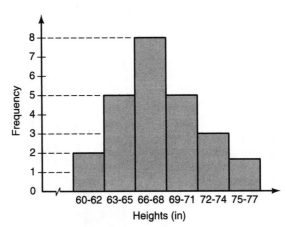

(2) 63–65 (3) 69–71   (c) 0.625

CHAPTER 16

Section 16.1

1. 72   **2.** 90   **3.** 5   **4.** (b) $\dfrac{3}{9}$ (c) $\dfrac{5}{9}$ (d) $\dfrac{3}{9}$

5. (b) (1) $\dfrac{6}{12}$ (2) $\dfrac{2}{12}$ (3) 0   **6.** (a) $\dfrac{1}{9}$ (b) $\dfrac{5}{9}$ (c) $\dfrac{4}{9}$ (d) $\dfrac{3}{9}$

Section 16.2

1. (1)   **2.** (3)   **3.** (3)   **4.** $\dfrac{11}{87}$ **5.** $\dfrac{5}{9}$ **6.** $\dfrac{472}{756}$

7. $\dfrac{29}{35}$ **8.** (a) $\dfrac{6}{72}$ (b) $\dfrac{36}{72}$

9. (a) $x = \dfrac{4}{9}$, $y = \dfrac{5}{9}$, $z = 0$ (b) (1) $\dfrac{20}{90}$ (2) $\dfrac{30}{90}$

10. (a) 18 cherry, 6 lemon, 16 orange   (b) $\dfrac{576}{1560}$

11. (a) $\dfrac{18}{56}$ (b) $P(\text{total} = \$6) = \dfrac{24}{56}$ $P(\text{total} \geq \$10) = \dfrac{26}{56}$

12. (a) $\dfrac{1}{12}$ (b) $\dfrac{6}{12}$ (c) $\dfrac{6}{12}$ **13.** (a) $\dfrac{108}{240}$ (b) $\dfrac{132}{240}$

14. (a) $\dfrac{8}{12}$ (b) $\dfrac{22}{132}$ (c) $\dfrac{72}{132}$

Section 16.3

1. (3)   **2.** (1)   **3.** (2)   **4.** 48   **5.** 144   **6.** 280

7. (a) 45,360 (b) $\dfrac{1}{3}$ **8.** $\dfrac{36}{60}$ **9.** $\dfrac{240}{6720}$ **10.** $\dfrac{6}{120}$

11. (a) $\dfrac{48}{720}$ (b) $\dfrac{192}{720}$ **12.** 7.98×10^6

Section 16.4

1. (2)   **2.** (1)   **3.** (3)   **4.** (2)   **5.** (1)   **6.** 10

7. $\dfrac{840}{2002}$ **8.** 490   **9.** 1540   **10.** (a) $\dfrac{1}{252}$ (b) $\dfrac{18}{252}$ (c) 0

11. $\dfrac{36}{792}$ **12.** (a) $\dfrac{33}{495}$ (b) 1   **13.** (a) $\dfrac{5}{210}$ (b) $\dfrac{155}{210}$

14. (a) $\dfrac{18}{126}$ (b) $\dfrac{81}{126}$ **15.** (a) $\dfrac{6}{10}$ (b) $\dfrac{4}{10}$ (c) 6

16. (a) $\dfrac{60}{495}$ (b) $\dfrac{1}{495}$ (c) 0

GLOSSARY OF MATH A TERMS

abscissa See *x*-coordinate.

absolute value The distance from 0 to *x* of a number *x* on the number line; denoted as $|x|$. The absolute value sign is removed from the number as follows:

$$|-3| = -(-3) = 3 \text{ and } |+3| = 3$$

acute angle An angle whose degree measure is greater than 0 but less than 90.

acute triangle A triangle that contains three acute angles.

additive inverse The opposite of a number. The sum of a number and its additive inverse is 0. Thus, the additive inverse of +3 is –3 since (+3) + (–3) = 0.

adjacent angles Two angles that have the same vertex, share one side, but do not overlap.

alternate interior angles A pair of angles with different vertices that lie between two lines and on opposite sides of a transversal.

altitude A line segment that is perpendicular to the side of the figure to which it is drawn.

angle The union of two rays that have the same endpoint.

angle of depression The angle formed by a horizontal line of vision and the line of sight when an object is viewed beneath the horizontal line of vision.

angle of elevation The angle formed by a horizontal line of vision and the line of sight when an object is viewed above the horizontal line of vision.

area The number of unit squares that a region encloses.

associative property The mathematical law that states the order in which three numbers are grouped when added together or multiplied together does not matter.

average See **mean**.

axis of symmetry The line of symmetry of a parabola that contains the turning point. For the parabola $y = ax^2 + bx + c$, an equation of the axis of symmetry is $x = -\dfrac{b}{2a}$.

base angles of an isosceles triangle The congruent angles that lie opposite the congruent sides, called the *legs*, of an isosceles triangle.

base of a power The number that is used as the factor when a product is expressed in exponential form. In 3^4, the *exponent*, 4, tells the number of times that the base, 3, is used as a factor in the expanded product.

binomial A polynomial with two unlike terms, for example, $2x + 3y$.

bisector A line, ray, or line segment that divides an angle or a line segment into two parts that have the same measure.

box-and-whisker plot A diagram that shows how spread out data are about the median using five key values: the lowest score, the lower quartile, the median, the upper quartile, and the highest data value.

circle The set of points (x, y) in a plane that are the same distance *r*, called the *radius*, from a fixed point (h, k), called the *center*. Thus:

$$(x - h)^2 + (y - k)^2 = r^2$$

circumference The distance around a circle. The circumference *C* of a circle with radius *r* is given by the formula $C = 2\pi r$.

closure property The mathematical law that states that a set is closed under an operation if performing that operation on members of that set always produces another member of the same set. The set of integers is closed under addition, but is not closed under division.

coefficient The number by which the variable factor(s) in a monomial are multiplied. For example: the coefficient of $-5xy^2$ is -5.

collinear points Points that lie on the same line.

combination A selection of objects from a set without regard to order, as in choosing people from a committee to join a subcommittee.

combination notation The notation $_nC_r$ means the combination of n objects taken r at a time ($r \le n$) and can be evaluated using the formula

$$_nC_r = \frac{n!}{r!(n-r)!}$$

commutative property The mathematical law that states that the order in which two numbers are added or multiplied does not matter.

complementary angles Two angles whose degree measures add up to 90.

compound loci The set of points that satisfy two or more locus conditions.

conditional statement A statement of the form "If statement p, then statement q," which is always true except in the single instance when p is true and q is false.

congruent figures Figures that have the same size and the same shape. Two angles or two sides are congruent when they have the same measures. The symbol $\cong$ means "is congruent to."

congruent triangles Two triangles for which any one of the following conditions is true: (1) the sides of one triangle are congruent to the corresponding sides of the other triangle (SSS $\cong$ SSS); two sides and the included angle of one triangle are congruent to the corresponding parts of the other triangle (SAS $\cong$ SAS); two angles and the included side of one triangle are congruent to the corresponding parts of the other triangle (ASA $\cong$ ASA); two angles and a nonincluded side of one

triangle are congruent to the corresponding parts of the other triangle (AAS $\cong$ AAS). Two right triangles are congruent if the hypotenuse and a leg of one right triangle are congruent to the corresponding parts of the other right triangle (HL $\cong$ HL).

conjugate pair The sum and difference of the same two terms, as in $a + b$ and $a - b$.

conjunction Two statements joined by the logical connective *AND*. The conjunction of statements p and q is true only when statements p and q are true at the same time.

constant A quantity that does not change in value. In the formula $A = \pi r^2$, A represents the area of a circle whose radius is r and π is a numerical constant.

contrapositive The conditional statement formed by switching and negating both parts of another conditional. Thus, the contrapositive of "If p, then q" is "If not q, then not p." A conditional and its contrapositive always have the same truth values.

converse The conditional statement formed by switching both parts of another conditional. The converse of "If p, then q" is "If q, then p."

coordinate plane A plane divided into four rectangular regions, called *quadrants*, by a horizontal number line and a ventricular number line intersecting at their zero points, called the *origin*.

corresponding angles A pair of angles consisting of one interior angle and one exterior angle with different vertices on the same side of a transversal.

cosine ratio In a right triangle, the ratio of the length of the leg that is adjacent to a given acute angle to the length of the hypotenuse.

counting principle The fact that, if event A can occur in m ways and event B can occur in n ways, then the number of ways in which both events can occur is m times n.

cumulative frequency The sum of all frequencies from a given data point up to and including another data point.

cumulative frequency histogram A histogram whose bar heights represent the cumulative frequencies at stated intervals.

degree A unit of angle measure representing $\frac{1}{360}$ of one complete circular rotation of a ray about its endpoint.

degree of a monomial The sum of the exponents of the variables in the monomial. The degree of $7xy^2$ is 3 since 1 (the exponent of x) plus 2 (the exponent of y) equals 3.

degree of a polynomial The greatest degree of the terms of the polynomial. The degree of $y = x^2 + 5x + 11$ is 2.

dilation A transformation in which a figure is enlarged or reduced in size without changing its shape. The numerical factor by which the dimensions of the original figure are multiplied is called the *scale factor* or *constant of dilation*.

direct variation Variation in which the ratio of two variable quantities remains the same.

disjunction Two statements joined by the logical *OR*. The disjunction of statements p and q is true when p is true, q is true, or p and q are both true.

distance formula The formula for the distance d between points $A(x_A, y_A)$ and $B(x_B, y_B)$:

$$d = \sqrt{(x_B - x_A)^2 + (y_B - y_A)^2}$$

distributive property The mathematical law that states that

$$a(b + c) = ab + ac$$

and

$$(b + c)a = ba + ca$$

where a, b, and c are real numbers.

equation A statement formed by placing an equal symbol between two mathematical expressions or numbers.

equilateral triangle A triangle whose three sides have the same length.

equivalent equations Two or more equations that have the same solution; for example, $2x = 6$ and $x + 2 = 5$ are equivalent equations.

event A particular set of outcomes of a probability experiment.

exponent A number written to the right and a half line above another number, called the *base*, that tells the number of times the base is used as a factor in the expanded product. Thus,

$$5^3 = 5 \times 5 \times 5 = 125$$

factor A number, variable, or other expression that is being multiplied to form a product. In the product $3 \times 2 = 6$, numbers 3 and 2 are factors of 6.

factorial *n* For any positive integer n, the product of the consecutive whole numbers from n to 1; denoted by $n!$. By definition, $0! = 1$. Thus, $4! = 4 \times 3 \times 2 \times 1 = 24$.

factoring The process by which an integer is written as the product of two or more other integers, or a polynomial is written as the product of two or more lower degree polynomials.

factoring completely Factoring an integer or polynomial so that each of the factors cannot be factored further.

favorable outcomes The set of outcomes for which a specified event occurs.

FOIL An acronym for the rule for multiplying two binomials horizontally by forming the sum of the products of the first terms (F), the outer terms (O), the inner term (I), and the last terms (L) of the binomial factors; for example:

$$(x + 5)(x - 2)$$

$$= \overbrace{x \cdot x}^{F} + \overbrace{-2x}^{O} + \overbrace{5x}^{I} + \overbrace{-2 \cdot 5}^{L}$$

$$= x^2 + 3x - 10$$

frequency The number of times that a particular data value appears in a set of data values.

greatest common factor (GCF) The monomial that, when compared to a given set of monomials, has the greatest numerical coefficient and the greatest power of each variable that is common to all of the given monomials. Thus, the GCF of $8a^2b^2$ and $20ab^3$ is $4ab^2$ since 4 is the greatest whole number that divides evenly into 8 and 20, and a and b^2 are the greatest powers of the variable factors that are common to both monomials.

histogram A vertical bar graph in which the height of each rectangular bar shows an amount or frequency of a quantity that is indicated at the base of the bar.

hypotenuse The side of a right triangle that is opposite the right angle.

image The point or figure produced by a transformation. If, under a transformation, P' is the image of P, then the original point P is the preimage of P'.

inequality A statement formed by placing an inequality symbol between two mathematical expressions or numbers.

inequality symbols The symbol $<$ means "is less than"; $\leq$ means "is less than or equal to;" $>$ means "is greater than;" $\geq$ means "is greater than or equal to."

integer A number from the set $\{\ldots, -4, -3, -2, -1, 0, 1, 2, 3, 4, \ldots\}$.

inverse The statement formed by negating both parts of a conditional statement.

irrational number A number, such as $\sqrt{3}$ or π, that cannot be expressed as the quotient of two integers.

isosceles trapezoid A trapezoid in which the nonparallel sides, called *legs*, have the same length.

isosceles triangle A triangle in which two sides, called *legs*, have the same length.

least common denominator (LCD) The smallest positive expression into which all of the denominators of a set of fractions divide evenly. For example, the LCD of $\frac{1}{3}$ and $\frac{1}{4}$ is 12.

leg of a right triangle Either of the two sides of a right triangle that include the right angle.

like terms Terms that differ only in their numerical coefficients; for example, $3xy^2$ and $-5xy^2$.

line An undefined term in geometry that can be described as a continuous set of points that form a straight path, extending indefinitely in opposite directions.

linear equation An equation in which the greatest exponent of the variable is 1, as in $3x + 5 = 17$ or $y = -2x + 7$.

line reflection A transformation in which each point P of a figure is matched with point P', its image, so that the reflecting line is the perpendicular bisector of $\overline{PP'}$. If point P is on the line of reflection, then P' coincides with P.

line segment A part of a line that consists of two different points on the line, called *endpoints*, and the set of points on the line that are between them.

line symmetry A figure has line symmetry when a line can be drawn that divides the figure into two parts that can be made to coincide when "folded" along this line.

locus The set of points, and only those points, that satisfy a given condition.

logically equivalent statements Two statements that always agree in their truth values. A conditional statement and its contrapositive are logically equivalent statements.

mean The sum of the n data values in a set divided by n.

median The middle value when a set of data values are arranged in size order. If the set has an even number of values, then the median is the average of the two middle values.

median of a triangle A line segment drawn from a vertex of a triangle to the midpoint of the opposite side.

midpoint The point on a line segment that divides the segment into two shorter segments that have the same length.

midpoint formula A formula stating that the coordinates of the midpoint of a line segment are the averages of the corresponding coordinates of the endpoints. Thus, the midpoint of the line segment whose endpoints are $A(x_A, y_A)$ and $B(x_B, y_B)$ is $\left(\dfrac{x_A + x_B}{2}, \dfrac{y_A + y_B}{2} \right)$.

mode The value in a set of data values that has the greatest frequency.

monomial A single term that consists of a number, a variable, or the product of a number and one or more variables, for example 7, a, or ab.

multiplicative identity The number that when multiplied by any other expression results in that same expression. The multiplicative identity for real numbers is 1.

multiplicative inverse The reciprocal of a nonzero expression. Thus, the multiplicative inverse of $\frac{2}{3}$ is $\frac{3}{2}$. The product of a number and its multiplicative inverse is 1.

negation The statement that has the opposite truth value of a given statement. Thus, the negation of statement p is the statement "not p."

obtuse angle An angle whose degree measure is greater than 90 and less than 180.

obtuse triangle A triangle that contains an obtuse angle.

open sentence A sentence whose truth value cannot be determined until a variable or placeholder is replaced with a specific value. For example, "*He* was a President of the United States" is an open sentence whose truth value cannot be determined until *he* is replaced with the name of a particular person.

opposite rays Two rays that have the same endpoint and form a line.

ordered pair Two numbers that are written in a definite order.

ordinate See *y*-coordinate.

origin The zero point on a number line.

parabola The graph of a quadratic equation in which either x or y (but not both) is squared. The graph of

$$y = a^2 + bx + c \qquad (a \neq 0)$$

is a parabola that has a vertical axis of symmetry, an equation of which is

$$x = -\frac{b}{2a}$$

parallel lines Lines that lie in the same plane and do not intersect.

parallelogram A quadrilateral in which both pairs of opposite sides are parallel.

percentile See *p*th percentile.

perfect square A rational expression whose square root is also rational; for example,

$$4 \ (\sqrt{4} = 2) \text{ and } \frac{9}{25} \left(\sqrt{\frac{9}{25}} = \frac{3}{5} \right)$$

perimeter The distance around a figure.

permutation An ordered arrangement of objects.

permutation notation The notation $_nP_n$ represents the arrangement of n objects in n positions, so $_nP_n = n!$. The arrangement of n objects when fewer than n positions are available is denoted by $_nP_r$, where $r \leq n$. To evaluate $_nP_r$, multiply together the r greatest factors of n; for example, $_5P_3 = 5 \times 4 \times 3 = 60$.

perpendicular lines Two lines that intersect at right angles.

plane An undefined term in geometry that can be described as a flat surface that extends indefinitely in all directions.

point An undefined term in geometry that can be described as indicating location with no size.

point symmetry A figure has point symmetry about a point Q when any line drawn through Q that intersects the figure at point P will also intersect the figure at another point P' in such a way that $PQ = QP'$.

polygon A simple closed figure whose sides are line segments.

polynomial A monomial or the sum or difference of two or more monomials.

postulate A basic geometric assumption that is accepted without proof.

power A number written with an exponent; for example, 2^4, which is read "2 raised to the fourth power."

preimage See **image**.

prime number An integer greater than 1 whose only positive factors are itself and 1. The number 2 is the only even integer that is a prime number.

probability of an event If all the outcomes of an event are equally likely to occur, then the theoretical probability that the event will occur is the number of ways in which it can occur divided by the total number of possible outcomes.

proportion An equation that states that two ratios are equal.

pth percentile The value at or below which p percent of the values in a set of data values lie.

Pythagorean theorem A statement of the fact that in a right triangle the square of the length of the hypotenuse is equal to the sum of the squares of the lengths of the two legs.

quadrant One of four rectangular regions into which the coordinate plane is divided.

quadratic equation An equation in which the greatest exponent of a variable is 2.

quadratic polynomial A polynomial whose degree is 2.

quadrilateral A polygon with four sides.

quartiles Numbers that divide a set of data values into four equal groups. The *lower quartile* is the median of the set of data values below the overall median. The *upper quartile* is the median of the set of data values above the overall median.

radical sign The symbol $\sqrt{\ }$, which denotes one of two equal numbers whose product is the number underneath the radical. Thus, $\sqrt{49} = 7$ since $7 \times 7 = 49$.

radicand The number that appears underneath a radical sign.

radius of a circle Any segment whose endpoints are the center and a point on the circle. The radius of a circle also refers to the distance from the center to any point on the circle.

ratio A comparison of two numbers by division. The ratio of a to b ($a:b$) is the fraction $\frac{a}{b}$, provided that $b \neq 0$.

rational number A number that can be written as a fraction having an integer in the numerator and a nonzero integer in the denominator. Thus, the set of rational numbers includes decimals in which a set of digits endlessly repeats; for example, $0.25000\ldots\left(=\frac{1}{4}\right)$ and $0.333\ldots\left(=\frac{1}{3}\right)$.

ray The part of a line that consists of a point, called an *endpoint*, and the set of points on one side of the endpoint.

real number A number that is either rational or irrational.

reciprocal For a nonzero number x, the number that, when multiplied by x, produces 1. Thus, the reciprocal of x is $\frac{1}{x}$.

rectangle A parallelogram with four right angles.

reflection in the origin After a reflection in the origin, the image of point P is point P', whose coordinates have signs opposite to those of the corresponding coordinates of P. Thus, after a reflection in the origin, the image of point $(3, -2)$ is $(-3, 2)$.

regular polygon A polygon in which all of the sides have the same length and all of the angles have the same degree measure.

replacement set In an open sentence, the set of values that may be substituted for the variable.

rhombus A parallelogram with four sides that have the same length.

right angle An angle whose degree measure is 90.

right triangle A triangle that contains a right angle.

root A number from the replacement set for a variable that makes an equation a true statement.

rotation A transformation in which a point or figure is turned about a fixed point a specified number of degrees.

sample space The set of all possible outcomes of a probability experiment.

scalene triangle A triangle in which the three sides have different lengths.

scatter plot A collection of plotted points in which the x-value represents one set of measurements and the y-value represents another set of measurements.

scientific notation The expression of a number as the product of a number from 1 to 10 and a power of 10. Thus, in scientific notation, 81,000 is written as 8.1×10^4, and 0.0072 as 7.2×10^{-3}.

similar polygons Two polygons with the same number of sides are similar when corresponding angles have equal measures and corresponding sides have lengths that are in proportion.

similar triangles Two triangles are similar when two angles of one triangle

are congruent to the corresponding angles of the other triangle.

sine ratio In a right triangle, the ratio of the length of the leg that is opposite a given acute angle to the length of the hypotenuse.

slope formula The formula for slope m of a nonvertical line that contains points $A(x_A, y_A)$ and $B(x_B, y_B)$:

$$m = \frac{y_B - y_A}{x_B - x_A}$$

slope-intercept form An equation of a line that has the form $y = mx + b$, where m is the slope of the line and b is the y-intercept.

solution One or more numbers from the replacement set for a variable that make an equation a true statement.

square A rectangle whose four sides have the same length.

square root For a nonnegative number x, one of two identical factors whose product is x. Thus, $\sqrt{9} = 3$ since $3 \times 3 = 9$.

statement A sentence that can be judged as either true or false, but not both.

stem-and-leaf plot An organization of data in a table in which each data value is split into a "stem" and a "leaf." The stems are listed in the first vertical column; then, to the right of each stem, its leaf values are listed horizontally on the same line in ascending order.

successes See **favorable outcomes**.

supplementary angles Two angles whose degree measures add up to 180.

system of equations A set of equations whose solution is the set of values that make each of the equations in the set true at the same time.

tangent ratio In a right triangle, the ratio of the length of the leg opposite a given acute angle to the length of the leg adjacent to the same angle.

theorem A generalization that can be proved.

transformation The process of "moving" each point of a figure according to an established rule.

translation A transformation in which each point of a figure is "moved" the same distance and in the same direction.

transversal A line that intersects two other lines in two different points.

trapezoid A quadrilateral that has one pair of parallel sides and one pair of sides that are not parallel.

tree diagram A diagram whose branches describe all of the different possible outcomes in a probability experiment.

triangle A polygon with three sides.

trinomial A polynomial with three unlike terms.

truth value For a statement, a value of either true or false, but not both.

turning point of a parabola The point at which the axis of symmetry intersects a parabola. The turning point is the highest or lowest point on a parabola.

variable A symbol, usually a single letter, that is a placeholder for an unspecified member of the replacement set.

vertex of a parabola See **turning point of a parabola**.

vertical angles The opposite pairs of congruent angles that are formed when two lines intersect.

volume A measure of capacity that gives the number of unit cubes a solid figure can hold.

x-axis The horizontal axis in a rectangular coordinate plane.

x-coordinate The first member (abscissa) in an ordered pair of numbers that indicates the location of a point in the coordinate plane.

x-intercept The x-coordinate of the point at which the graph of an equation intersects the x-axis. To find the x-intercept without graphing, let $y = 0$ in the equation of the graph and solve for x.

y-axis The vertical axis in a rectangular coordinate plane.

y-coordinate The second member (ordinate) in an ordered pair of numbers that indicates the location of a point in the coordinate plane.

y-intercept The y-coordinate of the point at which the graph of an equation intersects the y-axis. To find the y-intercept without graphing, let $x = 0$ in the equation of the graph and solve for y.

BASIC GEOMETRIC CONSTRUCTIONS

Geometric constructions, unlike drawings, are made only with a straightedge (for example, an unmarked ruler) and compass. The point at which the sharp point of the compass is placed is called the **center**, and the fixed compass setting that is used to draw **arcs** is the **radius length**.

Construction 1: Copy a Segment

Objective: To construct a segment that has the same length as a given line segment $\overline{AB}$.

Step	Diagram
1. Draw any line, and choose any convenient point on it. Label the line as ℓ and the point as C.	
2. Using a compass, measure $\overline{AB}$ by placing the compass point on A and the pencil point to B.	
3. Using the same compass setting, place the compass point on C and draw an arc that intersects line ℓ. Label the point of intersection as D.	

Conclusion: $\overline{AB} \cong \overline{CD}$.

Construction 2: Copy an Angle

Objective: To construct an angle that has the same measure as a given angle, $\angle ABC$.

Step	Diagram
1. Draw any line and choose any point on it. Label the line as ℓ and the point as S.	
2. Using any convenient compass setting, place the compass point on B and draw an arc intersecting $\overrightarrow{BC}$ at X and $\overrightarrow{BA}$ at Y.	
3. Using the same compass setting, place the compass point at S and draw arc WT, intersecting line ℓ at T.	
4. Adjust the compass setting to measure the line segment determined by points X and Y by placing the compass point at X and the pencil at Y.	
5. Using the same compass setting, place the compass point at T and construct an arc intersecting arc WT at point R.	
6. Using a straightedge, draw $\overrightarrow{SR}$.	

Conclusion: $\angle ABC \cong \angle RST$.

Reason: The arcs were constructed so that $\overline{BX} \cong \overline{ST}$, $\overline{BY} \cong \overline{SR}$, and $\overline{XY} \cong \overline{TR}$. Therefore, $\triangle XYB \cong \triangle TRS$ by the SSS postulate. Since all of the corresponding pairs of parts of congruent triangles have equal measures, $m\angle ABC = m\angle RST$.

Construction 3: Bisect a Line Segment

Objective: To construct the perpendicular bisector of the segment determined by two given points.

Step	Diagram
1. Label points *A* and *B*, and draw $\overline{AB}$. Choose any compass setting (radius length) that is more than one-half the length of $\overline{AB}$. 2. Using this compass setting, and points *A* and *B* as centers, construct a pair of arcs above and below $\overline{AB}$. Label the points at which the pairs of arcs intersect as *P* and *Q*. 3. Draw $\overleftrightarrow{PQ}$, and label the point of intersection of $\overleftrightarrow{PQ}$ and $\overline{AB}$ as *M*.	

Conclusion: $\overleftrightarrow{PQ}$ is the perpendicular bisector of $\overline{AB}$.

Reason: The arcs were constructed in such a way that $AP = BP = AQ = BQ$. Since quadrilateral *APBQ* is equilateral, it is a rhombus. Since the diagonals of a rhombus are perpendicular bisectors, $\overline{AM} \cong \overline{BM}$ and $\overline{PQ} \perp \overline{AB}$.

486

Construction 4: Bisect an Angle

Objective: To construct the ray that bisects a given angle, $\angle ABC$.

Step	Diagram
1. Using *B* as a center, construct an arc, using any convenient radius length, that intersects $\overrightarrow{BA}$ at point *P* and $\overrightarrow{BC}$ at point *Q*. 2. Using points *P* and *Q* as centers and the same radius length, draw a pair of arcs that intersect. Label the point at which the arcs intersect as *D*. 3. Draw $\overrightarrow{BD}$.	

Conclusion: $\overrightarrow{BD}$ is the bisector of $\angle ABC$.

Reason: The arcs were constructed so that $BP = BQ$ and $PD = QD$, thus making $\triangle BPD \cong \triangle BQD$. Therefore, $m\angle 1 = m\angle 2$.

Construction 5: Construct a Perpendicular Line at a Point

Objective: To construct a line that is perpendicular to a given line, ℓ, at a point on the given line.

Step	Diagram
1. Using P as a center and any convenient radius length, construct an arc that intersects line ℓ at two points. Label these points as A and B. 2. Choose a radius length greater than one-half the length of $\overline{AB}$. Using points A and B as centers, construct, on either side of line ℓ, a pair of arcs that intersect at point Q. 3. Draw $\overleftrightarrow{PQ}$.	

Conclusion: $\overleftrightarrow{PQ} \perp \ell$

Reason: The arcs were constructed so that $AP = PB$ and $AQ = BQ$, thus making $\triangle APQ \cong \triangle BPQ$ by SSS $\cong$ SSS. This means that angles APQ and BPQ are both congruent and adjacent, so each must be a right angle, making $\overleftrightarrow{PQ} \perp \ell$.

Construction 6: Construct a Perpendicular Line from a Point

Objective: To construct a line that is perpendicular to a given line, ℓ, and through a point not on the given line.

Step	Diagram
1. Using *P* as a center and any convenient radius length, construct an arc that intersects line ℓ at two points. Label these points *A* and *B*. 2. Choose a radius length greater than one-half the length of $\overline{AB}$. Using points *A* and *B* as centers, construct a pair of arcs that intersect at point *Q*. 3. Draw $\overrightarrow{PQ}$, intersecting line ℓ at point *M*.	

Conclusion: $\overleftrightarrow{PQ}$ is perpendicular to line ℓ at point *M*.

Construction 7: Construct Parallel Lines

Objective: To construct a line that is parallel to another line, $\overline{AB}$, and passes through a given point, P.

Step	Diagram
1. Through P draw any convenient line, ℓ, extending it so that it intersects $\overleftrightarrow{AB}$. Label the point of intersection as Q. 2. Construct an angle with vertex at P, congruent to $\angle PQB$ (see Construction 2), as shown. 3. Draw $\overleftrightarrow{PS}$.	

Conclusion: $\overleftrightarrow{PS} \parallel \overleftrightarrow{AB}$

Reason: The arcs were constructed so that $\angle PQB$ and $\angle RPS$ are congruent corresponding angles, making $\overleftrightarrow{PS} \parallel \overleftrightarrow{AB}$.

Combining Constructions

Sometimes a construction requires that one or more basic constructions be performed.

Required Construction	What to Do
To construct a 45° angle	• Construct a line perpendicular to another line, thereby forming two adjacent right angles. • Bisect either right angle.
To construct the *median* to side $\overline{AC}$ of △*ABC*	• Locate the midpoint of $\overline{AC}$ by constructing its perpendicular bisector. • Draw a line segment connecting *B* and the midpoint of $\overline{AC}$.
To construct the *altitude* to side $\overline{AC}$ of △*ABC*	• Construct a line from *B* perpendicular to $\overline{AC}$. Label the point at which the perpendicular line intersects $\overline{AC}$ as *H*. • Draw $\overline{BH}$.
To locate the center of a circle that contains noncollinear points *A*, *B*, and *C*	• Construct the perpendicular bisector of $\overline{AB}$. • Construct the perpendicular bisector of $\overline{BC}$. • Label as *P* the point at which the two perpendicular bisectors intersect. Point *P* is the center of the desired circle.
To construct a triangle congruent to △*ABC* 	• Draw a line ℓ. At a point *D* on line ℓ, copy ∠*A* so that ∠*XDY* ≅ ∠*A*. • On ray *DX*, copy $\overline{AB}$ so that $\overline{AB}$ ≅ $\overline{DE}$. On ray *DY*, copy $\overline{AC}$ so that $\overline{AC}$ ≅ $\overline{DF}$. • Draw $\overline{EF}$. Triangles *ABC* and *DEF* are congruent by SAS ≅ SAS.

THE MATH A REGENTS EXAM

The Mathematics A Regents Exam is a three-hour exam that is divided into four parts with a total of 35 questions. No choice is permitted in any of the four parts. You are required to answer all 35 questions. Part I consists of 20 standard multiple-choice questions. You must record the answers to these 20 questions on a detachable answer sheet that is located at the end of the question booklet. Parts II, III, and IV each contain a set of five questions that must be answered directly in the question booklet. The directions for Parts II, III, and IV ask that you show or explain how you arrived at your answers. Since scrap paper is not permitted for any part of the exam, you may use the blank spaces in the question booklet as scrap paper. If you need to draw a graph, graph paper will be included in the question booklet.

How Is the Math A Exam Scored?

Each of your answers to the 20 multiple-choice questions in Part I is scored as either right or wrong. Solutions to questions in Parts II, III, and IV that are not completely correct may receive partial credit according to a special rating guide that is provided by the New York State Education Department. The answers and the work for the questions in Parts II, III, and IV must be written directly in the question booklet. The accompanying table shows how the Math A test breaks down.

Part	Number of Questions	Point Value	Points
I	20 multiple choice	2 points each	40
II	5 questions	2 points each	10
III	5 questions	3 points each	15
IV	5 questions	4 points each	20
Total = 35 questions		Total = 85	

In order to receive full credit for a correct answer to a question in Parts II, III, or IV, you must show or explain how you arrived at your answer by indicating the necessary steps you take, including appropriate formula substitutions, diagrams, graphs, and charts. A correct numerical answer with no work shown will receive only 1 credit. If you use the guess and check strategy to arrive at an answer for a problem, you must include at least one guess that does not work.

How Is Your Final Math A Regents Score Determined?

The raw scores for the four parts of the test are added together. The maximum total raw score for the Math A test is 85 points. Using a conversion table that is also provided by the New York State Education Department, your teacher will convert your total raw score into a final test score that falls within the usual 0 to 100 scale.

What Type of Calculator Is Needed?

Scientific calculators are required for the Math A Regents Exam. You will need to be able to use your scientific calculator to work with trigonometric functions of angles. Your scientific calculator may also be helpful in performing routine arithmetic calculations, and in evaluating permutations ($_nP_r$) and combinations ($_nC_r$). Graphing calculators are permitted, but not required.

What Is the *Core Curriculum*?

The New York State Department of Education publishes a *Core Curriculum* that lists the topics that may be tested by the Mathematics A Regents Exam. The *Core Curriculum* covers a wide range of mathematics skills and concepts. If you have Internet access, you can view the *Core Curriculum* at the New York State Education Department's web site at

http://www.emsc.nysed.gov/ciai/mst/pub/matha&b.pdf

NYS Regents Updates

- Unless otherwise directed by the question, use *all* of the digits in the calculator display window. Do not round intermediate values. Rounding, if required, should be done only after the *final* answer is reached.

- When using a "guess and check" method, show a developing pattern that progresses to the correct solution. If the first guess is the correct solution, you must also include one trial below the correct answer and another trial above the correct answer with appropriate checks.

Examination
August 2002
Math A

Answer all questions in this part. Each correct answer will receive 2 credits. No partial credit will be allowed. Record your answers in the spaces provided. [40]

1 On a map, 1 centimeter represents 40 kilometers. How many kilometers are represented by 8 centimeters?

(1) 5   (3) 280
(2) 48   (4) 320   1 ____

2 In the accompanying diagram of parallelogram $ABCD$, diagonals $\overline{AC}$ and $\overline{DB}$ intersect at E, $AE = 3x - 4$, and $EC = x + 12$.

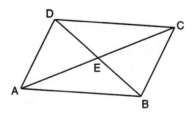

What is the value of x?

(1) 8   (3) 20
(2) 16   (4) 40   2 ____

3 What is the total number of points equidistant from two intersecting straight roads and also 300 feet from the traffic light at the center of the intersection?

(1) 1   (3) 3
(2) 2   (4) 4   3 ____

4 Juan has three blue shirts, two green shirts, seven red shirts, five pairs of denim pants, and two pairs of khaki pants. How many different outfits consisting of one shirt and one pair of pants are possible?

(1) 19   (3) 130
(2) 84   (4) 420   4 _____

5 Given the statement: "If two lines are cut by a transversal so that the corresponding angles are congruent, then the lines are parallel."

What is true about the statement and its converse?

(1) The statement and its converse are both true.
(2) The statement and its converse are both false.
(3) The statement is true, but its converse is false.
(4) The statement is false, but its converse is true.   5 _____

6 If the area of a square garden is 48 square feet, what is the length, in feet, of one side of the garden?

(1) $12\sqrt{2}$ (3) $16\sqrt{3}$
(2) $4\sqrt{3}$ (4) $4\sqrt{6}$ 6 _____

7 The sum of $\dfrac{3}{x}+\dfrac{2}{5}, x \neq 0$, is

(1) $\dfrac{1}{x}$ (3) $\dfrac{5}{x+5}$

(2) $\dfrac{2x+15}{5x}$ (4) $\dfrac{2x+15}{x+5}$ 7 _____

8 The number 0.14114111411114 . . . is

(1) integral   (3) irrational
(2) rational   (4) whole   8 _____

9 When $-2x^2 + 4x + 2$ is subtracted from $x^2 + 6x - 4$, the result is

(1) $-3x^2 - 2x + 6$ (3) $2x^2 - 2x - 6$
(2) $-x^2 + 10x - 2$ (4) $3x^2 + 2x - 6$ 9 _____

10 If 0.0347 is written by a scientist in the form 3.47×10^n, the value of n is
(1) -2 (3) 3
(2) 2   (4) -3 10 _____

11 If $x = -2$ and $y = -1$, which point on the accompanying set of axes represents the translation $(x, y) \rightarrow (x + 2, y - 3)$?

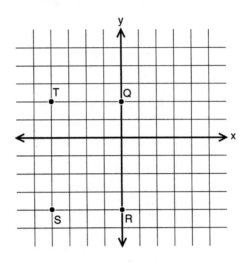

(1) Q (3) S

(2) R (4) T 11 _____

12 In the accompanying diagram, which transformation changes the solid-line parabola to the dotted-line parabola?

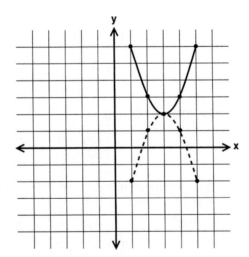

(1) translation   (3) rotation, only

(2) line reflection, only   (4) line reflection or rotation   12 _____

13 How many times larger than $\frac{1}{4}x$ is $5x$?

(1) 20   (3) $\frac{5}{4}$

(2) 9   (4) $\frac{4}{5}$ 13 _____

14 If the lengths of two sides of a triangle are 4 and 10, what could be the length of the third side?

(1) 6   (3) 14

(2) 8   (4) 16   14 _____

15 Which piece of paper can be folded into a pyramid?

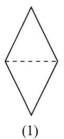

(1)

(3)

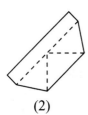

(2)

(4)

15 _____

16 What is the measure of the largest angle in the accompanying triangle?

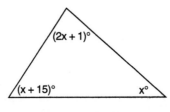

(1) 41   (3) 56

(2) 46.5   (4) 83   16 _____

17 M is the midpoint of $\overline{AB}$. If the coordinates of A are $(-1, 5)$ and the coordinates of M are $(3, 3)$, what are the coordinates of B?

(1) $(1, 4)$ (3) $(7, 1)$

(2) $(2, 8)$ (4) $(-5, 7)$ 17 _____

18 If $2m + 2p = 16$, p equals

(1) $8 - m$ (3) $16 + 2m$

(2) $16 - m$ (4) $9m$ 18 _____

19 If $2x + 5 = -25$ and $-3m - 6 = 48$, what is the product of x and m?

(1) -270 (3) 3

(2) -33 (4) 270   19 _____

20 In the graph of $y \leq -x$, which quadrant is completely shaded?

(1) I   (3) III

(2) II   (4) IV   20 _____

PART II

Answer all questions in this part. Each correct answer will receive 2 credits. Clearly indicate the necessary steps, including appropriate formula substitutions, diagrams, graphs, charts, etc. For all questions in this part, a correct numerical answer with no work shown will receive only 1 credit. [10]

21 In the accompanying diagram of $\triangle BCD$, $\triangle ABC$ is an equilateral triangle and $AD = AB$. What is the value of x, in degrees?

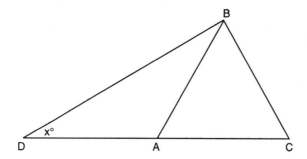

22 In the addition table for a subset of real numbers shown below, which number is the inverse of 3? Explain your answer.

$\oplus$	1	2	3	4
1	2	3	4	1
2	3	4	1	2
3	4	1	2	3
4	1	2	3	4

23 An image of a building in a photograph is 6 centimeters wide and 11 centimeters tall. If the image is similar to the actual building and the actual building is 174 meters wide, how tall is the actual building, in meters?

24 A doughnut shop charges $0.70 for each doughnut and $0.30 for a carryout box. Shirley has $5.00 to spend. At most, how many doughnuts can she buy if she also wants them in one carryout box?

25 In bowling leagues, some players are awarded extra points called their "handicap." The "handicap" in Anthony's league is 80% of the difference between 200 and the bowler's average. Anthony's average is 145. What is Anthony's "handicap"?

PART III

Answer all questions in this part. Each correct answer will receive 3 credits. Clearly indicate the necessary steps, including appropriate formula substitutions, diagrams, graphs, charts, etc. For all questions in this part, a correct numerical answer with no work shown will receive only 1 credit. [15]

26 In a telephone survey of 100 households, 32 households purchased Brand *A* cereal and 45 purchased Brand *B* cereal. If 10 households purchased both items, how many of the households surveyed did *not* purchase either Brand *A* or Brand *B* cereal?

27 Tamika could not remember her scores from five mathematics tests. She did remember that the mean (average) was exactly 80, the median was 81, and the mode was 88. If all her scores were integers with 100 the highest score possible and 0 the lowest score possible, what was the *lowest* score she could have received on any one test?

28 There are 28 students in a mathematics class. If $\frac{1}{4}$ of the students are called to the guidance office, $\frac{1}{3}$ of the remaining students are called to the nurse, and, finally, $\frac{1}{2}$ of those left go to the library, how many students remain in the classroom?

29 On a bookshelf, there are five different mystery books and six different biographies. How many different sets of four books can Emilio choose if two of the books must be mystery books and two of the books must be biographies?

30 On the accompanying grid, graph a circle whose center is at (0, 0) and whose radius is 5. Determine if the point (5, −2) lies on the circle.

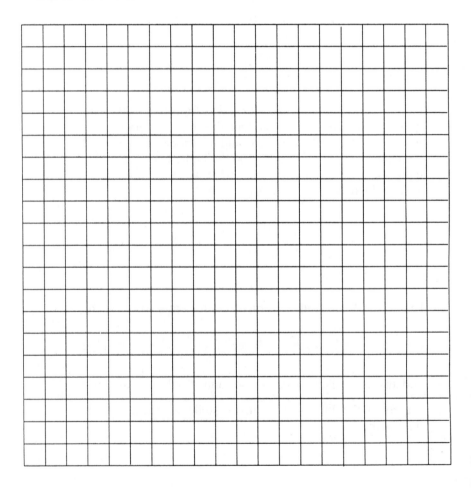

PART IV

Answer all questions in this part. Each correct answer will receive 4 credits. Clearly indicate the necessary steps, including appropriate formula substitutions, diagrams, graphs, charts, etc. For all questions in this part, a correct numerical answer with no work shown will receive only 1 credit. [20]

31 In the accompanying diagram, x represents the length of a ladder that is leaning against a wall of a building, and y represents the distance from the foot of the ladder to the base of the wall. The ladder makes a 60° angle with the ground and reaches a point on the wall 17 feet above the ground. Find the number of feet in x and y.

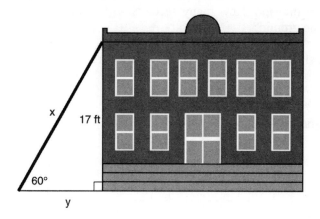

32 A rectangular park is three blocks longer than it is wide. The area of the park is 40 square blocks. If w represents the width, write an equation in terms of w for the area of the park. Find the length and the width of the park.

33 Tanisha and Rachel had lunch at the mall. Tanisha ordered
   three slices of pizza and two colas. Rachel ordered two slices
   of pizza and three colas. Tanisha's bill was $6.00, and
   Rachel's bill was $5.25. What was the price of one slice of
   pizza? What was the price of one cola?

34 Greg is in a car at the top of a roller-coaster ride. The distance, d, of the car from the ground as the car descends is
   determined by the equation $d = 144 - 16t^2$, where t is the
   number of seconds it takes the car to travel down to each
   point on the ride. How many seconds will it take Greg to
   reach the ground?

   For an algebraic solution show your work here.

   For a graphic solution show your work here.

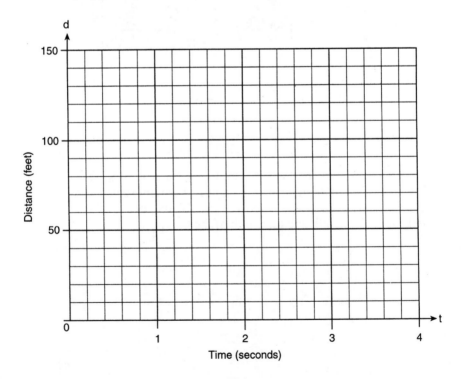

Time (seconds)

35 Determine the distance between point $A(-1, -3)$ and point $B(5, 5)$. Write an equation of the perpendicular bisector of $\overline{AB}$. [The use of the accompanying grid is optional.]

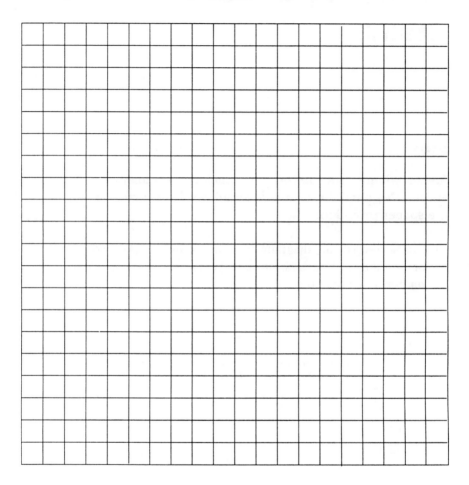

Examination
January 2003
Math A

PART I

Answer all questions in this part. Each correct answer will receive 2 credits. No partial credit will be allowed. Record your answers in the spaces provided. [40]

1 The accompanying diagram shows a box-and-whisker plot of student test scores on last year's Mathematics A midterm examination.

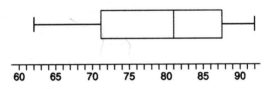

What is the median score?

(1) 62   (3) 81

(2) 71   (4) 92   1 ____

2 Triangle $A'B'C'$ is the image of $\triangle ABC$ under a dilation such that $A'B' = 3AB$. Triangles ABC and $A'B'C'$ are

(1) congruent but not similar
(2) similar but not congruent
(3) both congruent and similar
(4) neither congruent nor similar

2 _____

3 What is the inverse of the statement "If Mike did his homework, then he will pass this test"?

(1) If Mike passes this test, then he did his homework.
(2) If Mike does not pass this test, then he did not do his homework.
(3) If Mike does not pass this test, then he only did half his homework.
(4) If Mike did not do his homework, then he will not not pass this test.

3 _____

4 In which list are the numbers in order from least to greatest?

(1) $3.2, \pi, 3\frac{1}{3}, \sqrt{3}$ (3) $\sqrt{3}, \pi, 3.2, 3\frac{1}{3}$

(2) $\sqrt{3}, 3.2, \pi, 3\frac{1}{3}$ (4) $3.2, 3\frac{1}{3}, \sqrt{3}, \pi$

4 _____

5 The accompanying diagram shows a transformation.

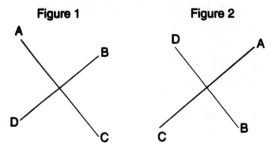

Which transformation performed on figure 1 resulted in figure 2?

(1) rotation   (3) dilation
(2) reflection   (4) translation   5 ____

6 The product of $3x^5$ and $2x^4$ is

(1) $5x^9$ (3) $6x^9$
(2) $5x^{20}$ (4) $6x^{20}$ 6 ____

7 There are 12 people on a basketball team, and the coach needs to choose 5 to put into a game. How many different possible ways can the coach choose a team of 5 if each person has an equal chance of being selected?

(1) $_{12}P_5$ (3) $_{12}C_5$
(2) $_5P_{12}$ (4) $_5C_{12}$ 7 ____

8 Given the true statement: "If a person is eligible to vote, then that person is a citizen."

Which statement must also be true?

(1) Kayla is not a citizen; therefore, she is not eligible to vote.
(2) Juan is a citizen; therefore, he is eligible to vote.
(3) Marie is not eligible to vote; therefore, she is not a citizen.
(4) Morgan has never voted; therefore, he is not a citizen.   8 _____

9 Line P and line C lie on a coordinate plane and have equal slopes. Neither line crosses the second or third quadrant. Lines P and C must

(1) form an angle of 45°
(2) be perpendicular
(3) be horizontal
(4) be vertical

9 _____

10 The equation $P = 2L + 2W$ is equivalent to

(1) $L = \dfrac{P - 2W}{2}$

(3) $2L = \dfrac{P}{2W}$

(2) $L = \dfrac{P + 2W}{2}$

(4) $L = P - W$ 10 _____

11 The sum of $\sqrt{75}$ and $\sqrt{3}$ is

(1) 15

(3) $6\sqrt{3}$

(2) 18

(4) $\sqrt{78}$ 11 _____

12 Which graph represents the solution set for $2x - 4 \le 8$ and $x + 5 \ge 7$?

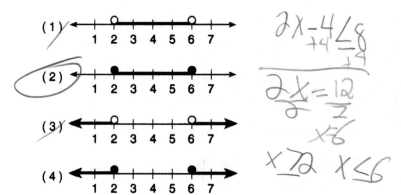

(1)

1 2 3 4 5 6 7

(2)

1 2 3 4 5 6 7

(3)

1 2 3 4 5 6 7

(4)

1 2 3 4 5 6 7

12 _____

$2x - 4 \le 8$

$+4 \quad +4$

$\dfrac{2x}{2} = \dfrac{12}{2}$

$x - 6$

$x \le 2 \quad x \le 6$

13 If the measure of an angle is represented by $2x$, which expression represents the measure of its complement?

(1) $180 - 2x$ (3) $90 + 2x$

(2) $90 - 2x$ (4) $88x$ 13 _____

14 Which equation illustrates the multiplicative identity element?

(1) $x + 0 = x$ (3) $x \cdot \dfrac{1}{x} = 1$

(2) $x - x = 0$ (4) $x \cdot 1 = x$ 14 _____

15 The ages of five children in a family are 3, 3, 5, 8, and 18. Which statement is true for this group of data?

(1) mode > mean   (3) median = mode

(2) mean > median   (4) median > mean   15 _____

$90 - 2x = 180$

Mean = 7.4
Med = 5
mode = 3

510

16 In the accompanying diagram of right triangle ABC, $AB = 8, BC = 15, AC = 17$, and $m\angle ABC = 90$.

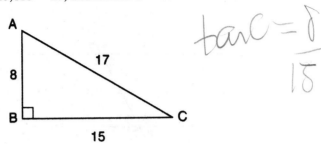

A

8

17

B

15

C

What is tan $\angle C$?

(1) $\frac{8}{15}$

(3) $\frac{8}{17}$

(2) $\frac{17}{15}$

(4) $\frac{15}{17}$

16 _____

17 The locus of points equidistant from two sides of an acute scalene triangle is

(1) an angle bisector

(3) a median

(2) an altitude

(4) the third side

17 _____

18 What are the factors of $x^2 - 10x - 24$?

(1) $(x - 4)(x + 6)$

(3) $(x - 12)(x + 2)$

(2) $(x - 4)(x - 6)$

(4) $(x + 12)(x - 2)$

18 _____

19 What is the value of $\dfrac{6.3 \times 10^8}{3 \times 10^4}$ in scientific notation?

(1) 2.1×10^{-2}

(3) 2.1×10^{-4}

(2) 2.1×10^2

(4) 2.1×10^4

19 _____

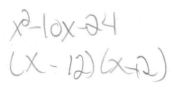

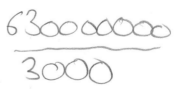

511

20 In the accompanying figure, what is one pair of alternate interior angles?

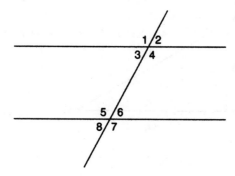

(1) ∠1 and ∠2
(2) ∠4 and ∠5

(3) ∠4 and ∠6
(4) ∠6 and ∠8

20 _____

PART II

Answer all questions in this part. Each correct answer will receive 2 credits. Clearly indicate the necessary steps, including appropriate formula substitutions, diagrams, graphs, charts, etc. For all questions in this part, a correct numerical answer with no work shown will receive only 1 credit.   [10]

21 If Laquisha can enter school by any one of three doors and the school has two staircases to the second floor, in how many different ways can Laquisha reach a room on the second floor? Justify your answer by drawing a tree diagram or listing a sample space.

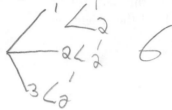

22 The world population was 4.2 billion people in 1982. The population in 1999 reached 6 billion. Find the percent of change from 1982 to 1999.

23 Six members of a school's varsity tennis team will march in a parade. How many different ways can the players be lined up if Angela, the team captain, is always at the front of the line?

$$1 \cdot 5 \cdot 4 \cdot 3 \cdot 2 \cdot 1 = 120$$

24 A fish tank with a rectangular base has a volume of 3,360 cubic inches. The length and width of the tank are 14 inches and 12 inches, respectively. Find the height, in inches, of the tank.

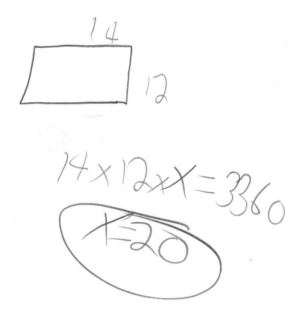

$$14 \times 12 \times X = 3360$$
$$X = 20$$

25 Mr. Smith's class voted on their favorite ice cream flavors, and the results are shown in the accompanying diagram. If there are 20 students in Mr. Smith's class, how many students chose coffee ice cream as their favorite flavor?

Favorite Ice Cream Flavors

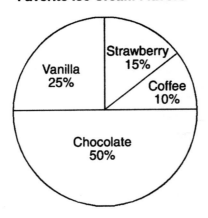

$$\frac{10\%}{100} = \frac{x}{20}$$

$$\boxed{2}$$

PART III

Answer all questions in this part. Each correct answer will receive 3 credits. Clearly indicate the necessary steps, including appropriate formula substitutions, diagrams, graphs, charts, etc. For all questions in this part, a correct numerical answer with no work shown will receive only 1 credit. [15]

26 Three brothers have ages that are consecutive even integers. The product of the first and third boys' ages is 20 more than twice the second boy's age. Find the age of *each* of the three boys.

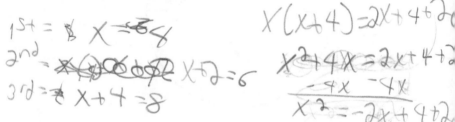

27 Arielle has a collection of grasshoppers and crickets. She has 561 insects in all. The number of grasshoppers is twice the number of crickets. Find the number of *each* type of insect that she has.

28 The graph of a quadratic equation is shown in the accompanying diagram. The scale on the axes is a unit scale. Write an equation of this graph in standard form.

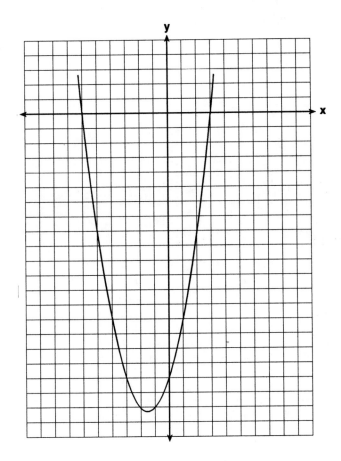

29 Currently, Tyrone has \$60 and his sister has \$135. Both get an allowance of \$5 each week. Tyrone decides to save his entire allowance, but his sister spends all of hers each week plus an additional \$10 each week. After how many weeks will they each have the same amount of money? [The use of the accompanying grid is optional.]

5 weeks

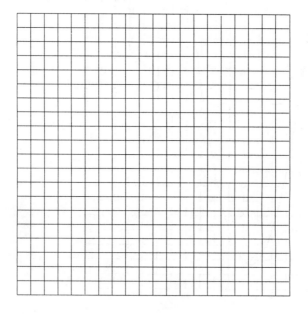

135
- 10
─────
125
- 10
─────
115
- 10
─────
105

165
- 10
─────
95
- 10
─────
85

518

30 A rectangular garden is going to be planted in a person's rectangular backyard, as shown in the accompanying diagram. Some dimensions of the backyard and the width of the garden are given. Find the area of the garden to the *nearest square foot*.

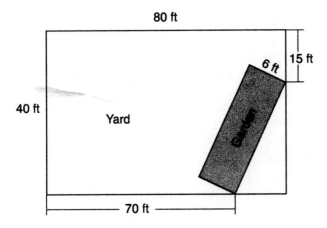

PART IV

Answer all questions in this part. Each correct answer will receive 4 credits. Clearly indicate the necessary steps, including appropriate formula substitutions, diagrams, graphs, charts, etc. For all questions in this part, a correct numerical answer with no work shown will receive only 1 credit. [20]

31 At the Phoenix Surfboard Company, $306,000 in profits was made last year. This profit was shared by the four partners in the ratio 3:3:5:7. How much *more* money did the partner with the largest share make than one of the partners with the smallest share?

$$3x + 3x + 5x + 7x = 306,000$$

$$18x = 306000$$

$$x = 1700$$

$$\begin{array}{r} 119000 \\ -51000 \\ \hline \$68000 \end{array}$$

32 Alexandra purchases two doughnuts and three cookies at a doughnut shop and is charged $3.30. Briana purchases five doughnuts and two cookies at the same shop for $4.95. All the doughnuts have the same price and all the cookies have the same price. Find the cost of one doughnut and find the cost of one cookie.

33 On the accompanying grid, draw and label quadrilateral
   ABCD with points $A(1,2)$, $B(6,1)$, $C(7,6)$, and $D(3,7)$. On
   the same set of axes, plot and label quadrilateral $A'B'C'D'$,
   the reflection of quadrilateral *ABCD* in the *y*-axis. Determine the area, in square units, of quadrilateral $A'B'C'D'$.

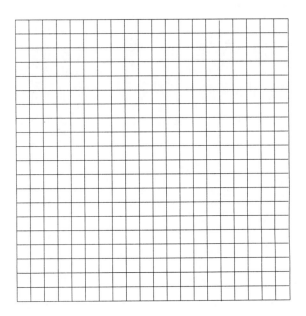

34 Sarah's mathematics grades for one marking period were
85, 72, 97, 81, 77, 93, 100, 75, 86, 70, 96, and 80.

   a Complete the tally sheet and frequency table below,
   and construct and label a frequency histogram for
   Sarah's grades using the accompanying grid.

Interval (grades)	Tally	Frequency
61–70		
71–80		
81–90		
91–100		

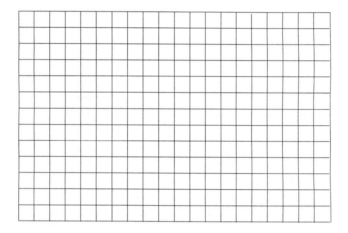

   b Which interval contains the 75th percentile (upper
   quartile)?

35 On the accompanying set of axes, graph and label the following lines:

$$y = 5$$
$$x = -4$$
$$y = \frac{5}{4}x + 5$$

Calculate the area, in square units, of the triangle formed by the three points of intersection.

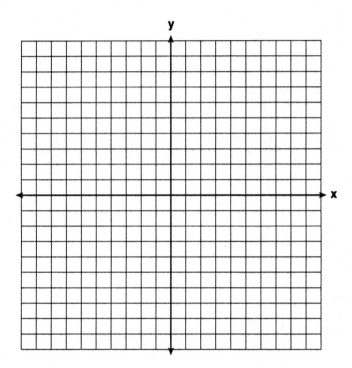

Answers to Regents Examinations

AUGUST 2002
PART I

1. (4)	**6.** (2)	**11.** (2)	**16.** (4)
2. (1)	**7.** (2)	**12.** (4)	**17.** (3)
3. (4)	**8.** (3)	**13.** (1)	**18.** (1)
4. (2)	**9.** (4)	**14.** (2)	**19.** (4)
5. (1)	**10.** (1)	**15.** (3)	**20.** (3)

PART II

21. 30   **24.** 6

22. 1   **25.** 44

23. 319

PART III

26. 33   **28.** 7   **30.** Correctly graphed circle; show that $(5, -2)$ does not lie

27. 63   **29.** 150   on the circle.

PART IV

31. $x = 19.62990915$ and $y = 9.814954576$

32. $w(w + 3) = 40$, width $= 5$, and length $= 8$

33. \$1.50 for one slice of pizza, and \$0.75 for one cola

34. 3

35. 10 and $y - 1 = -\dfrac{3}{4}(x - 2)$

Parts II–IV You are required to show how you arrived at your answers.

JANUARY 2003
PART I

1. (3)	**6.** (3)	**11.** (3)	**16.** (1)
2. (2)	**7.** (3)	**12.** (2)	**17.** (1)
3. (4)	**8.** (1)	**13.** (2)	**18.** (3)
4. (3)	**9.** (4)	**14.** (4)	**19.** (4)
5. (1)	**10.** (1)	**15.** (2)	**20.** (2)

PART II

21. 6
22. 42.85714286
23. 120
24. 20
25. 2

PART III

26. 4, 6, and 8
27. 374 grasshoppers and 187 crickets
28. $y = x^2 + 3x - 18$
29. 5
30. 162

PART IV

31. $68,000
32. $0.75 for one doughnut, and $0.60 for one cookie
33. The area of $A'B'C'D'$ is 24
34. **a.** Correctly completed frequency table and histogram
    b. The interval 91–100
35. All lines are graphed and labeled correctly and area = 10

Parts II–IV You are required to show how you arrived at your answers.

INDEX

solving algebraically, 188–189, 217, 222
solving graphically, 388–389
Quadratic-linear systems, 392–396
Quadrilateral, 124
Quartiles, 411

Radicals, 219, 221
Radius, 251
Ratio, 53–55
Rational numbers, 219
Ray, 106
Real numbers, 7–8
Rectangle, 164
Rectangular solid, 49
Reflection, 331, 337
Regular polygon, 124
Rhombus, 165
Right
    angle, 109
    triangle, 125, 234–235
Root, 24
Rotation, 332

Sample space, 339
SAS, 142
Scalene triangle, 124
Scatter plot, 324
Scientific notation, 16
Signed numbers, 2, 4
Similar figures, 266
Sine ratio, 272
Slope
    formula, 296–297
    of parallel and perpendicular lines, 300
Slope-intercept equation, 306
Solution set, 24
Sphere, 262
Square, 167
Square root, 219
SSS, 143
Statement, 92
Stem-and-Leaf plot, 417
Straight angle, 108

Successes, 400
Sum of angles
    of a polygon, 137
    of a triangle, 127–128
Supplementary angles, 112
Symmetry
    axis of, 379
    line, 328
    point, 329
Systems
    of linear equations, 352–354, 360–361, 366–368
    of linear inequalities, 355–357
    of linear-quadratic equations, 392–396

Tangent ratio, 272
Theorem, 107
Transformations
    types of, 331
    using coordinates, 337–340
Translation, 332, 338
Transversal, 116
Trapezoid, 116
Tree diagram, 431
Triangle(s)
    classifying, 124–125
    congruent, 142
    exterior angle theorem, 132
    inequalities, 152–154
    similar, 266
    special right triangle relationships, 235
    sum of angles, 127–128
Trigonometric ratios, 272, 275–276
Trinomial, 172
Turning point, 379

Undefined terms, 105

Variable, 1
Venn diagram, 89–91
Vertex
    of an angle, 106

529

NOTES

NOTES